GENETIC ASPECTS OF PHOTOSYNTHESIS

GENETIC ASPECTS OF PHOTOSYNTHESIS

Selected papers from the Symposium held on October 17-24, 1972, in Dushanbe, U.S.S.R.

Edited by
Yu. S. NASYROV (Dushanbe) & Z. ŠESTÁK (Prague)

Dr. W. JUNK B.V. Publishers, The Hague 1975

ISBN-13: 978-90-6193-027-3   e-ISBN-13: 978-94-010-1936-1
DOI: 10.1007/978-94-010-1936-1
© Dr. W. Junk B.V. Publishers, The Hague 1975

Cover design Charlotte van Zadelhoff

PREFACE

The Symposium 'Genetic Aspects of Photosynthesis' was sponsored by the Institute of Plant Physiology and Biophysics of the Tajik Academy of Sciences and the Scientific Councils of the Academy of Sciences of the U.S.S.R. on the problem of photosynthesis, genetics and selection, and also by the N.I. Vavilov All-Union Society of Geneticists and Selectionists. The Symposium took place in Dushanbe on October 17-24, 1972. 223 scientists attended, including 209 participants from different parts of the Soviet Union and 14 honoured guests from Australia, Czechoslovakia, Denmark, German Democratic Republic, Hungary and the U.S.A. They represented various fields of biological sciences: molecular biology, biochemistry, genetics, cytology, radiobiology, biophysics, plant physiology, and selection.

The chairmen of the ten symposial sessions were the leading Soviet scientists, B. L. ASTAUROV, A. A. NICHIPOROVICH, A. A. SHLYK, Yu. S. NASYROV, V. B. EVSTIGNEEV, O. V. ZALENSKIĬ, B. F. VANYUSHIN, N. P. VOSKRESENSKAYA, S. V. TAGEEVA and K. V. KVITKO. There were 57 speakers, and a further 40 experimental papers on the problems of genetic control of photosynthesis were displayed on an exhibition of scientific research. Two special round-table evening sessions were devoted to discussions on the mechanisms of the C_3 and C_4 pathways of photosynthetic carbon metabolism and the structure of photosynthetic membranes. The symposial sessions were concluded by a general discussion on the genetics of photosynthesis.

This proceedings volume presents the most important and interesting papers from Soviet participants of the Symposium and almost all papers from the foreign guests. Nevertheless, this selection gives — according to our opinion — a fairly good picture of recently solved problems of the genetics of photosynthesis.

The proceedings were prepared by a joint effort of co-workers of the Institute of Plant Physiology and Biophysics of the Tajik Academy of Sciences in Dushanbe, U.S.S.R., and of the Institute of Experimental Botany of the Czechoslovak Academy of Sciences, Prague, Czechoslovakia. The editors-in-chief express their thanks to all colleagues who helped in editorial work, namely: from Dushanbe, U.S.S.R., H. ABDULLAEV, K. A. ALIEV, RAISA I. CHERNER, Yu. E. GILLER, H. H. KARIMOV, DILBAR H. KADYROVA, ALLA G. KOLTUNOVA, LILIYA M. MAKHMADBEKOVA, GALINA I. NAGEL, KLAVDIYA P. RAKHMANINA, P. D. USMANOV; from Prague, Czechoslovakia, STANISLAVA FOUSOVÁ, DRAHOMÍRA TĚŽKÁ, J. VELEMÍNSKÝ and JARMILA VEVERKOVA. The careful linguistic revision was undertaken by G. J. KELLY from Australia.

The manuscripts were elaborated and their style, abbreviations, references drawings, etc., unified from the editorial point-of-view in Dushanbe and Prague. Nevertheless, the manuscripts were not reviewed and hence the original ideas of the authors need not agree with the views of the editors who present this material to all plant physiologists and geneticists interested in photosynthetic activity and productivity. The editors hope that this volume will stimulate further research in this important field of biology and agronomy.

Yu. S. NASYROV & Z. ŠESTÁK

CONTENTS

Preface . . . . by Yu. S. NASYROV & Z. ŠESTÁK		V
Authors' addresses		X
Foreword . . . . by A.A. NICHIPOROVICH		XV

Genome and Origin of Chloroplasts

M. S. TURISHCHEVA (Moscow): Chloroplast DNA-Membrane Complex in *Pisum sativum* 1

J. R. THOMAS, R. KOLODNER & K. K. TEWARI (Irvine, Cal.): Molecular Size and the Information Content of Chloroplast DNAs from Higher Plants 9

V. V. PINEVICH, S. B. IVANOVA & A. A. LIPSKAYA (Leningrad): Characteristics of Chloroplast DNA-Protein Complex 31

M. S. ODINTSOVA & N. P. YURINA (Moscow): Ribosomal Proteins and the Origin of Plastids 37

V. G. LADYGIN, G. A. SEMYONOVA & S. V. TAGEEVA (Pushchino-na-Oke): Electron-microscopic Study of Chloroplasts in Zygotes of *Chlamydomonas reinhardi* 43

Genetic Control of Biosynthesis in Chloroplasts

L. BOGORAD (Cambridge, Mass.): Genetic and Evolutionary Relationships between Plastids and the Nuclear-cytoplasmic System 51

J. A. SCHIFF (Waltham, Mass.): Interactions among Cellular Compartments in *Euglena* during Chloroplast Development 63

K. A. ALIEV, S. MUZAFAROVA, H. RADZHABOV, M. KHOLMATOVA, G. ULUGBEKOVA & Yu. S. NASYROV (Dushanbe): Chloroplast Messenger RNA from Etiolated Pea Seedlings 93

I. I. FILIPPOVICH, B. A. ALINA, I. N. BEZSMERTNAYA, A. M. TONGUR & A. I. OPARIN (Moscow): Relationship between the Protein-Synthesizing System and the Structure of Chloroplasts 105

R. HAGEMANN, F. HERRMANN & T BÖRNER (Halle/S.): The Use of Plastid and Gene Mutants of Higher Plants in Studying the Genetic Control of Plastid Functions 115

A. A. SHLYK, I. V. PRUDNIKOVA, G. E. SAVCHENKO, N. G. AVERINA, N. N. KOSTYUK, L. K. KAMYSHENKO, L. I. VLASENOK, V. I. GAPONENKO, E. F. BALEVA, T. K. PARAMONOVA, T. V. LOSITSKAYA & A. Yu. VEZITSKII (Minsk): The Relationship of Chlorophyll Biosynthesis to Protein and RNA Synthesis in Greening and Green Leaves 119

Yu. S. NASYROV, Yu. E. GILLER & P. D. USMANOV (Dushanbe): Genetic Control of Chlorophyll Biosynthesis and Formation of Its Forms in vivo 133

Genetic Control of Photosynthetic CO_2 Assimilation

M. GIBBS, E. LATZKO, L. J. LABER & G. HINES (Waltham, Mass.): Photosynthetic CO_2 Assimilation in Maize and Spinach Leaves and Chloroplasts 149

Z. ŠESTÁK, J. ČATSKÝ, J. SOLÁROVÁ, H. STRNADOVÁ & I. TICHÁ (Praha): Carbon Dioxide Transfer and Photochemical Activities as Factors of Photosynthesis during Ontogenesis of Primary Bean Leaves 159

N. P. VOSKRESENSKAYA (Moscow), N. M. POYARKOVA (Moscow), A. KHODZHIEV (Dushanbe) & I. S. DROZDOVA (Moscow): Regulatory Action of Blue Light on the Activity of Carboxylating Enzymes and Enzymes of Glycollate Pathway in Broad Bean and Maize Plants 167

Yu. S. KARPILOV, T. A. AVDEEVA & V. M. PERSANOV (Pushchino-na-Oke): Localization of Carbon Metabolism in Two Assimilation Tissues of Maize Leaf . 177

Genetic Nature of Photosynthetic Mutations and Their Phenogenesis

P. D. USMANOV, H. A. ABDULLAEV, O. V. USMANOVA & Sh. SOKHIBNAZAROV (Dushanbe): Mutation Variability of Chloroplasts in *Arabidopsis thaliana* (L.) HEYNH. 189

B. T. MUKHAMADIEV (Dushanbe) & K. V. KVITKO (Leningrad): The Relation of Pigment Composition in Algae Mutant Cells to Their Resistance to Inhibitors of Photophosphorylation 203

S. I. DEMCHENKO, V. A. AVETISOV & R. G. BUTENKO (Moscow): Nature of Chlorophyll Chimeras in the M_1 of *Arabidopsis* by means of Tissue Culture. I. Production of Regenerants from Chlorophyll-deficient Tissues. 209

A. V. STOLBOVA (Leningrad): Genetic Analysis of the Light-sensitive Mutants of *Chlamydomonas reinhardi* 217

K. V. KVITKO, V. V. TUGARINOV, PHAM THAN HO, A. S. CHUNAEV (Leningrad), E. E. TEMPER (Blagoveshchensk) & B. T. MUKHAMADIEV (Dushanbe): Mutation Analysis as a Method of Studying Genotype Structure of Green Algae . 225

Models for Studying the Mechanism of Photosynthesis

Á. FALUDI-DÁNIEL (Szeged): Pigment Synthesis and Photosynthetic Activity in Carotenoid Deficient Mutants of Maize 239

H. SAGROMSKY (Gatersleben): Mutants as Objects for Investigations on the Function of Chlorophylls 247

N. V. KARAPETYAN (Moscow), V. V. KLIMOV (Moscow), F. LÁNG (Budapest) & A. A. KRASNOVSKIĬ (Moscow): Fluorescence Induction of Normal and Mutant Maize Seedlings 255

M. M. YAKUBOVA (Dushanbe), A. B. RUBIN (Moscow), G. A KHRAMOVA (Dushanbe) & D. N. MATORIN (Moscow): Hill Reaction and Delayed Fluorescence in Mutants of *Gossypium hirsutum* 263

Yu. E. GILLER, L. R. VAKHIDOVA, L. M. ASOEVA, L. N. YUKHANANOVA, S. K. ABDULLAEVA, B. I. LIPKIND, G. V. KRASICHKOVA & G. A. YUSUPOVA (Dushanbe): Synthetic Pigment-protein-lipoid Complexes – Models of Molecular Organization and Functional Properties of the Pigment System of the Photosynthetic Apparatus 271

A. A. YASNIKOV, B. I. BERSHTEĬN, N. V. VOLKOVA, L. I. VASILENOK, O. I. VOLOVIK, N. A. ZAĬTSEVA, N. P. KANIVETS, L. S. MUSHKETIK, A. S. OKANENKO, L. K. OSTROVSKAYA, S. S. PETRENKO, A. I. POLISHCHUK, T. A. REĬNGARD & I. I. SEMENYUK (Kiev): Regulation by Pyruvate Kinase and Phosphatase of Inorganic Phosphate Incorporation during Photophosphorylation . . 287

V. L. KALER (Minsk): The Metabolic and Epigenetic Control of Chlorophyll Biosynthesis 295

Genetic Basis of Optimization of Plant Photosynthetic Activity

R. M. SMILLIE, D. G. BISHOP & J. CONROY (Sydney): The Implications of Genetic, Physiological and Environmental Control of Chloroplast Ultrastructure to the Optimization of Photosynthetic Activity 305

A. A. NICHIPOROVICH (Moscow): The Genetics of Photosynthesis and Rational Means of Breeding Highly Productive Plants 315

N. AVRATOVŠČUKOVÁ & S. FOUSOVÁ (Praha): Genetic Variation of Photosynthetic Rate in Leaf Discs of *Zea mays* L. 343

A. SHNEYOUR, R. M. SMILLIE & J. K. RAISON (Sydney): Biochemical and Genetical Basis for the Temperature Sensitivity of Photosynthesis and Growth in Chilling-sensitive Plants 349

M. DÉVAY & S. RAJKI (Martonvásár): Photosynthetic Adaptation in the Autumnization Process 357

I. A. TARCHEVSKIĬ, V. I. CHIKOV, A. Yu. SULEĬMANOVA, A. P. IVANOVA & Yu. E. ANDRIYANOVA (Kazan): Photosynthesis, Assimilation Numbers, and Photosynthate Translocation in Wheats of Different Stalk Length 363

Conclusions

Actual Problems of Genetics of Photosynthesis ... by Yu. S. NASYROV, K. V. KVITKO and Z. ŠESTÁK 369

Authors' Index 375
Subject Index 385
Plant Index 391

AUTHORS' ADDRESSES

ABDULLAEV, H. A., Institute of Plant Physiology and Biophysics, Academy of Sciences of the Tajik S.S.R., Dushanbe GSP, Ordzhonikidzeabadskoe shosse, 8-km, Akademgorodok, U.S.S.R.
ABDULLAEVA, S. K., Institute of Plant Physiology and Biophysics, Academy of Sciences of the Tajik S.S.R., Dushanbe GSP, Ordzhonikidzeabadskoe shosse, 8-km, Akademgorodok, U.S.S.R.
ALIEV, K. A., Institute of Plant Physiology and Biophysics, Academy of Sciences of the Tajik S.S.R., Dushanbe GSP, Ordzhonikidzeabadskoe shosse, 8-km, Akademgorodok, U.S.S.R.
ALINA, B. A., A. N. Bakh Institute of Biochemistry, Academy of Sciences of the U.S.S.R., Moscow V-71, Leninskiï prospekt 33, U.S.S.R.
ANDRIYANOVA, Yu. E., Kazan State University and Tatar Research Agricultural Institute, Kazan, U.S.S.R.
ASOEVA, L. M., Institute of Plant Physiology and Biophysics, Academy of Sciences of the Tajik S.S.R., Dushanbe GSP, Ordzhonikidzeabadskoe shosse, 8-km, Akademgorodok, U.S.S.R.
AVDEEVA, T. A., Institute of Photosynthesis, Academy of Sciences of the U.S.S.R., Pushchino-na-Oke, Moscow region, U.S.S.R.
AVERINA, N. G., Institute of Photobiology, Academy of Sciences of the B.S.S.R., 220733 Minsk-72, Akademicheskaya 27, U.S.S.R.
AVETISOV, V. A., Institute of Chemical Physics, Academy of Sciences of the U.S.S.R., Moscow V-344, Vorob'evskoe shosse 2-b, U.S.S.R.
AVRATOVŠČUKOVÁ, N., Department of Genetics, Faculty of Natural Sciences, Charles University, Viničná 5, 120 00 Praha 2, Czechoslovakia
BALEVA, E. F., Institute of Photobiology, Academy of Sciences of the B.S.S.R., 220733 Minsk-72, Akademicheskaya 27, U.S.S.R.
BERSHTEÏN, B. I., Institute of Organic Chemistry, Ukrainian Academy of Sciences, Kiev 252094, Murmanskaya 5, U.S.S.R.
BEZSMERTNAYA, I. N., A. N. Bakh Institute of Biochemistry, Academy of Sciences of the U.S.S.R., Moscow V-71, Leninskiï prospekt 33, U.S.S.R.
BISHOP, D. G., Plant Physiology Unit, C.S.I.R.O., Division of Food Research and School of Biological Sciences, Macquarie University, North Ryde 2113, Sydney, Australia
BOGORAD, L., Harvard University, The Biological Laboratories, Cambridge, Massachusetts 02138, U.S.A.
BÖRNER, T., Department of Genetics, Section of Biosciences, The University, DDR-402 Halle/S., G.D.R.
BUTENKO, R. G., K.A. Timiryazev Institute of Plant Physiology, Academy of Sciences of the U.S.S.R., Moscow, I-273, Botanicheskaya 35, U.S.S.R.
ČATSKÝ, J., Institute of Experimental Botany, Czechoslovak Academy of Sciences, 160 00 Praha 6, Flemingovo n.2, Czechoslovakia
CHIKOV, V. I., Kazan State University and Tatar Research Agricultural Institute, Kazan, U.S.S.R.
CHUNAEV, A. S., Leningrad State University, Department of Genetics and Selection, Leningrad V-164, Universitetskaya naberezhnaya 7/9, U.S.S.R.
CONROY, J., Plant Physiology Unit, C.S.I.R.O., Division of Food Research and School of Biological Sciences, Macquarie University, North Ryde 2113, Sydney, Australia
DEMCHENKO, S. I., Institute of Chemical Physics, Academy of Sciences of the U.S.S.R., Moscow V-344, Vorob'evskoe shosse 2-b, U.S.S.R.
DÉVAY, M., Agricultural Research Institute of the Hungarian Academy of Sciences, Martonvásár, Hungary

DROZDOVA, I. S., K.A. Timiryazev Institute of Plant Physiology, Academy of Sciences of the U.S.S.R., Moscow I-273, Botanicheskaya 35, U.S.S.R.
FALUDI-DÁNIEL, Á., Institute of Plant Physiology, Biological Research Center, Hungarian Academy of Sciences, Szeged, Hungary
FILIPPOVICH, I. I., A.N. Bakh Institute of Biochemistry, Academy of Sciences of the U.S.S.R., Moscow V-71, Leninskiï prospekt 33, U.S.S.R.
FOUSOVÁ, S., Institute of Experimental Botany, Czechoslovak Academy of Sciences, Flemingovo n.2, 160 00 Praha 6, Czechoslovakia
GAPONENKO, V. I., Institute of Photobiology, Academy of Sciences of the B.S.S.R., 220733 Minsk-72, Akademicheskaya 27, U.S.S.R.
GIBBS, M., Department of Biology, Brandeis University, Waltham, Massachusetts 02154, U.S.A.
GILLER, Yu. E., Institute of Plant Physiology and Biophysics, Academy of Sciences of the Tajik S.S.R., Dushanbe GSP, Ordzhonikidzeabadskoe shosse, 8-km, Akademgorodok, U.S.S.R.
HAGEMANN, R., Department of Genetics, Section of Biosciences, The University, DDR-402 Halle/S., G.D.R.
HERRMANN, F., Department of Genetics, Section of Biosciences, The University, DDR-402 Halle/S., G.D.R.
HINES, G., Department of Biology, Brandeis University, Waltham, Massachusetts 02154, U.S.A.
IVANOVA, A. P., Kazan State University and Tatar Research Agricultural Institute, Kazan, U.S.S.R.
IVANOVA, S. V., Biological Research Institute of Leningrad State University, 198904 Leningrad, Petergof, Oranienbaumskoe shosse 2, U.S.S.R.
KALER, V. L., Institute of Experimental Botany, Academy of Sciences of the B.S.S.R., 220733, Minsk-72, Akademicheskaya 27, U.S.S.R.
KAMYSHENKO, L. K., Institute of Photobiology, Academy of Sciences of the B.S.S.R., 220733, Minsk-72, Akademicheskaya 27, U.S.S.R.
KANIVETS, N. P., Institute of Organic Chemistry, Ukrainian Academy of Sciences, Kiev 105, 252094, Murmanskaya 5, U.S.S.R.
KARAPETYAN, N. V., A.N. Bakh Institute of Biochemistry, Academy of Sciences of the U.S.S.R., Moscow V-71, Leninskiï prospekt 33, U.S.S.R.
KARPILOV, Yu. S., Institute of Photosynthesis, Academy of Sciences of the U.S.S.R., Pushchino-na-Oke, Moscow region, U.S.S.R.
KHODZHIEV, A., K.A. Timiryazev Institute of Plant Physiology, Academy of Sciences of the U.S.S.R., Moscow I-273, Botanicheskaya 35, U.S.S.R.
KHOLMATOVA, M., Institute of Plant Physiology and Biophysics, Academy of Sciences of the Tajik S.S.R., Dushanbe GSP, Ordzhonikidzeabadskoe shosse, 8-km, Akademgorodok, U.S.S.R.
KHRAMOVA, G. A., V.I. Lenin Tajik State University, Department of Biology, Dushanbe, Ostrovskogo 2, U.S.S.R.
KLIMOV, V. V., A.N. Bakh Institute of Biochemistry, Academy of Sciences of the U.S.S.R., Moscow V-71, Leninskiï prospekt 33, U.S.S.R.
KOLODNER, R., Department of Molecular Biology and Biochemistry, University of California, Irvine, Cal. 92664, U.S.A.
KOSTYUK, N. N., Institute of Photobiology, Academy of Sciences of the B.S.S.R., 220733, Minsk-72, Akademicheskaya 27, U.S.S.R.
KRASICHKOVA, G. V., Institute of Plant Physiology and Biophysics, Academy of Sciences of the Tajik S.S.R., Dushanbe GSP, Ordzhonikidzeabadskoe shosse, 8-km, Akademgorodok, U.S.S.R.
KRASNOVSKIÏ, A. A., A.N. Bakh Institute of Biochemistry, Academy of Sciences of the U.S.S.R., Moscow V-71, Leninskiï prospekt 33, U.S.S.R.
KVITKO, K. V., Leningrad State University, Department of Genetics and Selection, Lenin-

grad V-164, Universitetskaya naberezhnaya 7/9, U.S.S.R.
LABER, L. J., Department of Biology, Brandeis University, Waltham, Massachusetts 02154, U.S.A.
LADYGIN, V. G., Institute of Biological Physics, Academy of Sciences of the U.S.S.R., Pushchino-na-Oke, Moscow region, U.S.S.R.
LÁNG, F., Department of Plant Physiology, Eötvös Loránd University, Budapest, VIII, Múzeum körút 4/a, Hungary.
LATZKO, E., Department of Biology, Brandeis University, Waltham, Massachusetts 02154, U.S.A.
LIPKIND, B. I., Institute of Plant Physiology and Biophysics, Academy of Sciences of the Tajik S.S.R., Dushanbe GSP, Ordzhonikidzeabadskoe shosse, 8-km, Akademgorodok, U.S.S.R.
LIPSKAYA, A. A., Biological Research Institute of Leningrad State University, 198904 Leningrad, Petergof, Oranienbaumskoe shosse 2, U.S.S.R.
LOSITSKAYA, T. V., Institute of Photobiology, Academy of Sciences of the B.S.S.R., 220733, Minsk-72, Akademicheskaya 27, U.S.S.R.
MATORIN, D. N., M.V. Lomonosov State University, Moscow, U.S.S.R.
MUKHAMADIEV, B. T., Institute of Plant Physiology and Biophysics, Academy of Sciences of the Tajik S.S.R., Dushanbe GSP, Ordzhonikidzeabadskoe shosse, 8-km, Akademgorodok, U.S.S.R.
MUSHKETIK, L. S., Institute of Organic Chemistry, Ukrainian Academy of Sciences, Kiev 252094, Murmanskaya 5, U.S.S.R.
MUZAFAROVA, S., Institute of Plant Physiology and Biophysics, Academy of Sciences of the Tajik S.S.R., Dushanbe GSP, Ordzhonikidzeabadskoe shosse, 8-km, Akademgorodok, U.S.S.R.
NASYROV, Yu. S., Institute of Plant Physiology and Biophysics, Academy of Sciences of the Tajik S.S.R., Dushanbe GSP, Ordzhonikidzeabadskoe shosse, 8-km, Akademgorodok, U.S.S.R.
NICHIPOROVICH, A. A., K.A. Timiryazev Institute of Plant Physiology, Academy of Sciences of the U.S.S.R., Moscow I-273, Botanicheskaya 35, U.S.S.R.
ODINTSOVA, M. S., A.N. Bakh Institute of Biochemistry, Academy of Sciences of the U.S.S.R., Moscow V-71, Leninskiï prospekt 33, U.S.S.R.
OKANENKO, A. S., Institute of Organic Chemistry, Ukrainian Academy of Sciences, Kiev 252094, Murmanskaya 5, U.S.S.R.
OPARIN, A. I., A.N. Bakh Institute of Biochemistry, Academy of Sciences of the U.S.S.R., Moscow V-71, Leninskiï prospekt 33, U.S.S.R.
OSTROVSKAYA, L. K., Institute of Organic Chemistry, Ukrainian Academy of Sciences, Kiev 252094, Murmanskaya 5, U.S.S.R.
PARAMONOVA, T. K., Institute of Photobiology, Academy of Sciences of the B.S.S.R., 220733, Minsk-72, Akademicheskaya 27, U.S.S.R.
PERSANOV, V. M., Institute of Photosynthesis, Academy of Sciences of the U.S.S.R., Pushchino-na-Oke, Moscow region, U.S.S.R.
PETRENKO, S. S., Institute of Organic Chemistry, Ukrainian Academy of Sciences, Kiev 252094, Murmanskaya 5, U.S.S.R.
PHAM THAN HO, Leningrad State University, Department of Genetics and Selection, Leningrad V-164, Universitetskaya naberezhnaya 7/9, U.S.S.R.
PINEVICH, V. V., Biological Research Institute of Leningrad State University, 198904 Leningrad, Petergof, Oranienbaumskoe shosse 2, U.S.S.R.
POLISHCHUK, A. I., Institute of Organic Chemistry, Ukrainian Academy of Sciences, Kiev 252094, Murmanskaya 5, U.S.S.R.
POYARKOVA, N. M., K.A. Timiryazev Institute of Plant Physiology, Academy of Sciences of the U.S.S.R., Moscow I-273, Botanicheskaya 35, U.S.S.R.
PRUDNIKOVA, I. V., Institute of Photobiology, Academy of Sciences of the B.S.S.R., 220733, Minsk-72, Akademicheskaya 27, U.S.S.R.

RADZHABOV, H., Institute of Plant Physiology and Biophysics, Academy of Sciences of the Tajik S.S.R., Dushanbe GSP, Ordzhonikidzeabadskoe shosse, 8-km, Akademgorodok, U.S.S.R.
RAISON, J. K., Plant Physiology Unit, C.S.I.R.O., Division of Food Research and School of Biological Sciences, Macquarie University, North Ryde 2113, Sydney, Australia
RAJKI, S., Agricultural Research Institute, Hungarian Academy of Sciences, Martonvásár, Hungary
REÏNGARD, T. A., Institute of Organic Chemistry, Ukrainian Academy of Sciences, Kiev 252094, Murmanskaya 5, U.S.S.R.
RUBIN, A. B., M.V. Lomonosov State University, Moscow, U.S.S.R.
SAGROMSKY, H., Zentralinstitut für Genetik und Kulturpflanzenforschung der Akademie der Wissenschaften der DDR, 4325 Gatersleben, DDR
SAVCHENKO, G. E., Institute of Photobiology, Academy of Sciences of the B.S.S.R., 220733, Minsk-72, Akademicheskaya 27, U.S.S.R.
SCHIFF, J. A., Department of Biology, Brandeis University, Waltham, Massachusetts 02154, U.S.A.
SEMENYUK, I. I., Institute of Organic Chemistry, Ukrainian Academy of Sciences, Kiev 252094, Murmanskaya 5, U.S.S.R.
SEMYONOVA, G. A., Institute of Biological Physics, Academy of Sciences of the U.S.S.R., Pushchino-na-Oke, Moscow region, U.S.S.R.
ŠESTÁK, Z., Institute of Experimental Botany, Czechoslovak Academy of Sciences, 160 00 Praha 6, Flemingovo n.2, Czechoslovakia
SHLYK, A. A., Institute of Photobiology, Academy of Sciences of the B.S.S.R., 220733, Minsk-72, Akademicheskaya 27, U.S.S.R.
SHNEYOUR, A., Plant Physiology Unit, C.S.I.R.O., Division of Food Research and School of Biological Sciences, Macquarie University, North Ryde 2113, Sydney, Australia
SMILLIE, R. M., Plant Physiology Unit, C.S.I.R.O., Division of Food Research and School of Biological Sciences, Macquarie University, North Ryde 2113, Sydney, Australia
SOKHIBNAZAROV, Sh., Institute of Plant Physiology and Biophysics, Academy of Sciences of the Tajik S.S.R., Dushanbe GSP, Ordzhonikidzeabadskoe shosse, 8-km, Akademgorodok, U.S.S.R.
SOLÁROVÁ, J., Institute of Experimental Botany, Czechoslovak Academy of Sciences, 160 00 Praha 6, Flemingovo n.2, Czechoslovakia.
STOLBOVA, A. V., The Leningrad State University, Department of Genetics and Selection, Leningrad, V-164, Universitetskaya naberezhnaya 7/9, U.S.S.R.
STRNADOVÁ, H., Institute of Experimental Botany, Czechoslovak Academy of Sciences, 160 00 Praha 6, Flemingovo n.2, Czechoslovakia
SULEÏMANOVA, A. Yu., Kazan State University and Tatar Research Agricultural Institute, Kazan, U.S.S.R.
TAGEEVA, S. V., Institute of Biological Physics, Academy of Sciences of the U.S.S.R., Pushchino-na-Oke, Moscow region, U.S.S.R.
TARCHEVSKIÏ, I. A., Kazan State University and Tatar Research Agricultural Institute, Kazan, U.S.S.R.
TEMPER, E. E., Department of Microbiology, Medical Institute, Blagoveshchensk, U.S.S.R.
TEWARI, K. K., Department of Molecular Biology and Biochemistry, University of California, Irvine, Cal. 92664, U.S.A.
THOMAS, J. R., Department of Molecular Biology and Biochemistry, University of California, Irvine, Cal. 92664, U.S.A.
TICHÁ, I., Institute of Experimental Botany, Czechoslovak Academy of Sciences, 160 00 Praha 6, Flemingovo n.2, Czechoslovakia
TONGUR, A. M., A.N. Bakh Institute of Biochemistry, Academy of Sciences of the U.S.S.R., Moscow V-71, Leninskiï prospekt 33, U.S.S.R.
TUGARINOV, V. V., Leningrad State University, Department of Genetics and Selection, Leningrad V-164, Universitetskaya naberezhnaya 7/9, U.S.S.R.

TURISHCHEVA, M. S., A.N. Bakh Institute of Biochemistry, Academy of Sciences of the U.S.S.R., Moscow V-71, Leninskiï prospekt 33, U.S.S.R.
ULUGBEKOVA, G., Institute of Plant Physiology and Biophysics, Academy of Sciences of the Tajik S.S.R., Dushanbe GSP, Ordzhonikidzeabadskoe shosse, 8-km, Akademgorodok, U.S.S.R.
USMANOV, P. D., Institute of Plant Physiology and Biophysics, Academy of Sciences of the Tajik S.S.R., Dushanbe GSP, Ordzhonikidzeabadskoe shosse, 8-km, Akademgorodok, U.S.S.R.
USMANOVA, O. V., Institute of Plant Physiology and Biophysics, Academy of Sciences of the Tajik S.S.R., Dushanbe GSP, Ordzhonikidzeabadskoe shosse, 8-km, Akademgorodok, U.S.S.R.
VAKHIDOVA, L. R., Institute of Plant Physiology and Biophysics, Academy of Sciences of the Tajik S.S.R., Dushanbe GSP, Ordzhonikidzeabadskoe shosse, 8-km, Akademgorodok, U.S.S.R.
VASILENOK, L. I., Institute of Organic Chemistry, Ukrainian Academy of Sciences, Kiev 252094, Murmanskaya 5, U.S.S.R.
VEZITSKII, A. Yu., Institute of Photobiology, Academy of Sciences of the B.S.S.R., 220733, Minsk-72, Akademicheskaya 27, U.S.S.R.
VLASENOK, L. I., Insitute of Photobiology, Academy of Sciences of the B.S.S.R., 220733, Minsk-72, Akademicheskaya 27, U.S.S.R.
VOLKOVA, N. V., Institute of Organic Chemistry, Ukrainian Academy of Sciences, Kiev 252094, Murmanskaya 5, U.S.S.R.
VOLOVIK, O. I., Institute of Organic Chemistry, Ukrainian Academy of Sciences, Kiev 252094, Murmanskaya 5, U.S.S.R.
VOSKRESENSKAYA, N. P., Institute of Plant Physiology, Academy of Sciences of the U.S.S.R., Moscow I-273, Botanicheskaya 35, U.S.S.R.
YAKUBOVA, M. M., V.I. Lenin Tajik State University, Department of Biology, Dushanbe, Ostrovskogo 2, U.S.S.R.
YASNIKOV, A. A., Institute of Organic Chemistry, Ukrainian Academy of Sciences, Kiev 252094, Murmanskaya 5, U.S.S.R.
YUKHANANOVA, L. N., Institute of Plant Physiology and Biophysics, Academy of Sciences of the Tajik S.S.R., Dushanbe GSP, Ordzhonikidzeabadskoe shosse, 8-km, Akademgorodok, U.S.S.R.
YURINA, N. P., A.N. Bakh Institute of Biochemistry, Academy of Sciences of the U.S.S.R., Moscow V-71, Leninskiï prospekt 33, U.S.S.R.
YUSUPOVA, G. A., Institute of Plant Physiology and Biophysics, Academy of Sciences of the Tajik S.S.R., Dushanbe GSP, Ordzhonikidzeabadskoe shosse, 8-km, Akademgorodok, U.S.S.R.
ZAITSEVA, N. A., Institute of Organic Chemistry, Ukrainian Academy of Sciences, Kiev 252094, Murmanskaya 5, U.S.S.R.

FOREWORD

Knowledge of the genetic nature of photosynthesis implies the knowledge of the basic mechanisms which govern the formation of chloroplast structure and of the systems determining its functional activity.

A highly complicated structural, biochemical, biophysical and physiological organization of the chloroplast and a large number of mutually dependent, specifically interacting systems in plant cells make photosynthesis a complex process for genetic analysis. This process is even more complex due to a great variety of forms of photosynthetic activity in plants which, in turn, depend on different paths of biochemical, physiological and ecological evolution of representatives of the flora.

The genetic apparatus is also responsible for the complicated systems strongly relating the photosynthetic function to the vital systems of the plant as a whole through forward and feedback mechanisms.

The role of the photosynthetic apparatus is equally important with respect to its capability of adaptation to reactions which makes its functioning well-adjusted both to environmental conditions and to the internal states of the plants. No other process establishes such specific, complex and important ties between a plant and a set of most essential environmental factors (light, heat, CO_2, O_2, H_2O, mineral elements) as photosynthesis.

These complex functions and relations make photosynthesis dependent on the two types of genetic apparatus: a nuclear genetic apparatus and an internal semi-independent apparatus of chloroplasts. Possibly, this dual dependence (whatever its origin, symbiotic or otherwise) is what is required for such an important and complex function; it is possible that the nuclear genetic apparatus relates the photosynthetic activity with the properties and vital functions of a given plant species as a whole, while the internal apparatus of chloroplasts promotes photosynthesis in interactions of plants with such intricate and essential factors as sources of radiant energy and CO_2.

The complex and so far little-studied genetic nature of photosynthesis makes genetic studies both difficult and tempting and, perhaps, the most important field of photosynthesis research. Indeed, in order to understand the genetics of photosynthesis, we must have a detailed knowledge of features of the components, its structure and functional activity. Only if such knowledge is available, can we proceed to study the 'gene-character' pairs and systems connecting the components of these pairs and their complexes by actual bonds.

The ultimate task of studying the genetic nature of photosynthesis is, first, to control photosynthesis at a phenotype level, though within a genetically determined potential range, and second, to improve the photosynthetic apparatus at a genetic level.

The very statement and essence of the problem assert that the genetics of photosynthesis is a complex problem, the solution of which calls for joint efforts on the part of not only genetists, but also photochemists, biochemists, physiologists and those engaged in all fields of photosynthesis research.

With respect to the approach to the genetic aspects of photosynthesis, the method which has proved its value and requires further extensive development is the use of mutants as model objects. But equally important must be the trend of studying the genetic nature of the diversity and evolution of photosynthetic functions in the plant kingdom.

All these problems have been treated at the Symposium 'Genetic Aspects of Photosynthesis' in Dushanbe:

Genome and Origin of Chloroplasts;
Genetic Control of Biosynthesis in Chloroplasts;
Genetic Control of Photosynthetic CO_2 Assimilation;
Genetic Nature of Photosynthetic Mutations and their Phenogenesis;
Models for Studying the Mechanism of Photosynthesis;
Genetic Basis of Optimization of Plant Photosynthetic Activity.

We hope that the publication of selected papers from this Symposium will give a new impetus to further systematic and extensive investigation in this important field.

A. A. NICHIPOROVICH

In the auditorium. First row (from left): M. GIBBS, D. VON WETTSTEIN, R.M. SMILLIE, J.A. SCHIFF. Second row: Mr. (X), L. BOGORAD, Z. ŠESTÁK. Third row: O.V. ZALENSKII, N.S. MAMUSHINA, M.V. CHULANOVSKAYA. (Photo YU.I. PINKHASOV.)

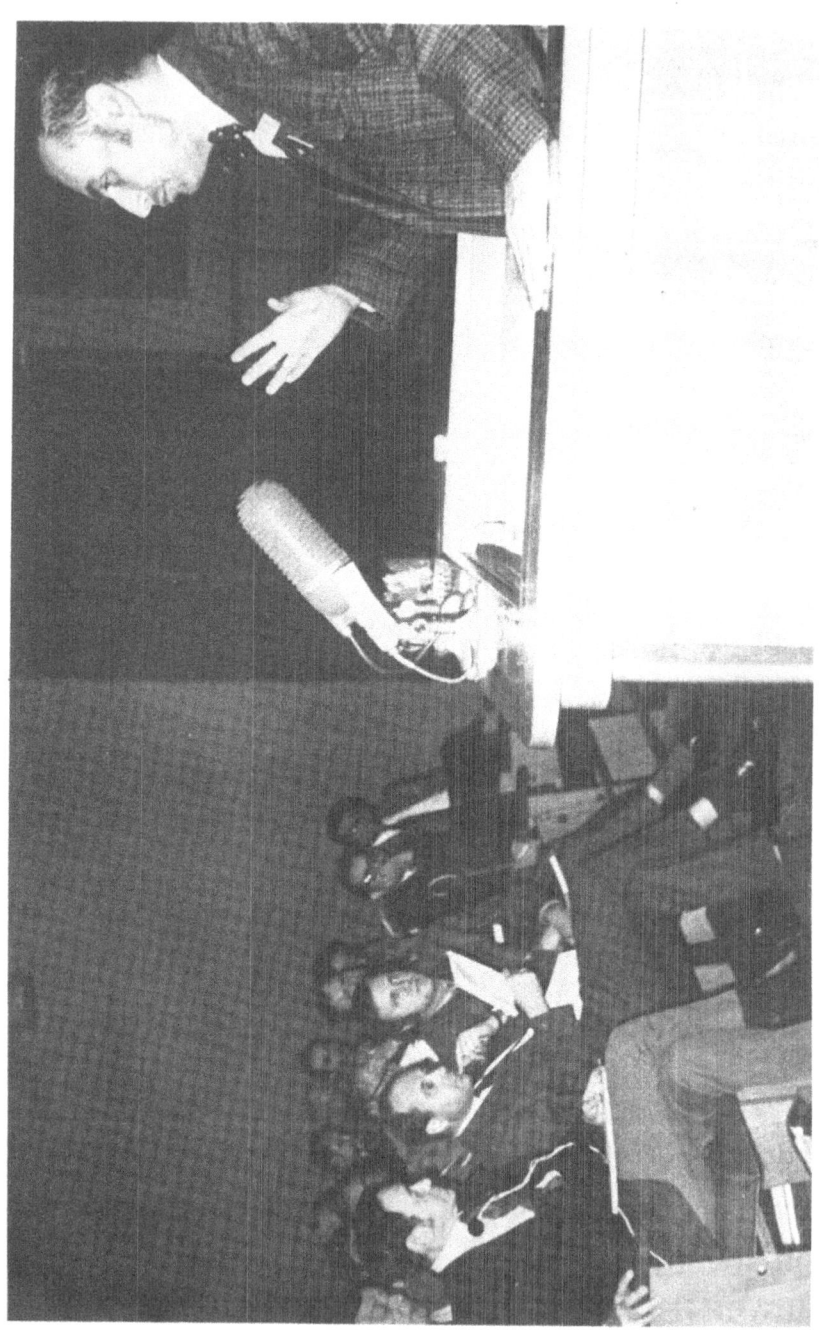

Lecture by J.A. SCHIFF. (Photo YU.I. PINKHASOV.)

Scientific discussion (from left) A.A. NICHIPOROVICH, K. TEWARI, YU.S. NASYROV and R.M. SMILLIE. (Photo YU.I. PINKHASOV.)

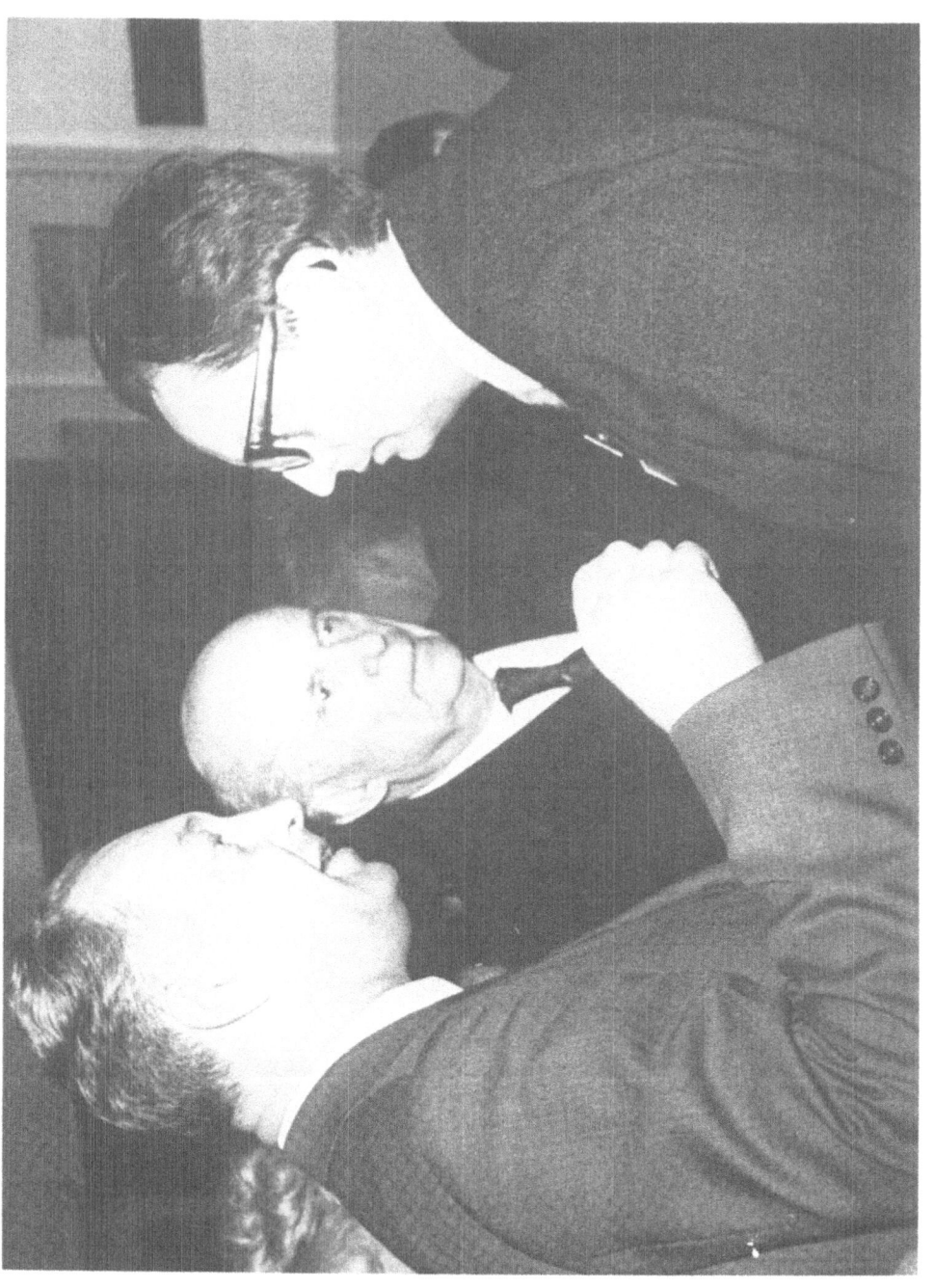

Scientific discussion (from left) A.A. SHLYK, A.A. NICHIPOROVICH and Z. ŠESTÁK. (Photo YU.I. PINKHASOV.)

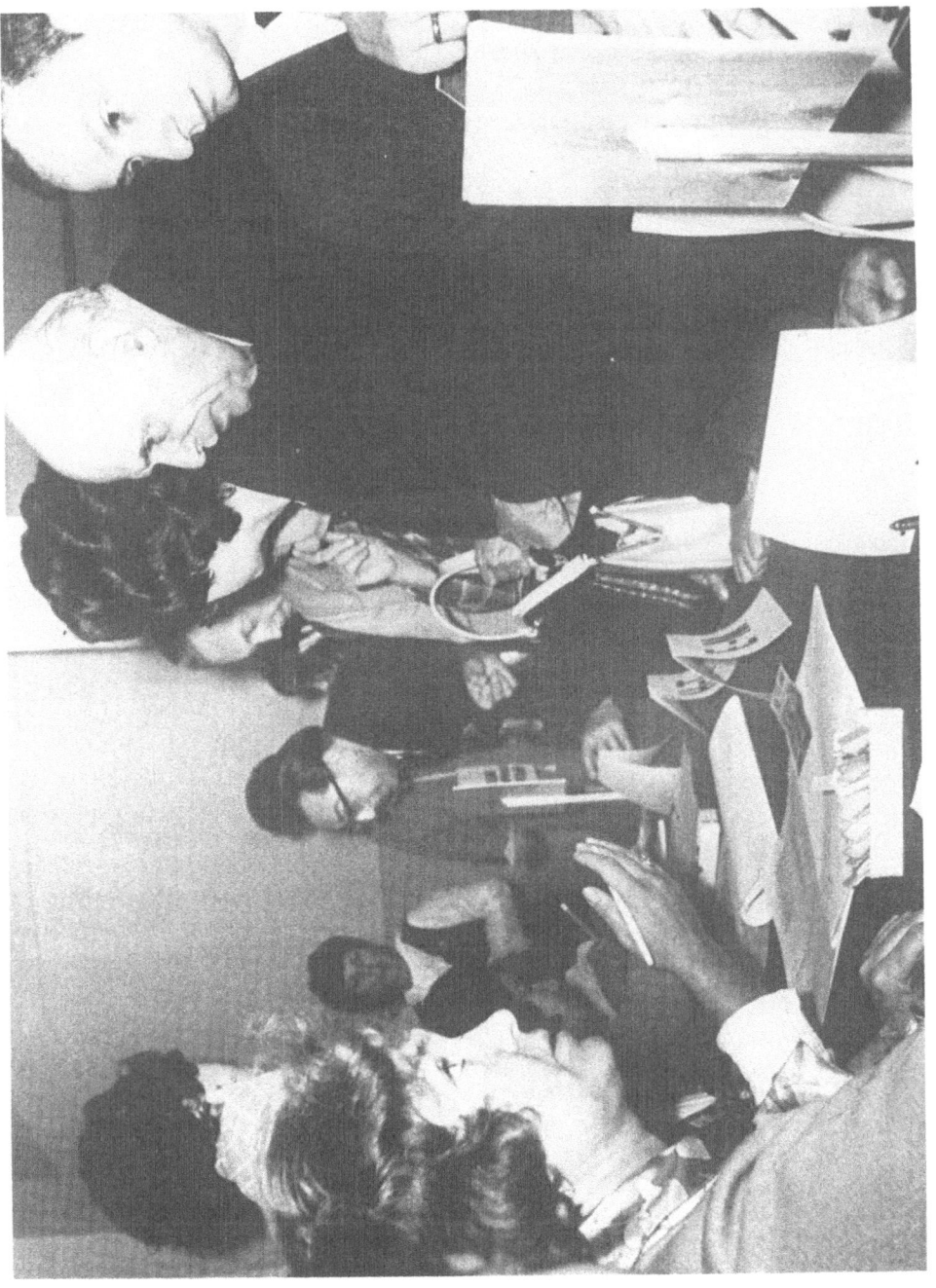
At the registration desk. (Photo YU.I. PINKHASOV.)

Excursion to the Varzob mountain botanical station Kondara. From left: A.A. SHLYK, L. BOGORAD, Mrs. BOGORAD, J.A. SCHIFF, YU.S. NASYROV, M. GIBBS and S.V. TAGEEVA. (Photo YU.I. PINKHASOV.)

GENOME AND ORIGIN OF CHLOROPLASTS

CHLOROPLAST DNA-MEMBRANE COMPLEX IN PISUM SATIVUM

MARINA S. TURISHCHIVA

A. N. Bakh Institute of Biochemistry, Academy of Sciences of the U.S.S.R., Moscow, U.S.S.R.

During recent years it has become well established that chloroplasts contain DNA which differs in a number of respects from nuclear DNA. The presence of DNA in plastids of many plant species has been demonstrated by cytochemical, biochemical, radioautographic and electron microscopic methods.

RIS & PLAUT (1962) were the first to detect electron-transparent areas in the chloroplast matrix; they found in *Chlamydomonas* chloroplasts fibrils 2.5-3.0 nm thick which could be removed by DNase treatment. Later, DNA-containing electron-transparent areas were found in chloroplasts of many plant species as well as in mitochondria of animal and plant origin. The electron-transparent areas within the chloroplast matrix of any one species differed in number, form and size and were usually situated at random (ODINTSOVA *et al.* 1970).

Apparently, in chloroplasts and mitochondria the DNA molecules are associated with the membrane system in the same way as in bacterial cells. After differential centrifugation of chloroplasts exposed to osmotic shock, practically all the DNA was in the lamellar fraction and could not be removed by washing. Different attempts were made to separate DNA from chloroplast membranes: incubation of chloroplasts in hypotonic medium, freezing and thawing, fragmentation in an homogenizer, treatment by enzymes. The only successful method was mechanical degradation and lysis of membranes with detergents (KUNG & WILLIAMS 1969, TEWARI & WILDMAN 1969).

The association of DNA with membranes of intracellular organelles can be seen in sections of chloroplasts and mitochondria by means of electron microscopy. This association is easily detected following complete digestion of the chloroplast matrix with proteases (trypsin or pronase which separate membranes and DNA fibrils on sections of cell organelles) or spreading of osmotically shocked organelles in a protein monolayer.

We have studied the association of DNA with the membrane system of chloroplasts of *Pisum sativum* L. isolated by means of differential centrifugation and purified on a sucrose density gradient.

Electron microscopic investigations of isolated chloroplasts exposed to osmotic shock revealed that approximately 30% of the chloroplast fragments were in association with the fine fibrils or aggregates of fibrils which could be completely removed by DNase treatment. The DNA filaments released from osmotically disrupted pea chloroplasts are essentially identical in diameter (Fig. 1).

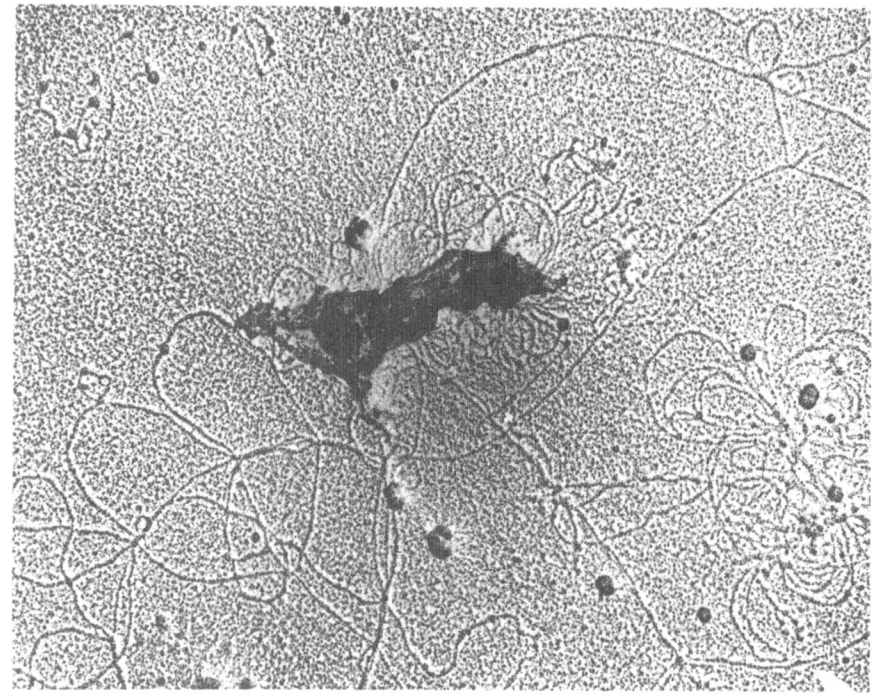

Fig. 1. Electron micrograph of pea chloroplast DNA released by osmotic shock. Rotary shadowing with Pt-Pd. The length of DNA filaments is *ca.* 100 μm. Magnification × 66 000.

They scatter from all sides of the chloroplast fragment and often form loops. The entire length of DNA molecules is 100 μm, corresponding to a molecular weight of 2×10^8 dalton. Sometimes even larger 'displays' of DNA connected with a chloroplast fragment can be observed (Fig. 2); unfortunately it is impossible to say how many DNA molecules are present in these cases. According to present literature data the maximum length of DNA fibrils associated with a membrane fragment in spinach chloroplast preparations is 150 μm (WOODCOCK & FERNÁNDEZ-MORÁN 1968) and in chloroplasts of *Acetabularia* 419-450 μm (GREEN & BURTON 1970, WOODCOCK & BOGORAD 1970). Thus, from electron microscopic data it follows that the amount of DNA per plastid is of the same range as that of a bacterial cell and much greater than that of mitochondria of animal cells.

Similar electron microscope pictures were obtained for DNA of proplastids and etioplasts lysed by osmotic shock (Fig. 3). These DNA threads radiate from membrane or organelle fragments; the threads often form multiple loops and clusters. It is difficult to find a single site where they attach to the fragments of these

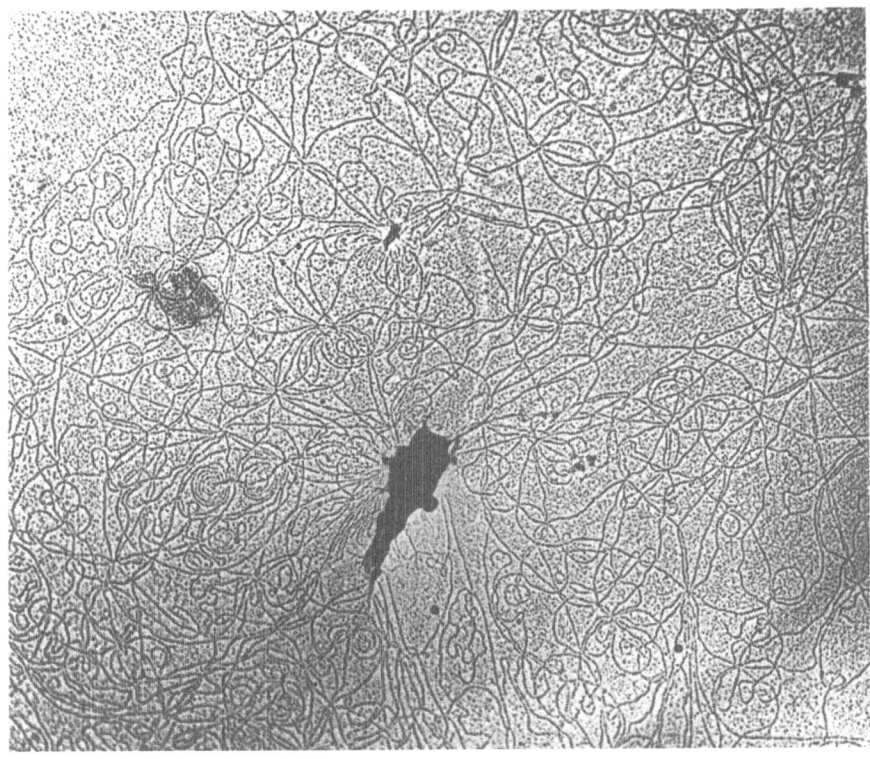

Fig. 2. Large 'display' of pea chloroplast DNA released by osmotic shock. Rotary shadowing with Pt-Pd. Magnification × 30 000.

organelles. In most cases it is impossible to measure the entire length of the DNA fibrils because they are interlaced and sometimes pass over or under the membranes.

The association of DNA fibrils and the membrane system of plastids can be demonstrated in a more direct way with a membrane fraction isolated from pea chloroplasts. The isolated chloroplasts, purified in a sucrose density gradient, were exposed to osmotic shock by suspending in 0.02 M Tris-HCl buffer (pH 7.6) containing 0.001 M EDTA. The suspension was centrifuged at 12 000 × g for 15 min. The sediment of membrane was suspended in initial buffer, frozen and thawed many times and then purified with a *Ficoll* gradient (5 to 20%; 0.02 M Tris-HCl buffer, pH 7.6; 0.25 M sucrose; 0.001 M EDTA) or in a linear (5 to 30%) sucrose density gradient. Strands associated with the membrane fragment and removed by DNase are seen in Fig. 4. The entire length of DNA fibrils connected with the membrane fragment is *ca.* 10 μm (average of 25 measurements), i.e. much less

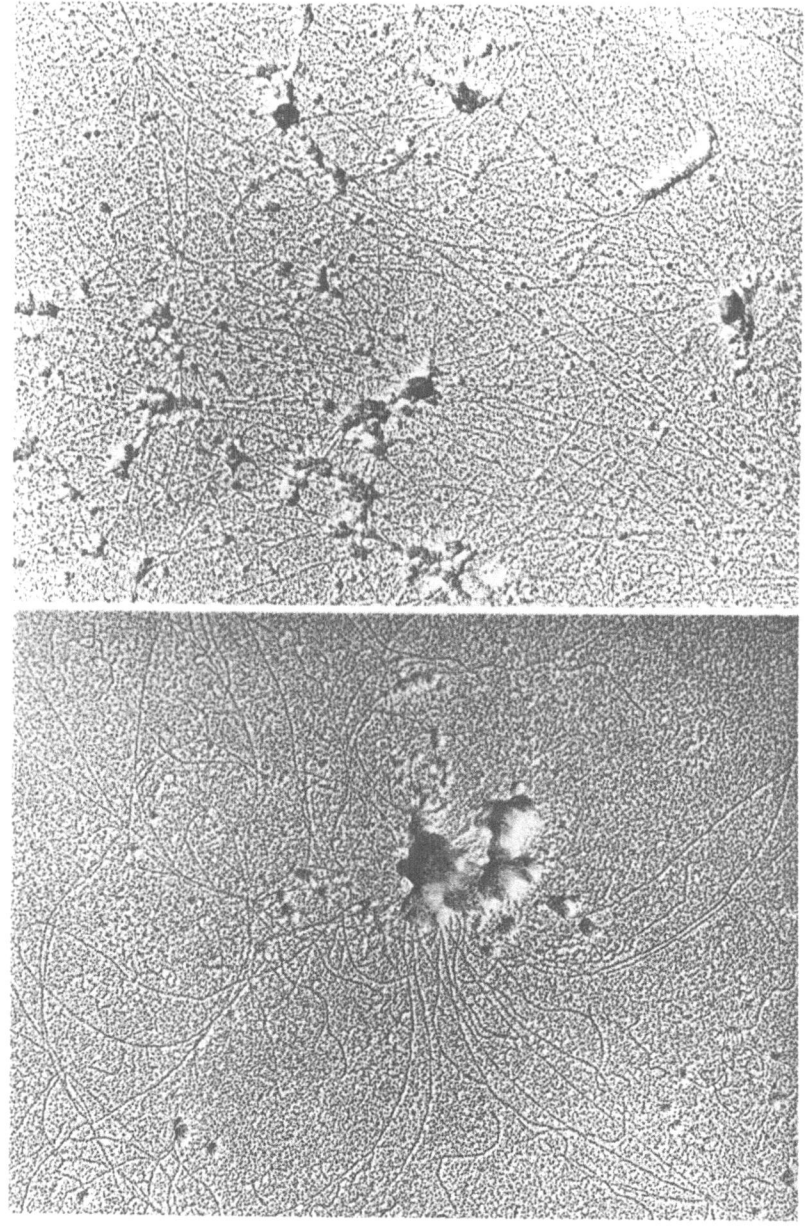

Fig. 3. Electron micrograph of pea etioplast DNA (top) and corn proplastid DNA (bottom) released by osmotic shock. Rotary shadowing with Pt-Pd. Magnification x 45 000.

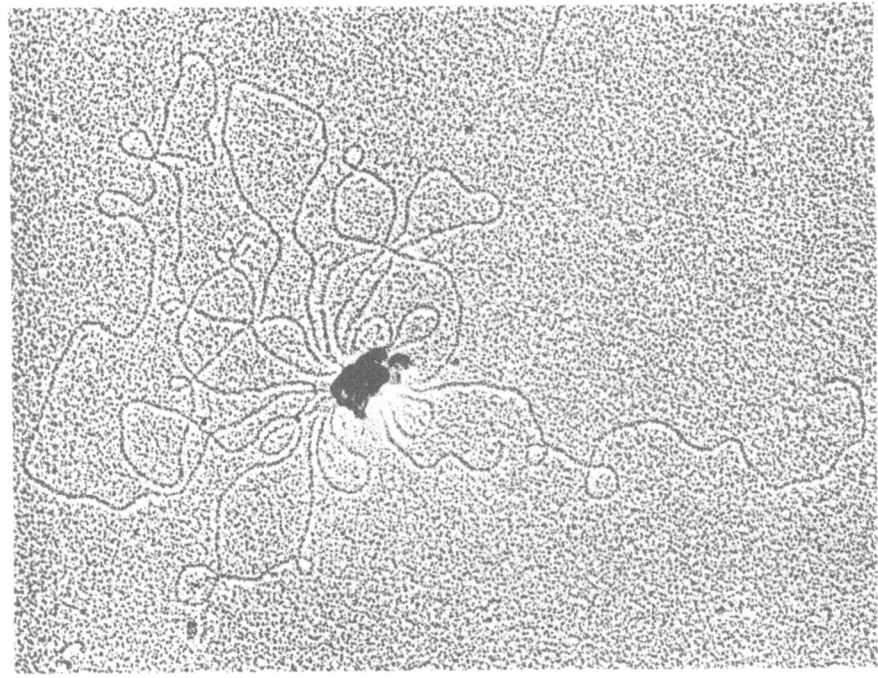

Fig. 4. DNA-membrane complex isolated from pea chloroplasts. The total length of the DNA molecule is 17.0 μm. Rotary shadowing with Pt-Pd. Magnification × 100 000.

than was found in osmotically disrupted membranes. Hence the degradation of DNA threads by nucleases and hydrodynamic forces probably takes place during the isolation of the membrane fraction.

Some electron micrographs of the membrane fraction of pea chloroplasts show the original regular packing of DNA threads into complex structures (Fig. 5). Careful analysis of DNA strands associated with the membrane fragments on preparations rotary-shadowed with platinum-palladium or stained with uranyl acetate has shown that single loops are the most frequently found DNA configurations (70-80% of all the attachments recognized). We often observed a great number of single loops attached in a row to the edge of the membrane segment. Sometimes two separate DNA loops (or a linear fragment and a single loop) were attached to a common point or to two proximal points within the membrane (Fig. 6,7). Such double loop DNA configurations are regarded as replicated DNA forms. We attempted to measure the length of the DNA loops with actual points of attachment to membrane fragments. The average length of 318 individual DNA loops ranged mostly from 0.1 to 5.0 μm and did not exceed 1.0 μm in the majority

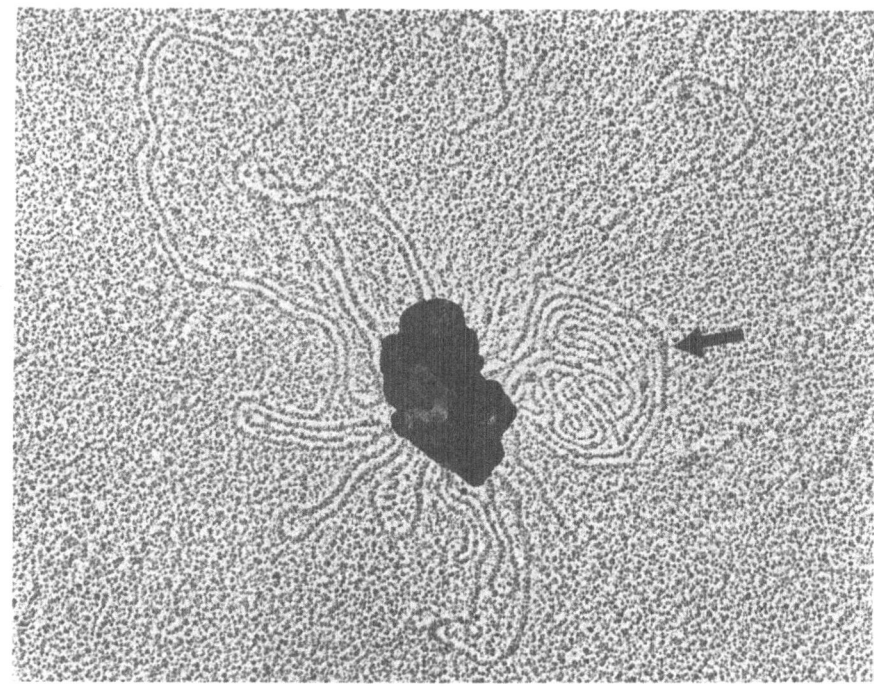

Fig. 5. Regular packing of DNA strands (arrow) isolated from pea chloroplasts. Rotary shadowing with Pt-Pd. Magnification x 45 000.

of DNA loops. BISALPUTRA & BURTON (1970) observed similar electron microscope pictures of DNA-membrane complexes in chloroplasts of lower plants but according to their data the average length of the loop replicons was 13.8 μm.

Up to the present, very little is known of the chemical origin of the DNA-membrane associations. According to our preliminary data treatment of the DNA-membrane complexes with pronase, 8 M urea or 5 M NaCl leads to separation of the DNA. RNase and phospholipases A and D do not affect the DNA-membrane association. The DNA is probably attached to the membrane via a protein which causes the specificity of this association.

Morphological studies of ultra-thin sections together with genetic and biochemical experiments have demonstrated an association between DNA and cell membranes in bacterial cells. As a consequence it has been generally assumed that membranes may play an important role in the replication and segregation of bacterial chromosomes. A similar DNA-membrane association has also come to light in autonomous cell organelles such as mitochondria and chloroplasts. As with the bacterial cell membrane, the possible role of the plastid membrane is participation in the replication and segregation of chloroplast DNA which occurs during the division

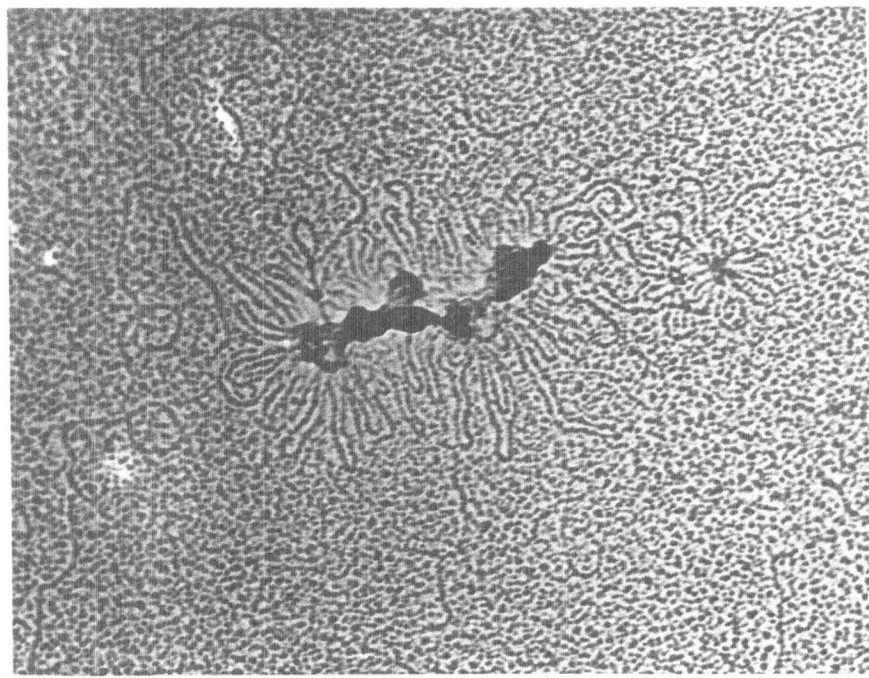

Fig. 6. Multitude of single DNA loops attached to the edge of a membrane fragment of pea chloroplast. Rotary shadowing with Pt-Pd. Magnification x 90 000.

of the organelle. The attachment to the membrane may be the physical basis for certain functions of extranuclear DNA. Such association may also be essential for the function and growth of chloroplast membranes.

Summary

Electron microscopic investigations of the membrane fraction of chloroplasts obtained from young pea seedlings by means of differential centrifugation, osmotic shock and density gradient purification, revealed numerous DNA strands connected with membrane fragments. The DNA strands, essentially identical in diameter, are scattered from all sides of membrane fragments. Rather often, regularly arranged DNA strands can be seen, probably reflecting the peculiarities of DNA packing in chloroplasts. Sometimes the attachment of DNA as a double loop, a single loop or linear fragment arising from a common point occur. The double loop configurations are regarded as replicated forms of DNA.

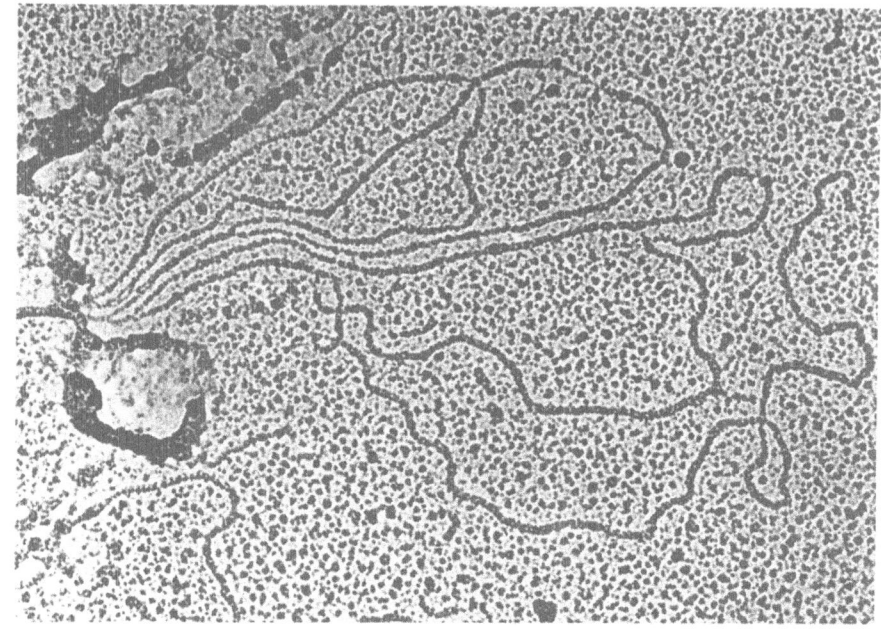

Fig. 7. Different configurations of DNA strands isolated from pea chloroplasts: a double loop, a linear molecule and a single loop. Rotary shadowing with Pt-Pd. Magnification × 66 000.

REFERENCES

BISALPUTRA, T. & BURTON, H., *J. Microscopie* 9: 661, 1970.
GREEN, B. R. & BURTON, H., *Science* 168: 981, 1970.
KUNG, S. D. & WILLIAMS, J. P., *Biochim. biophys. Acta* 195: 434, 1969.
ODINTSOVA, M. S., MIKUL´SKA, E. & TURISHCHEVA, M. S., *Mol. Biol. (Moskva)* 4: 889, 1970.
RIS, H. & PLAUT, W., *J. Cell Biol.* 13: 383, 1962.
TEWARI, K. K. & WILDMAN, S. G., *Biochim. biophys. Acta* 186: 358, 1969.
WOODCOCK, C. L. F. & BOGORAD, L., *J. Cell Biol.* 44: 361, 1970.
WOODCOCK, C. L. F. & FERNÁNDEZ-MORÁN, H., *J. mol. Biol.* 31: 627, 1968.

MOLECULAR SIZE AND THE INFORMATION CONTENT OF CHLOROPLAST DNAs FROM HIGHER PLANTS

J.R. THOMAS, R. KOLODNER & K.K. TEWARI

Department of Molecular Biology and Biochemistry, University of California, Irvine, Cal. 92664, U.S.A.

Introduction

Chloroplast of microorganisms through higher organisms have been shown to contain DNA which can be readily distinguished from the nuclear (n-)DNA by its base composition, buoyant density, and kinetic complexity (for review see TEWARI 1971). We have recently studied the molecular size and conformation of pea chloroplast (ct-)DNA by renaturation kinetics and electron microscopy (KOLODNER & TEWARI 1972). The ct-DNA was found to denature with a sharp T_m of 84 °C with no evidence of intramolecular base compositional heterogeneity. Denatured and sheared ct-DNA was shown to renature as a single kinetic class with no suggestion of repeating sequences. The molecular weight of the ct-DNA was calculated to be 95×10^6 by renaturation kinetics. Electron microscopic studies on ct-DNA showed it to exist in a circular conformation with the molecules ranging in contour lengths of 37-40 μm. Using ØX174 RF II DNA and circularized phage λ DNA as internal standards, the molecular weight of ct-DNA was calculated to be 91.0×10^6. We have now extended these studies to other higher plants, *viz.*, spinach, lettuce, corn and oats. The ct-DNAs from these plants were found to band at a density of 1.698 ± 0.001 g cm^{-3} and denature with a T_m of 84 ± 0.5 °C in a homogeneous manner. Fragmented and denatured ct-DNAs were found to renature as a single kinetic class with no indication of repeating sequences. Furthermore, the molecular weight of the ct-DNAs from these higher plants were found to range from $90-95 \times 10^6$ by renaturation rates. The ct-DNAs from these plants have also been isolated in covalently closed circular molecules. The contour lengths of circular molecules ranged from 37.1 to 41.5 μm which corresponds to the molecular size of $86-92 \times 10^6$ dalton. From these results, it is concluded that DNA in chloroplasts of higher organisms exists as circular molecules with a molecular weight of about $85-90 \times 10^6$. The ct-DNA of these organisms consist of unique base sequences and, within limits of analytical tools, intra- and intermolecular heterogeneity does not appear to exist between ct-DNA molecules inside an organelle.

The information content in the ct-DNA can be studied using molecular DNA-RNA hybridization. We have analyzed hybridization of ct-DNA with chloroplast ribosomal (r-)RNA in great detail (THOMAS 1972). The *in vivo* labeled ct-r-RNA from pea leaves was hybridized with ct-DNA from pea leaves. The hybridization with total ct-RNA and ct-DNA was found to be 4.4%. Ribosomal RNA subunits were fractionated in sucrose gradients, checked for purity by acrylamide gel electrophoresis, and also hybridized with ct-DNA. Twentythree S-RNA formed specific hybrids with 2.6% of the ct-DNA and 16S RNA similarly hybridized with 1.6% of the ct-DNA. Competition hybridization showed that 25S and 18S RNA from cytoplasmic 80S ribosomes did not compete for the sites of 23S and 16S RNA in the ct-DNA. The level of r-RNA hybridization with the ct-DNA accounts for two cistrons of ct-r-RNA in the ct-DNA. These studies have also been extended to other plants. The labeled ct-r-RNA from pea leaves was hybridized with ct-DNAs from beans, lettuce, spinach and corn (THOMAS & TEWARI 1973). The amount of ct-DNA hybridized in each case was found to be about 4%. Competition hybridization and thermal stability experiments involving homologous and heterologous DNA-r-RNA combinations have shown that ct-DNAs from higher plants contain two cistrons for each ct-r-RNA species and that the base sequences of these cistrons are very similar.

I. Molecular Size and Conformation of Chloroplast DNAs

A. Renaturation Kinetics

1. Purity of DNA: Chloroplast DNAs were obtained from different plants in large quantities by the method described for pea leaves (KOLODNER & TEWARI 1972). This method consisted of making a cell-free homogenate, centrifuging at 100 x g for 10 min to remove nuclei, and collecting the crude chloroplast fraction by centrifuging the 100 x g supernatant for 15 min at 1 020 x g. The crude chloroplast fraction was treated with DNase and the organelle DNA isolated by phenol extractions followed by alcohol precipitation. Contaminating RNA and proteins were removed by treating the DNA solution with RNase and RNase T_1, followed by pronase. The purified ct-DNA was banded in analytical CsCl density gradient centrifugation. The photoelectric scans of the DNAs banded in CsCl (Fig. 1) show that chloroplast DNAs from pea (*a*), lettuce (*b*), spinach (*c*), bean (*d*), and corn (*e*) band at a density of 1.698 ± 0.001 g cm^{-3} (Fig. 1, top). On denaturation, the densities of all the different ct-DNAs increased by 0.013 g cm^{-3} (Fig. 1, middle). Incubations of denatured ct-DNAs at a Cot of about 2.0 resulted in the formation of native ct-DNA molecules (Fig. 1, bottom), where all the ct-DNAs have regained their original buoyant density. In comparison, n-DNA of pea leaves banded at a density of 1.695 g cm^{-3}, and after renaturation at a Cot of about 5, the DNA renatured only to 20% (Fig. 1, bottom, *f, g*). N-DNAs from the corresponding plants studied showed the same pattern. Thus, ct-DNAs used in the following experiments are characteristically chloroplastal (TEWARI 1971) and are free of any nuclear contamination.

2. Melting pattern of ct-DNAs: The melting profile of pea ct-DNA performed in

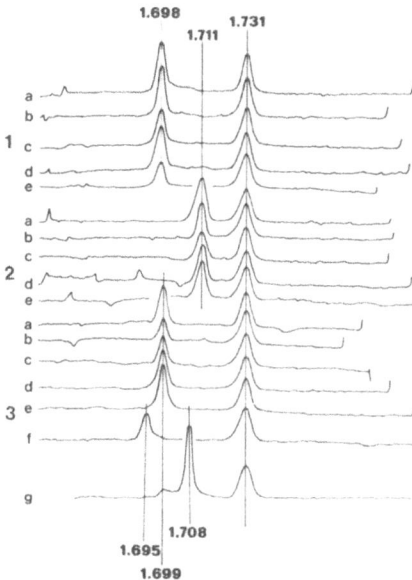

Fig. 1. Photoelectric scans of DNA centrifuged in CsCl of density 1.710 g cm^{-3} for 18 h at 18 °C in a Beckman Model E analytical ultracentrifuge. *Micrococcus lysodeikticus* DNA of density 1.731 g cm^{-3} has been used as a marker. 1: Native ct-DNAs; 2: Denatured ct-DNAs; 3: Renatured ct-DNAs; a: pea; b: lettuce; c: spinach; d: bean; e: corn; f: pea n-DNA; g: renatured pea n-DNA.

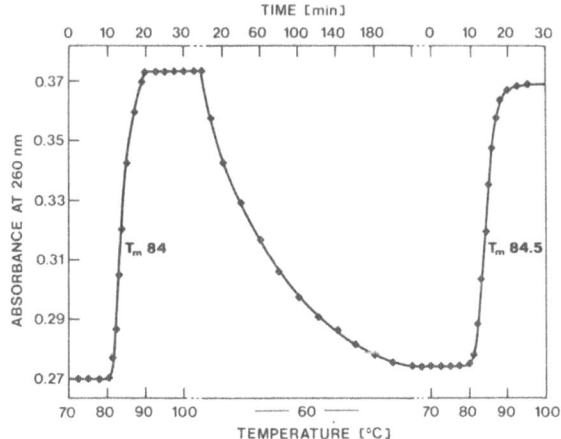

Fig. 2. The absorbance of ct-DNA at 260 nm as function of time and temperature. The isolated pea ct-DNA was used without fragmentation.

SSC (Fig. 2) was the same as profiles of all other ct-DNAs. The ct-DNAs showed a sharp transition with a T_m of 84.0 ± 0.5 °C, maximal hyperchromicity of 0.36 ($h_{max} = A_{max}/A_{20} - 1$) and a dispersion of ($\sigma 2/3$) 6.6 °C. ($A_{max}$ = absorbance at 260 nm of denatured DNA; A_{20} = absorbance of native DNA at 20°C; A_T = temperature range in which the increase in absorbance takes place.) There was no evidence of any other component in the melting pattern (cf. the melting patterns of ct-DNA and T_4-DNA plotted in a quadratic derivative form in Fig. 3).

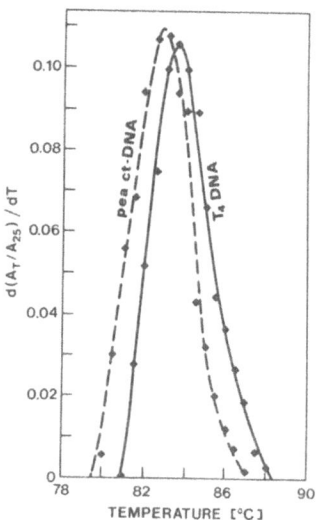

Fig. 3. Quadratic derivative of the melting curves of pea ct-DNA, T_4 DNA. d(A_T/A_{25})/dT is the derivative of A_T/A_{25} with respect to temperature where A_T is the absorbance at 260 nm of DNA at a particular temperature and A_{25} is the absorbance at 260 nm at 25 °C.

B. Renaturation Rates and Size of ct-DNAs

Ct-DNAs were broken to fragments of 2.8×10^6 dalton. The fragmented DNA was alkali denatured, neutralized, and the kinetics of renaturation monitored at 60 °C by the decrease in absorbance at 260 nm. The kinetics of renaturation using three different concentrations of pea ct-DNA (Fig. 4) shows a typical second order rate and the presence of a single kinetic class. The mean molecular weight of pea ct-DNA has been found to be 95×10^6 based on the molecular weight of 106×10^6 for T_4-DNA (see Table 1). Similarly ct-DNAs from spinach, beans, lettuce and corn renatured as a single component with a molecular weight of about $90-95 \times 10^6$. In mixed renaturation kinetics experiments, equal amounts of ct-DNAs from different sources were mixed in equal concentrations and renaturation studied. This solution showed a renaturation pattern which was indistinguishable

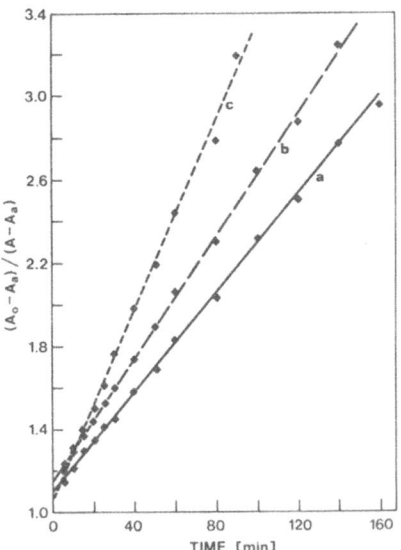

Fig. 4. Second order rate plots of pea ct-DNA. The concentration of a, b, and c were 10, 12 and 18 μg ml^{-1}, respectively. DNA was fragmented by passing through a No. 27 gauge needle ten times. It was made alkaline to pH 12, and after 15 min was neutralized to pH 7 and the salt concentration adjusted to SSC. Renaturation was carried out at T_m 26 °C.

Table 1. Molecular weight of pea chloroplast DNA. Renaturation rates of T4-DNA and ct-DNA were calculated according to KOLODNER & TEWARI (1972).

Preparation No.	Renaturation rate constant [mol^{-1} s^{-1}]		Mol. weight of ct-DNA (x 10^{-6}) calculated from T4-DNA assuming a mol. weight of	
	T4-DNA	ct-DNA	130 x 10^6	106 x 10^6
1	12.4	13.1	123.0	100.3
2	11.8	12.9	118.9	96.9
3	11.4	13.9	106.6	86.9
4	12.3	13.8	115.8	94.5
5	12.0	13.5	115.5	94.2
Mean			116.3	94.6

from the rate of renaturation of a single component. These results point to the similar size and sequences for the ct-DNAs from different higher plants.

C. *Electron Microscopy*

The method for the isolation of covalently closed circular molecules (see the legend of Fig. 5) has been used for the isolation of ct-DNAs from the other higher plants studied. A typical pattern of the final CsC1-EtBr density gradient centrifugation of the ^{32}P-labeled ct-DNA from pea leaves (Fig. 5) shows that the ct-DNA fractionates into forms I and II, and there is a considerable amount of DNA between these bands. 30% of the total DNA was found to band as covalently closed circles (form I) and about 60% as nicked circles (form II) and the rest of the DNA accounted for the bands between the two forms. In conformation with the chemical characteristics a typical microscopic picture of the DNA from band position I (shaded area in Fig. 5) shows a supercoiled molecule (Fig. 6), and a typical picture from the position II (shaded area in Fig. 5) shows an open circle (Fig. 7). Catenated dimers (Fig. 8), circular dimers (Fig. 9) and replicating molecules (Fig. 10) were found throughout the gradient. However, catenated dimers were enriched in the band III (shaded area in Fig. 5). Similar pattern was also obtained from lettuce, spinach, corn and oats.

The distribution profile of the lengths of circular ct-DNA molecules of pea leaves (Fig. 11) presents fourteen ct-DNA molecules ranging in length from 35.6 to 38.2 μm (s.d. ± 0.7 μm) compared to $\emptyset$X174 RF II DNA of 1.25 to 1.52 μm

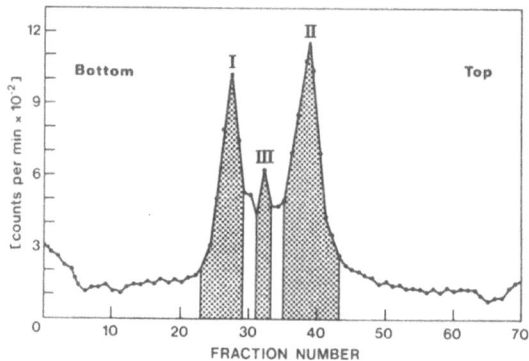

Fig. 5. Equilibrium density centrifugation of ct-DNA in CsC1-EtBr density gradient. DNase treated chloroplasts (KOLODNER & TEWARI 1972) from 100 g of pea leaves were suspended in 8.0 ml of buffer ST (0.3 M sucrose, 0.05 M Tris-base, 0.02 M Na$_2$-EDTA, pH 8.0). The chloroplasts were lysed by the addition of 16.0 ml of LS buffer (3% Na-sarkosyl, 0.05 M Tris-base, 0.02 M Na$_2$-EDTA, pH 8.0) and the suspension allowed to stand at room temperature for 30 min. 8 ml of a solution containing 4 M CsCl, 0.05 M Tris-base, 0.02 M Na$_2$-EDTA, pH 8.0 was added, the lysate was allowed to stand on ice for 90 min and centrifuged for 30 min at 10 000 × g. To each 8.0 ml of the clear lysate, 3.2 g CsCl and 200 μg ml^{-1} of EtBr was added in an SW-40 tube. This was underlayered with 3 ml of CsCl (p = 1.74 g cm^{-3}), 0.05 M Tris-base, 0.01 M Na$_2$EDTA, 200 μg ml^{-1} EtBr, pH 8.0. The tubes were overlayered with mineral oil and centrifuged for 12-16 h at 25 000 r.p.m. in a Spinco SW-40 rotor at 17 °C. The two DNA bands (readily seen after centrifugation) were collected. 5 ml of CsCl (p = 1.57 g cm^{-3}), 0.05 M Tris-base, 0.01 M Na$_2$-EDTA, 200 μg ml^{-1} EtBr, pH 8.0, was added and the solution further centrifuged for 36-44 h at 32 000 r.p.m. The labeled ct-DNA was isolated by this method and the fractions collected and counted.

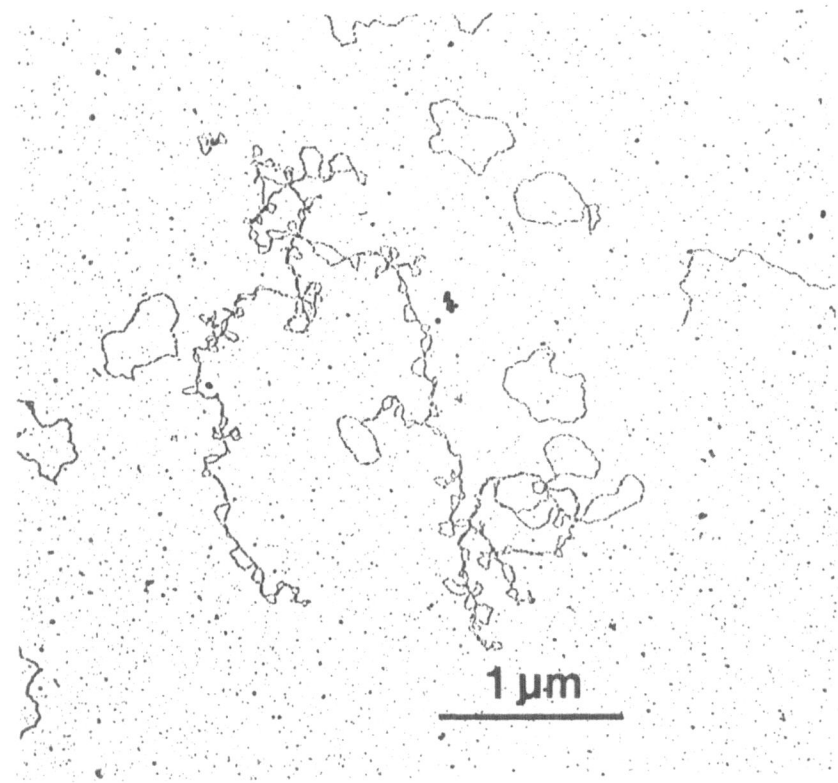

Fig. 6. A covalently closed circular DNA molecule from lettuce chloroplasts.

(s.d. ± 0.09 μm). The ct-DNA was also co-spread with circularized λ DNA. Nineteen molecules were found to range from 13.0 to 14.0 μm (s.d. ± 0.2 μm), and fourteen ct-DNA molecules ranged from 37.6 to 39.6 μm (s.d. ± 0.6 μm). The molecular weight of ct-DNA was calculated in relation to the molecular weights of ØX174 RF II DNA and λ DNA. Using the values of 3.4×10^6 and 10.0×10^6 dalton, respectively, for the two DNAs, the ct-DNA was found to be about 90.0×10^6 dalton.

We have measured the contour lengths of ct-DNAs from spinach, lettuce, oats and corn co-spread with ØX174 RF II DNA (Fig. 12). The ratio of lengths of ct-DNAs and ØX174 RF II DNA was found to be 25.9, 27.2, 26.0, and 25.4, respectively for spinach, lettuce, oats and corn. The calculated molecular weights for these ct-DNAs range from 86.3×10^6 to 92.5×10^6. Thus, the data obtained from electron microscopy are reasonably consistent with the molecular size of $90\text{-}95 \times 10^6$ dalton obtained by renaturation kinetics.

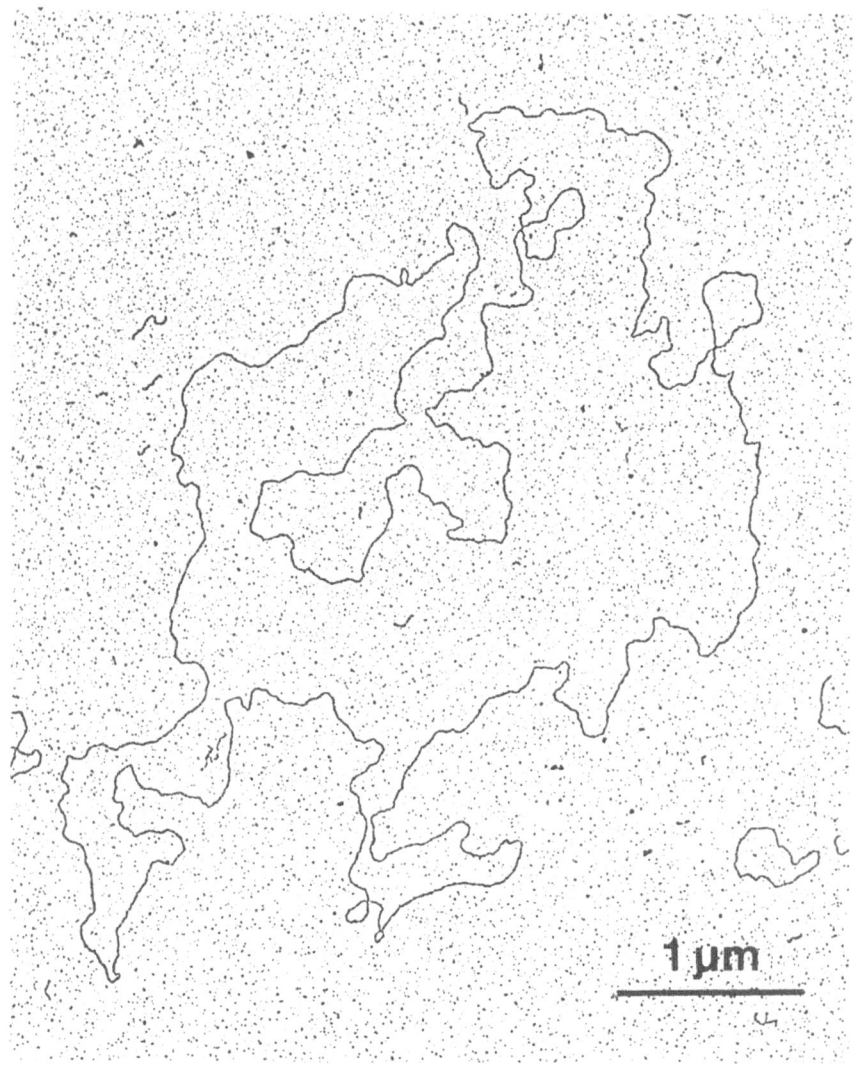

Fig. 7. An open circular DNA molecule from oat chloroplasts.

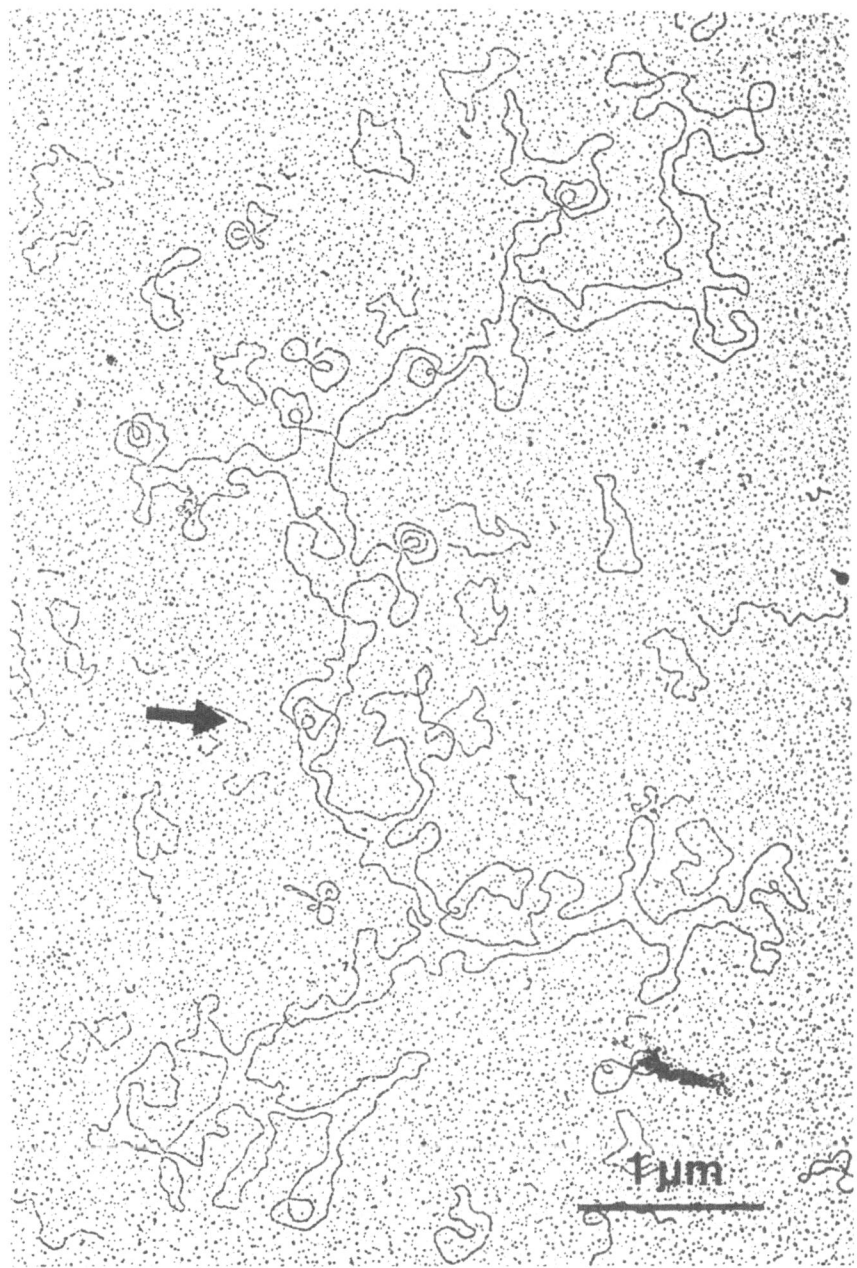

Fig. 8. A catenated dimer from pea chloroplasts. The arrow marks the point of catenation.

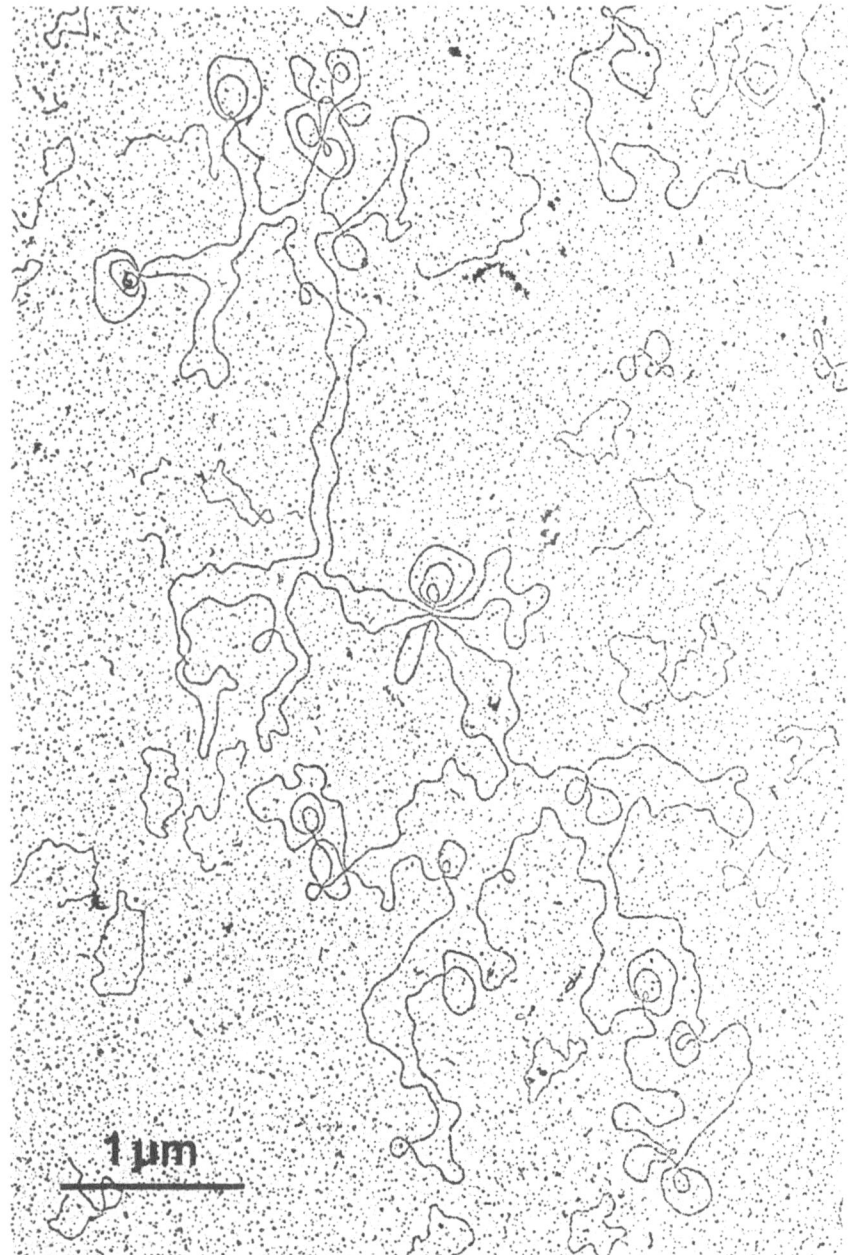

Fig. 9. A circular dimer from pea chloroplasts.

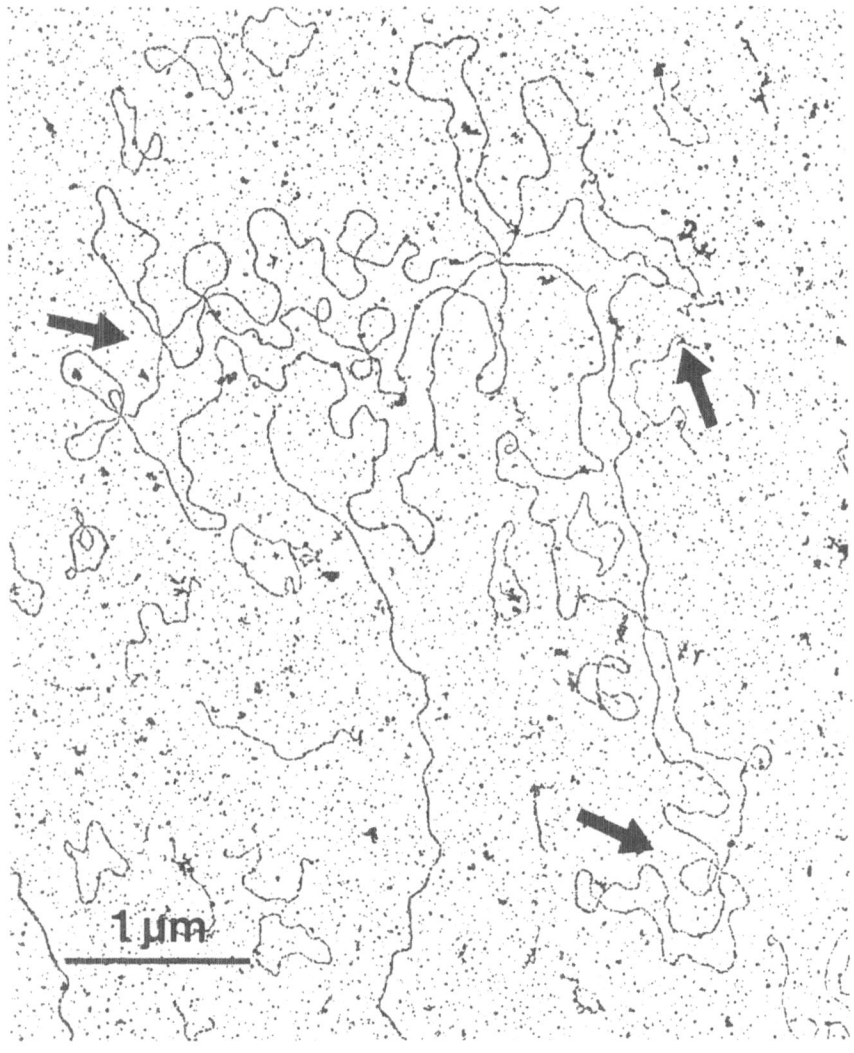

Fig. 10. A replicating pea chloroplast DNA molecule from the middle band of a CsCl-EtBr density gradient. The arrows mark the 2 branch points of the double stranded replicative fork as well as the position of a 'D'-loop.

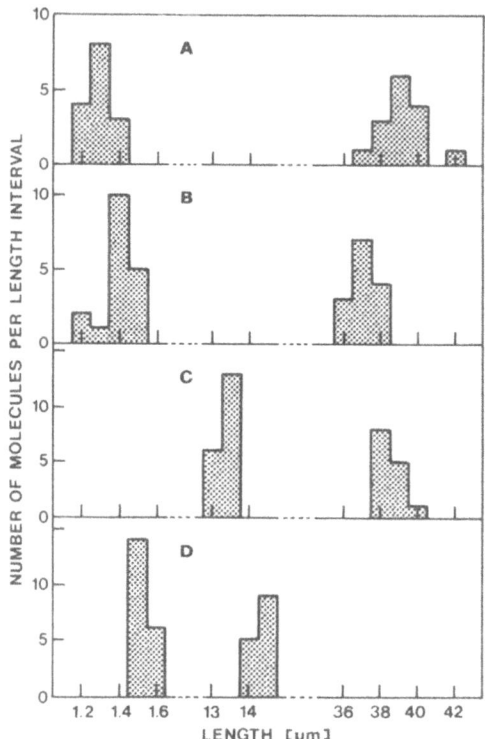

Fig. 11. Length distribution of DNAs. (A) Chloroplast lysate co-spread with ØX174 RF II DNA. Ct-DNA is found to have a contour length of 39.2 µm (s.d. = ± 1.1 µm) and ØX174 RF II DNA a contour length of 1.48 µm (s.d. = ± 0.05 µm). The ratio between the lengths of these DNAs is 26.5. (B) Isolated ct-DNA co-spread with ØX174 RF II DNA. Ct-DNA is found to be 37.1 µm (s.d. = ± 0.7 µm) compared to 1.39 µm (s.d. = ± 0.07 µm) for ØX174 RF II DNA. The ratio between the lengths of these two DNAs is 26.8. (C) Isolated ct-DNA co-spread with circularized lambda DNA. Ct-DNA is found to be 38.6 µm (s.d. = ± 0.6 µm), and lambda is found to be 13.6 µm (s.d. = ± 2.85 µm). The ratio between the lengths of two DNAs is 2.85. (D) ØX174 RF II DNA co-spread with circularized lambda DNA. Lambda DNA is found to have contour length of 14.4 µm (s.d. = ± 0.2 µm) and ØX174 RF II DNA is 1.53 µm (s.d. = ± 0.03 µm). The ratio between the lengths of two DNA molecules is 9.11.

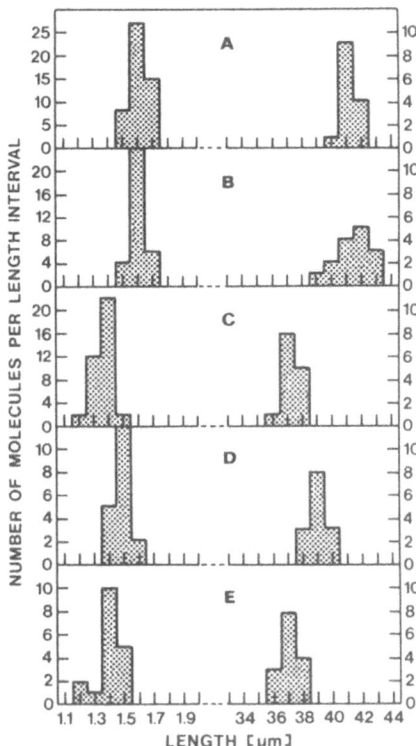

Fig. 12. Length distribution of chloroplast DNAs. (A) Corn ct-DNA of contour length 41.3 μm and ØX174 RF II DNA of 1.62 μm. The ratio between the lengths of the two DNAs is 25.4. (B) Oat ct-DNA of contour length 41.5 μm and ØX174 RF II DNA of 1.60 μm. The ratio between the lengths of the two DNAs is 26.0. (C) Lettuce ct-DNA of contour length 37.3 μm and ØX174 RF II DNA of 1.37 μm. The ratio between the lengths of the two DNAs is 27.2. (D) Spinach ct-DNA of contour length 39.4 μm and ØX174 RF II DNA of 1.52 μm. The ratio between the lengths of the two DNAs is 25.9. (E) Pea ct-DNA of contour length 37.1 μm and ØX174 RF II DNA of 1.39 μm. The ratio between the lengths of the two DNAs is 26.8.

II. Information Content in the Ct-DNAs

The information content in the ct-DNA for the ct-ribosomal (r-)RNA has been analyzed by DNA-RNA hybridization using labeled RNA.

A. *Number of r-RNA Cistrons*

Purified 70S ribosomes of the chloroplast were obtained from *in vivo* labeled plants. The r-RNA was isolated from these ribosomes by the method described in the legend of Fig. 13. The fractionation of r-RNA in linear sucrose gradients (15-30%) yielded 23S and 16S RNAs in the ratio of 2:1, respectively. The sizes of r-RNA were confirmed by co-electrophoresing them with labeled r-RNA from *Escherichia coli*, and the purity of the fractionated 23S and 16S r-RNA was analyzed by electrophoresis in acrylamide gels. In order to obtain 23S and 16S r-RNA free of contamination from each other, the fractions corresponding to the 23S and 16S r-RNA were collected and refractionated twice in linear sucrose gradients (15-30%).

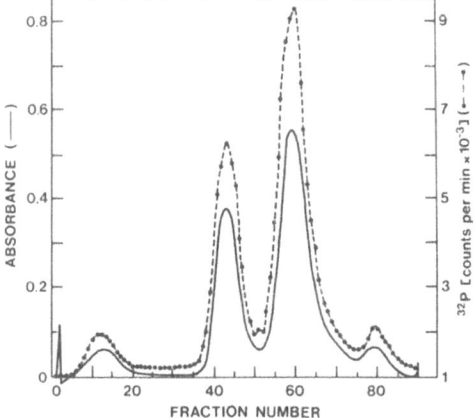

Fig. 13. Banding of ct-r-RNA in a sucrose gradient. Ct-r-RNA was isolated as follows: The labeled pea plants were chopped in SM buffer (0.5 M sucrose, RNase free, 0.05 Tris-base, 0.01 M $MgCl_2$, pH 8.0) containing 0.1% DEP (diethyl pyrocarbonate, Eastman). The homogenate was filtered through four layers of cheesecloth and centrifuged at 1500 xg for 15 min. The pellet was washed twice and suspended in TM buffer (0.01 M Tris-base, 0.01 M $MgCl_2$, pH 8.0) containing 0.1% DEP and 1% Triton X-100, then centrifuged at 17 000 xg for 20 min. The supernatant was centrifuged at 48 000 r.p.m. in a Spinco 50 rotor. The ribosomal pellet was suspended in EB buffer (0.025 M Tris-base, 0.025 M $MgCl_2$, 0.025 M KCl, 2% DSD, and 2% Triton X-100, pH 7.5) and extracted with water saturated phenol. The RNA was precipitated by the addition of 2 vol. of ethanol collected by centrifugation at 17 000 xg for 15 min. The RNA was dissolved in MgB buffer (0.03 M Tris-base, 0.001 M Mg-acetate, 0.000 1 M Na_4-EDTA, pH 8.0). ^{32}P-labeled r-RNA was isolated as described and centrifuged for 16 h in a 15-30% linear sucrose gradient. The fractions were collected and counted. The density increases from left to right.

The total ct-r-RNA from pea leaves was hybridized with 1.0 µg of pea ct-DNA on the filter. All hybridizations were carried out at the optimum conditions of time, salt concentration, and temperature with respect to specific hydrogen bond formation. The level of hybridization increases with increasing RNA to DNA ratio and reaches a maximum of 4.2% when the ratio has reached a level of 5.0 (Fig. 14). The ct-r-RNA sequences in the pea ct-DNA accounted for about 4% of the DNA in a number of different experiments. Taking the molecular weight of pea ct-DNA as 90×10^6, this level of hybridization indicates the presence of two r-RNA cistrons in the ct-DNA.

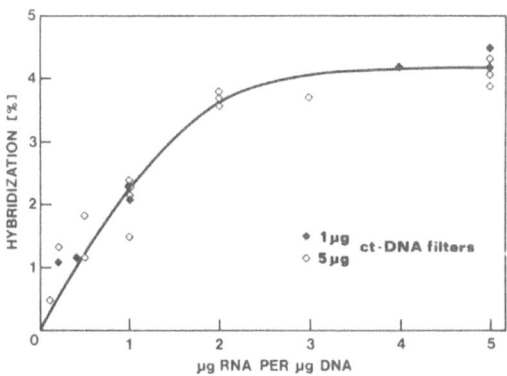

Fig. 14. Hybridization of pea ct-DNA to pea ct-r-RNA. The hybridization was carried out in 2.0 ml of 4 X SSC for 16 h. After hybridization, the filters were treated with 5.0 ml of 20 µg ml^{-1} RNase in 2 X SSC at 25 °C for 2 h and washed with 60 ml of 2 X SSC on each side.

We have also studied DNA-RNA hybridization using hydroxyapatite chromatography to fractionate the hybrids formed. For such an experiment, 30 µg of pea ct-DNA was sheared to 2.8×10^6 dalton fragments, denatured in 0.1 M NaOH, and neutralized with 0.2 M NaH_2PO_4. The DNA solution was incubated at 65 °C with 18 µg of pea ct-r-RNA. The aliquots were taken at various time intervals, treated with RNase and RNase T_1 and fractionated on hydroxyapatite columns as described by BRITTEN & KOHNE (1968). The hybridization is essentially complete in 2 h and the maximum hybrids formed amount to 4% of the ct-DNA (Fig. 15).

The hybridization experiments were carried out with 23S and 16S r-RNA to find out whether there were distinct cistrons for these RNAs. 23S r-RNA formed specific hybrids with 2.6% of ct-DNA and 16S r-RNA similarly hybridized to 1.6% of ct-DNA. These levels of hybridization add up to 4.2% for the total r-RNA, a value very close to that obtained using unfractionated r-RNA. The hybrids formed between ct-DNA and 23S r-RNA were specific as shown by competition with cold 23S r-RNA and these results were not appreciably affected by the presence of

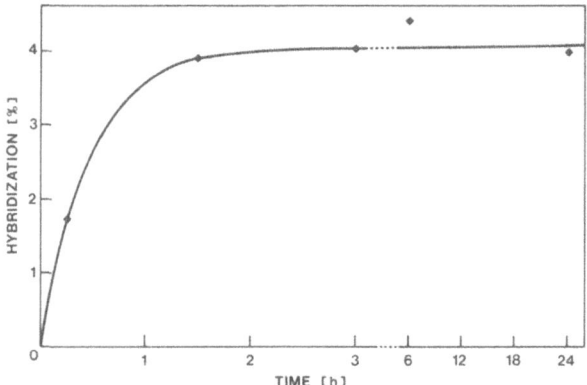

Fig. 15. Solution hybridization between pea ct-DNA and ct-r-RNA. Denatured and fragmented ct-DNA was incubated with RNA in 0.15 M NaH_2PO_4 and a final volume of 1.0 ml. The aliquots were fractionated on hydroxyapatite columns after treatment with RNase and RNase T_1.

cold 16S r-RNA. Similarly, the hybrids between ct-DNA and 16S r-RNA were unaffected by the presence of 23S r-RNA. These data show that there are two 23S and 16S r-RNA cistrons in the ct-DNA, and these cistrons do not have overlapping base sequences.

B. *Ribosomal RNA Cistrons in the ct-DNAs of Higher Plants*

Pea ct-DNA has been found to have two r-RNA cistrons (THOMAS & TEWARI 1973). We have now studied the divergence of r-RNA cistrons in higher plants. Purified ct-DNAs from bean, lettuce, spinach and corn were hybridized with ^{32}P-labeled ct-r-RNA from peas. The amount of hybridization for each of these DNAs was found to be about 4% (Fig. 16). The experiments were also carried out using purified 23S and 16S r-RNA components. The data were consistent in showing the presence of two non-overlapping cistrons in each ct-DNA. In order to find out the specificity of the DNA-r-RNA hybrids, competition experiments were carried out. Labeled ct-r-RNA from peas was hybridized with bean ct-DNA in competition experiments using cold lettuce ct-r-RNA. Theoretical competition was observed demonstrating specific hybrid formation between pea ct-r-RNA and bean ct-DNA (Fig. 17). Similarly, corn ct-DNA was hybridized with labeled pea r-RNA and competed with cold spinach ct-r-RNA (Fig. 17). In another experiment pea ct-DNA was hybridized with labeled pea r-RNA and competed with cold corn ct-r-RNA (Fig. 17). In all such experiments, cold ct-r-RNA from each plant competed with the labeled pea ct-r-RNA for the specific sites on the ct-DNAs. The experiments were extended to subunits: Pea ct-DNA was hybridized to labeled 23S r-RNA from peas and competed with spinach 23S r-RNA (Fig. 18). It is clear

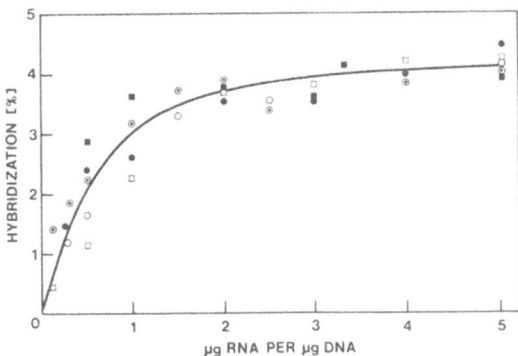

Fig. 16. Hybridization of labeled pea ct-r-RNA with ct-DNAs from different plants. Each filter contained 1.0 μg ct-DNA. (■ = peas, ○ = bean, ● = lettuce, ⊙ = spinach, □ = corn.)

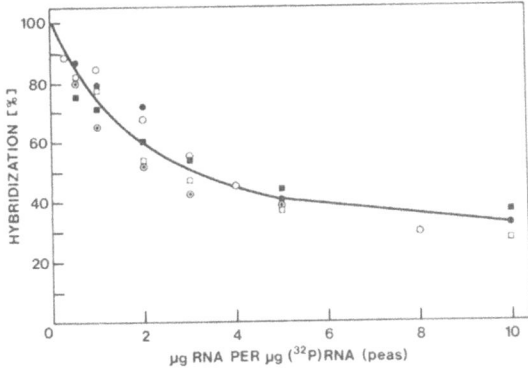

Fig. 17. Competition hybridization experiments. Ct-DNAs from different plants were hybridized with labeled pea ct-r-RNA in the presence of increasing cold r-RNA from different plants. Each filter contained 1.0 μg ct-DNA. ● = hybridization with pea ct-DNA in the presence of cold ct-r-RNA from oats; ■ = hybridization with pea ct-DNA in the presence of cold pea ct-r-RNA from oats; ○ = hybridization with bean ct-DNA in the presence of cold r-RNA from lettuce chloroplasts; □ = hybridization with pea ct-DNA in the presence of ct-r-RNA from corn; ⊙ = hybridization with corn ct-DNA in the presence of ct-r-RNA from spinach.

from the figure that spinach 23S r-RNA and pea 23S r-RNA have similar base sequences.

The specificity of the DNA-r-RNA hybrids was further studied by measuring the thermal stability of the complex. A thermal stability profile of pea ct-DNA with pea r-RNA (Fig. 19) indicates that the hybrid formed between pea ct-DNA and pea r-RNA is quite stable up to 70 °C, after which it begins to melt with a

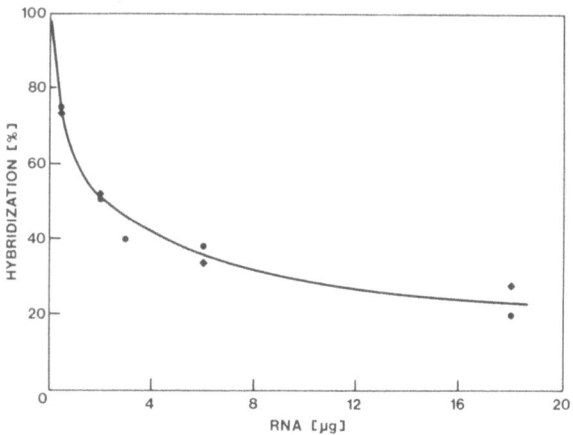

Fig. 18. Competition hybridization with subunits of RNA. Each filter contained 1.0 μg ct-DNA. Labeled pea 23S r-RNA was hybridized with pea ct-DNA in the presence of cold 23S r-RNA from peas (●) or from spinach (♦).

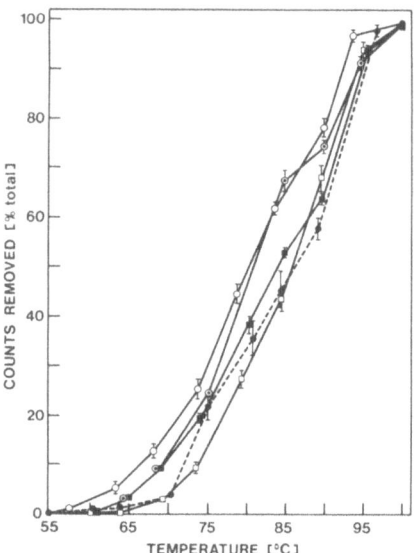

Fig. 19. Thermal stability of DNA-r-RNA complexes. DNA-r-RNA hybrids were studied by incubating the filters at the indicated temperature in 1 X SSC and analyzing the released counts. □ = pea ct-DNA, ⊙ = lettuce ct-DNA, ■ = spinach ct-DNA, ● = corn ct-DNA, ○ = bean ct-DNA.

T_{mi} (temperature of half dissociation) of about 85 °C. The thermal stability curve does not reveal heterogeneity in the hybrids and high T_{mi} points to the presence of stable DNA-r-RNA hybrids. The similar experiments have been carried out with pea r-RNA hybrids with ct-DNAs of spinach, bean, lettuce, and corn. The thermal stability of the DNA-r-RNA hybrids between such heterologous systems follows closely that of the homologous system (Fig. 19). These results again confirm the stability of the heterologous DNA-r-RNA hybrids. Thus, the data reported here show that ct-DNAs from higher plants contain two cistrons for each ct-r-RNA species and that the base sequences of these cistrons are very similar.

Discussion

We have modified our previous isolation procedure (KOLODNER & TEWARI 1972) as described in the text, and using this method, it has been possible for us to routinely isolate 75% of the ct-DNA from pea leaves in circular form. These observations show that practically all the ct-DNA exists as circles. As these molecules are extremely large, the chemical agents used to bring about the detachment of DNA from the thylakoids would have to be gentle and the following isolation procedures careful to maintain the stability of circular conformation. We have come across this problem particularly with the isolation of ct-DNA from corn leaves where we had to introduce minor variations in the isolation procedure for obtaining large proportions of intact ct-DNA. The isolated ct-DNA centrifuged in CsCl-EtBr density gradients bands at three distinct positions equivalent to form I (lower band), form II (upper band) and a middle band. We have analyzed the proportions of circular and catenated dimers in these three bands obtained from the ct-DNA of pea leaves. The lower band which amounted to 29% of the total DNA was found to contain 4.1% circular dimers and 1.7% catenated dimers. The middle band contained 20% of the total DNA and of this amount, 3.5% were circular dimers and 3.6% catenated dimers. It was difficult to decide between catenated dimers and replicative intermediates in the molecules composed of two circles. However, circular dimers could be unambigously identified. The top band of the gradient containing the rest of the DNA (51%) had 3.6% circular dimers and 2.0% catenated dimers.

We have discussed at length the necessity of including an internal standard in molecular weight determinations using electron microscopy (KOLODNER & TEWARI 1972). Therefore, we have utilized length ratios between ct-DNA and internal standards to determine the molecular weight of ct-DNAs. The length ratio between ct-DNA and λ DNA has been found to be 2.85 yielding a molecular weight of 85.5×10^6 for pea ct-DNA. The relative lengths between pea ct-DNA and $\emptyset$X174 RF II DNA has been found to be 26.8, yielding a molecular weight of 91.1×10^6 assuming the corresponding values of 3.4×10^6 for $\emptyset$X174 RF II DNA.

It has been possible to isolate ct-DNAs in circular conformation from bean, spinach, lettuce, corn and oats. In all these plants, it can be shown by rigorous isolation conditions that 75% of the isolated DNA can be found in circular conformation. The ct-DNA from these plants also have circular dimers and catenated dimers in about the same proportion as found in pea leaves. For example, in the case of lettuce, 5.6% of the ct-DNA from the lower band was found to exist as circular dimers and 2.4% as catenated dimers. We have calculated the molecular size of these ct-DNAs utilizing $\emptyset$X174 RF II DNA as an internal standard. The length ratio between lettuce ct-DNA and $\emptyset$X174 RF II DNA was found to be 27.2. This ratio corresponds to a molecular weight of 92.4×10^6. In the case of spinach, a similar ratio was found to be 25.9 corresponding to a molecular weight of 88.1×10^6. The ct-DNAs from oats and corn leaves had a ratio of 26.0 and 25.4, respectively. The corresponding molecular weights for oats ct-DNA was 88.44×10^6 and for corn ct-DNA was 86.6×10^6. Thus, the molecular weights of ct-DNA from corn, oats, spinach, lettuce and peas range from 86.4 to 92.4×10^6. The molecular size of ct-DNAs from higher plants appear to be of the same order of magnitude and differ from each other by about 6×10^6 dalton. We would like to point out that, though 6×10^6 dalton difference in the molecular weights of ct-DNAs from such diverse sources does not appear to be significant, the confidence level of our experiments makes us feel that this difference might be real. Experiments are currently in progress to ascertain whether or not such differences are significant.

The renaturation rates of ct-DNAs from the higher plants studied have shown these DNAs to consist of unique base sequences with no evidence for the existence of intra- or intermolecular heterogeneity. The molecular weight of pea ct-DNA has been calculated to be 94.6×10^6 assuming a value of 106×10^6 for T_4 DNA. The other ct-DNAs studied have similarly shown a molecular weight range of $90-95 \times 10^6$. The sensitivity of this method is not adequate enough to support or refute the difference in the molecular weights observed using electron microscopy. Within experimental errors, however, these two methods do give excellent agreement in the size of the ct-DNAs from higher plants.

We have also studied expression of ct-DNA by hybridizing it with r-RNA from chloroplasts. It has been found that about 4% of the pea ct-DNA is complementary to ct-r-RNA. Thus ct-DNA equivalent to a size of 3.6×10^6 dalton contains base sequences similar to the ct-r-RNA. This level of complementarity shows that there are two cistrons for the ct-r-RNA in the ct-DNA. This is the maximum level of r-RNA hybridization obtained with ct-DNA (see TEWARI 1971). The explanation for this level of hybridization lies in choosing the optimum conditions of hybridization. Our ct-r-RNA preparations are relatively intact during hybridization and the loss of the filter bound ct-DNA does not amount to more than 5%. It may be pointed out that ct-DNA is unique in having two cistrons for r-RNA. In comparison, animal mt-DNA has been shown to have only one r-RNA cistron (BORST 1972). The number of cistrons found in the pea ct-DNA is a correct estimate as the hybrids formed between ct-DNA and r-RNA are the result of specific hybridization. This is shown by the temperature stability of the DNA-r-

RNA hybrid. The T_m of the DNA-r-RNA hybrid has been found to be about 85 °C ± 2 °C, and the melting profile indicates the presence of only one class of bound RNA sequences. Purified 23S and 16S r-RNAs have also been hybridized with pea ct-DNA. We have found 2.6% of the ct-DNA to be complementary to the 23S r-RNA and 1.6% to be complementary to the 16S r-RNA. When the two classes of RNA were mixed in vitro and hybridized with pea ct-DNA, a hybridization of 4.2% was obtained. These results showed that the cistrons for the two sizes of r-RNA are distinct on the ct-DNA.

The above experiments have been further extended to hybridization studies with r-RNAs and ct-DNAs of spinach, lettuce, bean, corn, and oats. Radioactively labeled r-RNA from pea chloroplasts was hybridized with ct-DNAs from spinach, beans, lettuce and corn. The amount of r-RNA hybridized in each case was found to be about 4% of the ct-DNA. These experiments were also carried out with purified 23S and 16S RNA. The results were consistent with those obtained utilizing total ct-r-RNA. In each case it was found that there were two non-overlapping cistrons in each ct-DNA. The specificity of the DNA-r-RNA hybrids has been confirmed by competition hybridization experiments involving labeled pea r-RNA and pea ct-DNA in the presence of cold ct-r-RNA from spinach, beans, corn, lettuce and oats. In such experiments, the same level of competition was found between heterologous systems as well as the homologous system. These experiments were extended to include thermal stability of DNA-r-RNA hybrids. The heterologous DNA-r-RNA hybrids were found to melt with a T_m of 83 ± 2 °C compared to the homologous DNA-r-RNA hybrids which melted at 84 ± 1 °C. Thus, the data obtained here show that ct-DNAs from higher plants contain two cistrons for each ct-r-RNA species and that the base sequences of these cistrons are very similar. The conservation of two r-RNA cistrons in the plants which have diverged through billions of years and have been frequently cultivated through all the civilizations is unique in molecular evolution. In comparison, n-DNA contains varying numbers of cytoplasmic r-RNA copies in the different plant species (MATSUDA et al. 1970, INGLE & SINCLAIR 1972). For example, the number of cytoplasmic r-RNA genes per haploid compliment varies from 260 in the artichoke to 6 650 in the onion. Similar studies in bacteria have also shown that the DNA-r-RNA hybrids formed between distantly related species were relatively unstable even though the extent of cross reaction was quite high (MOORE & McCARTHY 1967). In the experiments presented here, the thermal stability of the hybrids have failed to show any such unstable hybrids.

Conclusion

Ct-DNAs in higher plants exist as circular molecules of about 90 x 10° dalton. There is no intra- or intermolecular heterogeneity in these ct-DNAs. Ct-DNAs in higher plants contain two cistrons for the ct-r-RNA and these sequences have been remarkably preserved in the different higher plants.

Acknowledgement

We thank Dr. R.C. WARNER for his support and encouragement. J.R.T. is a predoctoral trainee of PHS (HD-00347-03) and R.K. is a predoctoral trainee of NIGMS (GM-02063-01). This work was supported by National Science Foundation Grant GB-20674. The authors are thankful to JUTA KEITHE for her help in preparing the photographs and to NANCY HOFFNER for her help in preparing the manuscript.

REFERENCES

BORST, P., *Annu. Rev. Biochem.* 41: 333, 1972.
BRITTEN, R. J. & KOHNE, D. E., *Science* 161: 529, 1968.
INGLE, J. & SINCLAIR, J., *Nature* 235: 30, 1972.
KOLODNER, R. & TEWARI, K. K., *J. biol. Chem.* 247: 6355, 1972.
MATSUDA, K., SIEGEL, A. & LIGHTFOOT, D., *Plant Physiol.* 46: 6, 1970.
MOORE, R. L. & McCARTHY, B. J., *J. Bacteriol.* 94: 1066, 1967.
TEWARI, K. K., *Annu. Rev. Plant Physiol.* 22: 141, 1971.
THOMAS, J. R., *Fed. Proc.* 31: 914, 1972.
THOMAS, J. R. & TEWARI, K. K., *Fed. Proc.* 32: 642, 1973

CHARACTERISTICS OF CHLOROPLAST DNA-PROTEIN COMPLEX

V.V. PINEVICH, SVETLANA B. IVANOVA & A. A. LIPSKAYA

Biological Research Institute of Leningrad University, Leningrad, U.S.S.R.

Introduction

In recent years much consideration has been given to the structure and function of the genetic apparatus of organisms of different phylogenetic levels. It may be presumed that DNA[1] evolution ran parallel to the evolution of DNA-bound proteins in the course of formation of the genetic apparatus from procaryotes to eucaryotes. DNA is thus bound to protein at each stage of phylogenesis. DNA-bound proteins have been found in prokaryotes of bacterial type (RAAF & BONNER 1968; TAKEUCHI & TSUGITA 1970) and in blue-green algae (MAKINO & TSUZUKI 1971), in mesokaryotic organisms (NETRAWALI 1970; RIZZO & NOODÉN 1972), and in eukaryotic algae (IWAI 1964), yeasts (KRASHENNINIKOV et al. 1970; DUFFUS 1971), fungi (MOHBERG & RUSCH 1969), higher plants and animals (GOFSHTEĬN 1971). However, the qualitative and quantitative characteristics of the DNA-protein complexes seem to be little understood.

As the literature contains a rather small amount of data on the nature of the proteins bound to chloroplast DNA, the present study attempts to detect a specificity of the interaction of DNA-protein complex in chloroplasts.

Material and Methods

Leaves of 10 d-old pea seedlings (*Pisum sativum* L. cv. Spartanets) were used for the experiments. The peas were grown under artificial illuminance (lamps LDS-80, 420 W m^{-2}). Cell fractions (nuclei and chloroplasts) were isolated by a slightly modified method of TEWARI & WILDMAN (1969).

DNA was extracted from chloroplasts and nuclei according to MARMUR (1961).

DNA nativity was determined using the coefficient $E(p)$ and the melting temperature (T_M). A thermostatic cuvette equipped with a thermocouple (MARMUR & DOTY 1959) was used to obtain the T_M of DNA and T_M of DNP. Renaturation capacity of DNA was determined under standard conditions (KUNG et al. 1971). Histone isolated from pea nuclei was separated on polyacrylamide gel by the method of BONNER et al. (1968). This procedure was followed by chromatin

1. Abbreviations: DNA — deoxyribonucleic acid; DNP — deoxyribonucleo-protein

extraction by our modification of the method by TEWARI & WILDMAN (1969).

The soluble DNP was isolated as follows: Chromatin suspension in bi-distilled water was treated for 20-30 s at $0°C$ in an MSE ultrasonic disintegrator using 70% of its maximum capacity. The sonicated suspension was stirred for several hours, then centrifuged (VAC-60) at 25 000 r.p.m. The supernatant obtained (soluble DNP) was purified by column gel filtration on Sepharose 4 B. DNA and protein were separated according to LOEB (1968); this method does not denature DNA or protein. DNA and DNP template activities were determined by a modification of the method of TEWARI & WILDMAN (1969); in these experiments 20 µg DNA or DNP were added to the sample.

Formation of complexes of DNA and histones was studied by the method of BÚTLER & JOHNS (1964): various volumes of histone solution in 0.15 M NaCl were added to the solution of 100-150 µg DNA. The sample, of total volume 3 ml, was stirred for 30 min and afterwards centrifuged at 2 000 x g. DNA and protein amounts were determined from the difference between the initial sample and the supernatant.

DNA was determined spectrophotometrically with diphenyl amine (BURTON 1956), RNA spectrophotometrically according to SPIRIN (1958), proteins by the method of LOWRY et al. (1951), and chlorophyll according to ARNON (1949).

Results and Discussion

Nuclear and Chloroplast DNA

Melting curves of nuclear and chloroplast DNA were almost identical, with T_M 79-81 °C (Fig. 1). Under standard conditions an average of 90% of chloroplast DNA renaturated while nuclear DNA did not possess that capacity.

The samples of nuclear and chloroplast DNA and histones were used for complex formation, which revealed specific features of the DNA and histone interaction. This method has been used in experiments with animal tissues (BUTLER & JOHNS 1964; HOARE & JOHNS 1971), bacteria (ŠPONAR et al. 1967; ANSEVIN & BROWN 1971; PINEVICH & ZHDAN-PUSHKINA 1972), and higher plants with nuclear DNA only – (BONNER et al. 1968). Our results show that, under the same conditions, the chloroplast DNA precipitates by 15-18%, while the nuclear DNA precipitates by 70%. However, the chloroplast DNA is able to bind a larger amount (w/w) of histone (the ratio in precipitate was protein/DNA 4 to 5/1) than nuclear DNA (1.5 to 2.0/1).

It is known that addition of histone to DNA stabilizes the DNA structure, protects it from denaturation and increases its T_M. Nevertheless, in our experiments the addition of histone to chloroplast DNA in the ratio (w/w) 1:1 (this ratio avoids the formation of an insoluble DNA-histone complex) did not change the character of melting or the T_M of chloroplast DNA. These results are consistant with those reported by ŠPONAR et al. (1967): they showed that calf thymus histone stabilized the structure of calf thymus DNA but it did not influence the character of melting and T_M of DNA from some species of bacteria.

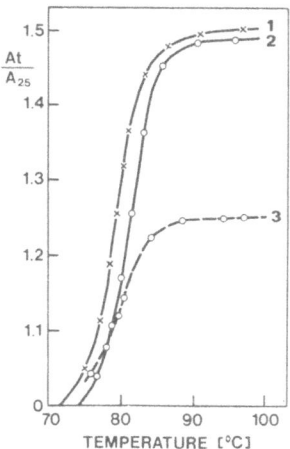

Fig. 1. Curves of melting of pea DNA and DNP preparations: 1 – nuclear DNA; 2 – chloroplast DNA; 3 – DNP of chloroplasts.

ŠPONAR et al. (1967) suggest that DNA might have two types of sites to bind histones and that the sites with tighter linkage require the presence of both pairs of bases, but in a definite ratio. The specificity of the interaction of DNA and protein has been proven by several investigators (LEWIN & ROSMARIN 1970; ANSEVIN & BROWN 1971) who indicate that besides non-specific ionic bonds, less tight hydrogen and hydrophobic bonds are typical of the interaction. ŠPONAR & ŠORMOVÁ (1972) reported that lysine-rich histones selectively link to A-T pairs of DNA.

Thus, our results show in vitro that the nuclear DNA and chloroplast DNA of pea differ significantly in structure and that the specificity of the interaction of histones with chloroplast and nuclear DNA is variable.

Chloroplast DNP

The soluble DNP which we isolated from pea chloroplasts to study the DNA-protein interaction in vivo contained about 90% of total DNA chromatin and was composed of DNA, RNA, protein and chlorophyll in the ratio 1 : 0.05 : 4 : 0.1. DNP purified on Sepharose 4B was composed of DNA, protein and chlorophyll in the ratio 1 : 1.5 : 0.03.

The specificity of DNP bonds was determined by means of T_M (Fig. 1). A DNP melting curve was similar to that of DNA, and melting temperatures of both DNA and DNP were 81°C. However, the chloroplast DNA exerted a greater hyperchromic effect. The literature data show that the histones of nuclear DNP protect DNA against thermal denaturation and hence the melting temperature of DNP is higher than that of DNA (BONNER et al. 1968). The absence of this effect in our experiments shows that the protein bound to chloroplast DNA does not have the properties of typical histone stabilizing DNA.

These considerations are in agreement with these by TEWARI & WILDMAN (1969) who reported that the protein bound to DNA in tobacco chloroplast chromatin does not increase the resistance of DNA to denaturation. According to TEWARI & WILDMAN (1969), tobacco chloroplasts contain DNA-dependent RNA-polymerase which is 30-40 times more active than nuclear enzyme. They assume that DNA binding to histones is responsible for the repression of nuclear activity and, therefore, the chloroplasts do not contain histones. However, BOTTOMLEY (1970) indicates that pea chloroplast RNA polymerase is less active than pea nuclear RNA polymerase and considers it to be a species characteristic.

Our experiments on the determination of chloroplast DNP template activity used DNP before purification on Sepharose 4B for comparison with standard calf thymus DNP (FRENSTER et al. 1963). It is known that soluble nuclear DNP is not capable of RNA synthesis in a standard system even when exogenous RNA polymerase is added (BONNER & HUANG 1963).

Table 1. Determination of pea chloroplast DNA and DNP template activities.

Sample	2-^{14}C-uridine incorporation [counts min^{-1} per 20μg DNA]
DNP I	2 773
DNP II	2 640
DNP III	1 772
DNA	449
DNP I + DNA	1 920

Our experiments (Table 1) show that the protein of DNA possesses the features of RNA-polymerase and that addition of exogenous DNA does not stimulate the incorporation of labelled uridine. Thus, it is highly probable that the protein examined does not inhibit the DNA template activity, but rather contributes to it. This confirms the belief that chloroplast DNA-bound protein does not have the properties of typical histone. Further experiments are required for ascertaining the template activity of purified chloroplast DNP as well as revealing the role of the protein isolated from chloroplast DNP in DNA template activity.

Further studies with proteins isolated from DNP confirm that the chloroplast protein in our material differs from typical histone. Fractions of total and acid-soluble proteins (30-40% of total DNP protein) were extracted from DNP. The ratio of acid-soluble protein to DNA was 0.5 to 0.6: 1. Samples containing 100 μg protein were subjected to electrophoretic separation on polyacrylamide gels. The acid-soluble DNP chloroplast protein appeared as one single band near the origin, in marked contrast to the pattern of six distinct bands obtained with pea nuclear

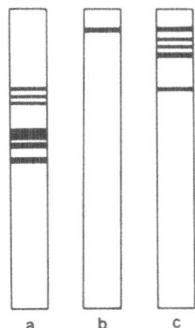

Fig. 2. Electropherograms of proteins in polyacrylamide gel: a – histone from cell nucleus in pea; b – acid-soluble protein from chloroplast DNP; c – total protein from chloroplast DNP.

histone (Fig. 2). The total DNP of the chloroplast gave five bands, differing in relative mobility from those of the nuclear histones (Fig. 2).

Hence our experiments show that the isolated chloroplast protein is not a histone, in agreement with the findings on tobacco chloroplast chromatin proteins by TEWARI & WILDMAN (1969). The proteins bound to chloroplast DNA are specific. Further comparative studies of these proteins and DNA-bound proteins from blue-green algae will certainly give new insights into the recent theory of symbiotic origin of chloroplasts and other cellular organelles.

Summary

DNA's were extracted from *P.sativum* nuclei and chloroplasts and their activities and heterogeneity were determined. Nuclear or chloroplast DNA form complexes with histone from pea nuclei. From the chromatin of pea chloroplasts a soluble deoxyribonucleo-protein (DNP) was extracted and purified on Sepharose 4B. A protein bound to DNA was extracted from DNP and separated electrophoretically on polyacrylamide gel. The electropherograms were compared with those of pea histone and calf thymus. The influence of proteins, bound to chloroplast DNA, and nuclear histones, on chloroplast DNA template activity was ascertained.

REFERENCES

ANSEVIN, A. T. & BROWN, B. W., *Biochemistry* 10 : 1133, 1971.
ARNON, D. I., *Plant Physiol.* 24 : 1, 1949.
BONNER, J., CHALKLEY, G. R., DAHMUS, M., FAMBROUGH, D., FUJIMURA, F., HUANG, R.-C. C., HUBERMAN, J., JENSEN, R., MARUSHIGE, K., OHLENBUSCH, H.,

OLIVERA, B. & WIDHOLM, J., in: COLOWICK, S. P. & KAPLAN, N. O. (ed.): Methods in Enzymology, Vol. XII, Nucleic Acids, Part B, p. 3, Academic Press, New York – London 1968.
BONNER, J. & HUANG, R.-C. C., *J. mol. Biol.* 6 : 169, 1963.
BOTTOMLEY, W., *Plant Physiol.* 45 : 608, 1970.
BURTON, K., *Biochem. J.* 62 : 315, 1956.
BUTLER, J. A. V. & JOHNS, E. W., *Biochem. J.* 91 : 156, 1964.
DUFFUS, J. H., *Biochim. biophys. Acta* 228 : 627, 1971.
FRENSTER, J. H., ALLFREY, V. G. & MIRSKY, A. E., *Proc. nat. Acad. Sci. U.S.A.* 50 : 1026, 1963.
GOFSHTEÏN, L. V., *Uspekhi biol. Khim.*, 12 : 72, 1971.
HOARE, T. A. & JOHNS, E. W., *Biochim. biophys. Acta* 247 : 408, 1971.
IWAI, K. in: BONNER, J. & TS'O, P. O. P. (ed.): The Nucleohistones, p. 59, Holden-Day Inc., San Francisco – London – Amsterdam 1964.
KRASHENNINIKOV, I. A., GOROZHANIN, P. P. & BELOZERSKIĬ, A. N., *Dokl. Akad. Nauk SSSR* 195 : 486, 1970.
KUNG, S. D., MOSCARELLO, M. A. & WILLIAMS, J. P., *Biochim. biophys. Acta* 232 : 614, 1971.
LEWIN, S. & ROSMARIN, M. N., *Biochem. J.* 119 : 62p, 1970.
LOEB, J. E., *Biochim. biophys. Acta* 157 : 424, 1968.
LOWRY, O. H., ROSEBROUGH, N. J., FARR, A. L. & RANDALL, R. J., *J. biol. Chem.* 193 : 265, 1951.
MAKINO, F. & TSUZUKI, J., *Nature* 231 : 446, 1971.
MARMUR, J., *J. mol. Biol.* 3 : 208, 1961.
MARMUR, J. & DOTY, P., *Nature* 183 : 1427, 1959.
MOHBERG, J. & RUSCH, H. P., *Arch. Biochem. Biophys.* 134 : 577, 1969.
NETRAWALI, M. S., *Exp. Cell Res.* 63 : 422, 1970.
PINEVICH, V. V. & ZHDAN-PUSHKINA, S. M., *Vestn. lening. gos. Univ., Ser. biol.* 9 (2) : 109, 1972.
RAAF, J. & BONNER, J., *Arch. Biochem. Biophys.* 125 : 567, 1968.
RIZZO, P. J. & NOODÉN, L. D., *Science* 176 : 796, 1972.
ŞPIRIN, A. S., *Biokhimiya* 23 : 656, 1958.
ŠPONAR, J., BOUBLÍK, M. & ŠORMOVÁ, Z., *Coll. czech. chem. Commun.* 32 : 4510, 1967.
ŠPONAR, J. & ŠORMOVÁ, Z., *Europe. J. Biochem.* 29 : 99, 1972.
TAKEUCHI, M. & TSUGITA, A., *J. Biochem.* 67 : 237, 1970.
TEWARI, K. K. & WILDMAN, S. G., *Biochim. biophys. Acta* 186 : 358, 1969.

RIBOSOMAL PROTEINS AND THE ORIGIN OF PLASTIDS

M. S. ODINTSOVA & N. P. YURINA

A. N. Bakh Institute of Biochemistry, Academy of Sciences of the U.S.S.R., Moscow, U.S.S.R.

Photosynthetic bacteria, which are the most primitive photosynthetic forms existing today, have a relatively simple lamellar photosynthetic apparatus. Blue-green algae occupy an intermediate position between photosynthetic bacteria and higher plants with respect to the organization of their photosynthetic apparatus. Nevertheless, their highly organized mechanism of photosynthesis draws them together with eukaryotic algae and higher plants and their more or less elaborate intracytoplasmic membrane system correspond functionally to the organelles of eukaryotic cells. At the same time, blue-green algae lack discrete plastids and other membrane-bound organelles.

At present numerous experimental data support the hypothesis of an endosymbiotic origin of plastids from free-living photosynthetic prokaryotes, ancient forms of blue-green algae and bacteria (MARGULIS 1970). Indeed, nucleic acids and the protein-synthesizing apparatus of chloroplasts are, to a certain extent, similar to those of blue-green algae. The DNAs have similar morphological organization and physico-chemical properties and they do not contain histones. The ribosomes of plastids resemble those of bacteria and blue-green algae by the sedimentation coefficients of monosomes, subunits and components of ribosomal RNA, use of N-formylmethionine as the initiating amino acid in protein synthesis, reaction to the ionic conditions of the medium and sensitivity to antibiotics. The present communication gives the results of a comparative study of proteins isolated from ribosomes of chloroplast, blue-green algae and photosynthetic bacteria.

Material and Methods

Ribosomes were isolated by means of differential and density gradient centrifugations from chloroplasts of *Pisum sativum* and *Chenopodium album*, from five species of blue-green algae of different systematic groups and from two species of purple bacteria. The isolated preparations had typical UV absorption spectra with A_{260}/A_{280} ratios about 1.9; A_{max}/A_{min} about 1.8 and, according to analytical ultracentrifugation data, they did not contain protein contaminations. The chemical composition of ribosomes was determined by isopicnic ultracentrifugation in CsCl density gradient.

Results and Discussion

Chloroplast and prokaryotic ribosomes are similar in many physico-chemical properties and hence the differences found in their relative protein content are very interesting. The buoyant density (Table 1) of ribosomes of photosynthetic bacteria is about 1.61-1.64 g cm^{-3}, corresponding to 40% protein and typical for ribosomes of bacterial type. Ribosomes of all species of blue-green algae studied are similar in RNA/protein ratio (1.5-1.8) and do not differ significantly from

Table 1. Buoyant densities in CsCl of formaldehyde-fixed bacterial and plant ribosomes and the relative content of protein and RNA calculated from the buoyant densities. (From YURINA et al. 1972.)

Organism	ς [g cm^{-3}]	Protein [%]
BACTERIA		
Escherichia coli	1.64	35.0
---"---	1.639±0.005	35.1
Rhodopseudomonas gelatinosa	1.623±0.007	37.9
Chromatium vinosum	1.616±0.004	39.0
BLUE-GREEN ALGAE		
Anacystis nidulans	1.615±0.007	39.2
Synechocystis aquatilis	1.612±0.002	39.7
Lyngbya sp.	1.622±0.006	38.0
Anabaena variabilis	1.634±0.004	36.0
Plectonema boryanum	1.618±0.015	38.7
Spirulina sp.	1.640	35.0
FUNGI		
Dictyostelium discoideum	1.58	45.0
Dictyostelium purpureum	1.58	45.0
GREEN ALGAE		
Chlorella pyrenoidosa cytoplasm	1.557	49.0
chloroplasts	1.570	46.7
HIGHER PLANTS		
Equisetum pratense	1.567±0.002	47.1
Triticum vulgare	1.553±0.003	49.5
---"---	1.565	47.5
Zea mays	1.553±0.003	49.5
Antirrhinum majus	1.568±0.002	47.0
Nicotiana tabacum	1.553±0.002	49.5
Phaseolus vulgaris	1.563	48.0
Pisum sativum cytoplasm	1.549	50.2
chloroplasts	1.576	45.7
cytoplasm	1.527±0.002	53.4
chloroplasts	1.555±0.003	49.2
seeds	1.55	50.0

ribosomes of bacteria. The buoyant density of cytoplasmic ribosomes of higher plants is 1.55-1.57 g cm^{-3} (RNA/protein ratio 1.0-1.14) and of chloroplast ribosomes about 1.57 g cm^{-3} (RNA/protein ratio 1.14), i.e. it differs significantly from that of the ribosomes of bacterial type.

The protein patterns obtained by disc-gel electrophoresis for ribosomes from photosynthetic bacteria (Fig. 1, left) show 28-30 bands. The electrophoretic patterns were consistently reproducible whether in different experiments or from different ribosome preparations. The number of bands as well as their distribution are nearly the same in two species of photosynthetic bacteria but differ from those of *Escherichia coli* (see middle section of gel columns).

With respect to the blue-green algae (Fig. 1, middle) several bands with identical electrophoretic mobility can be found in the protein patterns of ribosomes of two representatives of the unicellular *Chroococcales*, *Anacystis* and *Synechocystis*. Also, the protein electrophoretic patterns of ribosomes of the multicellular *Hormogonales*, *Lyngbya*, *Anabaena* and *Plectonema* are very similar, especially those of the closely related species *Anabaena* and *Plectonema* (12 common protein bands; 46% similarity). On the other hand, the high degree of heterology between ribosomal proteins of uni- and multicellular blue-green algae is assessed both by the single-gel and split-gel electrophoresis techniques.

The protein composition of the chloroplast ribosomes of two species of higher plants is very similar (Fig. 1, right). A certain difference may be found in the distribution of several minor components. Chloroplast ribosomes yield 21 stained bands migrating towards the cathode at pH 4.5.

The basic proteins derived from the cytoplasmic ribosomes of the same plant species showed 26-27 electrophoretic bands, distinctively different from those of chloroplast ribosomes. The interspecies differences in the composition of ribosomal proteins are more significant in chloroplasts than in the cytoplasm of higher plants.

The gel-electrophoretic patterns of the basic proteins of ribosomes of chloroplasts and photosynthetic bacteria are different (small number of common bands; different ratios between major and minor components). Protein electrophoretic patterns of chloroplast and blue-green algae ribosomes are also unlike in spite of the presence of about six similarly migrating bands. The differences are mainly in the number of bands and their electrophoretic mobility. VASCONCELOS & BOGORAD (1971) obtained similar electrophoretic data from ribosomal proteins from mung bean and maize chloroplasts and from the multicellular blue-green alga *Phormidium luridum* (species systematically close to *Lyngbya*). Among the several species of blue-green algae we studied, the protein patterns of chloroplast ribosomes and those of the unicellular blue-green alga *Anacystis nidulans* are the most similar as regards to both the staining properties and the electrophoretic mobility of 31-33% of protein bands.

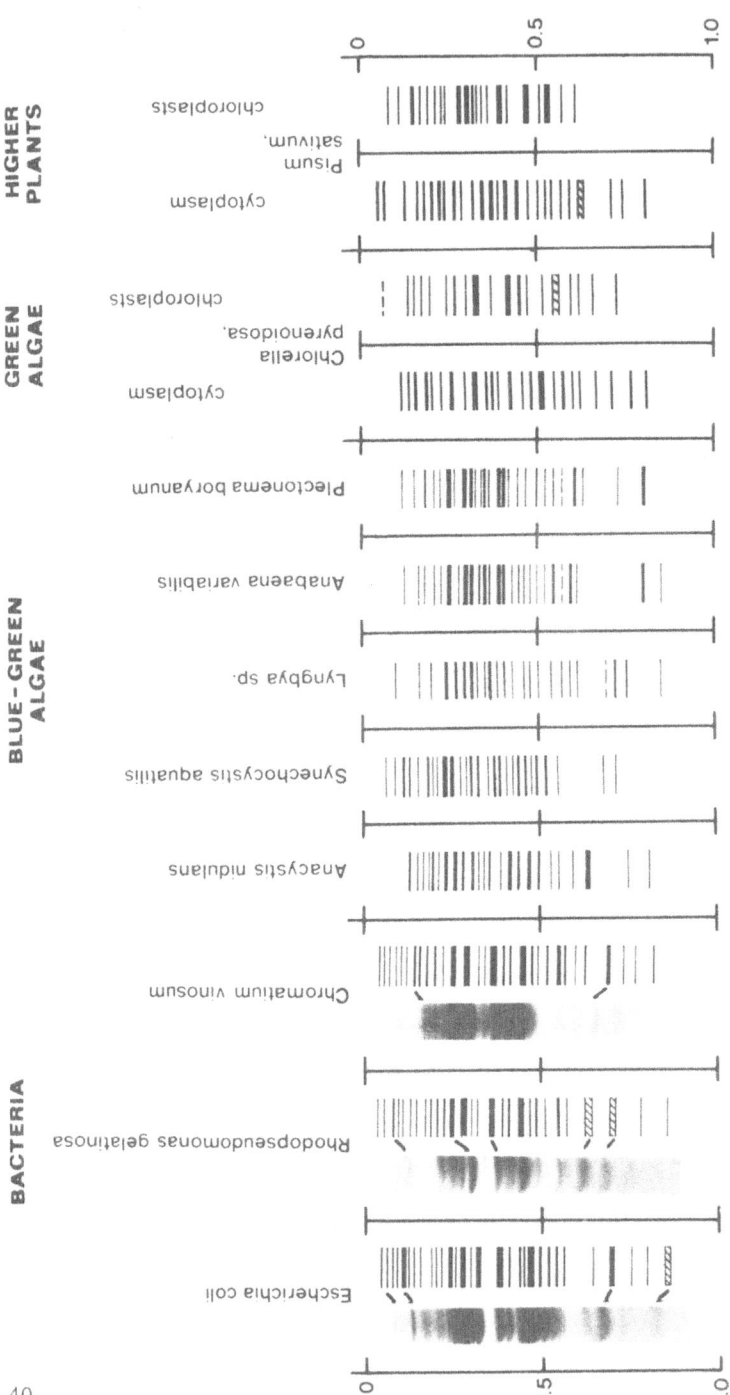

Fig. 1. Polyacrylamide gel electrophoresis patterns of basic ribosomal proteins of bacteria, blue-green and green algae, and higher plants.

Summary

Our results do not testify to the direct phylogenetic relation between chloroplasts, photosynthetic bacteria and blue-green algae, but they do not refute the possible evolutionary relationship of plastids and blue-green algae. These conclusions were confirmed recently by the hybridization experiments of PIGOTT & CARR (1972). Chloroplast DNA does not hybridize with rRNA of photosynthetic bacteria but does with rRNA of unicellular blue-green algae to a significant extent.

REFERENCES

MARGULIS, L., Origin of Eukaryotic Cells, Yale Univ. Press, New Haven 1970.
PIGOTT, G. H. & CARR, N. G., *Science* 175 : 1259, 1972.
VASCONCELOS, A. C. L. & BOGORAD, L., *Biochim. biophys. Acta* 228 : 492, 1971.
YURINA, N. P., ODINTSOVA, M. S. & OPARIN, A. I., *Dokl. Akad. Nauk SSSR* 205 : 997, 1972.

ELECTRON-MICROSCOPIC STUDY OF CHLOROPLASTS IN ZYGOTES OF CHLAMYDOMONAS REINHARDI

V. G. LADYGIN, GALINA A. SEMYONOVA & SOFIA V. TAGEEVA

Institute of Biological Physics, Academy of Sciences of the U.S.S.R., Moscow – Pushchino-na-Oke, U.S.S.R.

Introduction

The level of development of the lamellar system of chloroplasts and their functional ability are known to be genetically controlled characters. Many genes which control the structure and activity of chloroplasts were recently localized in the chromosome map, e.g. in *Chlamydomonas* (HASTINGS et al. 1965; LEVINE 1968; SURZYCKI et al. 1970; STOLBOVA 1971). In a majority of cases classical Mendelian inheritance was found. Attempts were also made to localize on the map extranuclear genes believed to be chloroplast genes (SAGER & RAMANIS 1970).

However, the data at present available does not permit a complete account of the genetic determination of the structure and function of the photosynthetic apparatus. It can only be suggested that the portion of genetic information present in the chloroplast itself (plastome) seems to be small as compared to the nuclear one (genome). It is difficult to evaluate the degree of chloroplast autonomy and to find the characters which are controlled by the plastome because the transmission mechanism of plastid genetic information from parents to progeny is as yet unknown. Some investigators think that, in the process of zygote formation, DNA of the paternal cell chloroplast is lyzed (SAGER & LANE 1969; SAGER & RAMANIS 1970) and consequently the progeny of one zygote obtains the plastom of the maternal chloroplast. Other scientists mentioned the presence of DNA of both gametes in the zygote chloroplast (GILLHAM 1969). Similar contradictions are also found while studying the chloroplast behaviour in the process of zygote formation. In some cases fusion of two gamete chloroplasts with the formation of one chloroplast of the zygote was observed (BASTIA et al. 1969; CAVALIER-SMITH 1970); in others, lysis of paternal cell chloroplasts was seen (GRANICK 1965, BRATEN 1971; FJELD 1971).

The present study follows the behaviour of two plastids of the gametes from copulation up to the formation of one chloroplast of the zygote in the unicellular green alga *Chlamydomonas reinhardi*.

Methods

To differentiate easily between the plastids of maternal and paternal gametes, cells of the initial wild type of the green strain (W.T.) with a well-developed system of thylakoids in the chloroplast were crossed with cells of yellow mutants (Y-3 and Y-4) in which the membrane organization of plastids was represented only by vesicles. Both mutants were obtained by γ-irradiation of wild type cells with doses which, in 95-99% of cases, produced lethal effects.

The composition of media and culture methods of W.T. and mutant cells were described by LADYGIN (1970). Vegetative cells were differentiated into gametes by transplanting them from solid acetate medium into nitrogen-free medium. The gametes were washed from the latter with bidistilled water, and an approximately equal number of the W.T. gametes, mating type minus, and Y-4, mating type plus, were poured together. Copulation and fusion of gametes with zygote formation began at once. Each hour zygote sediment was obtained by centrifugation. It was fixed with a 1% solution of osmic acid in phosphate buffer at pH 7.4 for 1-2 h at 0 °C. Samples were dehydrated in a series of alcohols of rising concentration and in anhydrous acetone. Then the material was embedded in epon-812. Ultrafine sections mounted on nets were studied with the electron microscope UEMV-100.

Results

The electron microscopic studies have shown that the chloroplast of the W.T. gamete has a great number of thylakoids, in some places grouped in stacks (Fig. 1a). The plastid of the Y-4 gamete has a strongly reduced membrane system (Fig. 1b). It has some vesicles and does not form thylakoids. Depending on the time of culture in nitrogen-free medium, partial destruction of lamellae and accumulation of a great number of starch grains is observed in gamete chloroplasts.

Fusion of two gametes proceeds for 20-30 min and is easily observed in a light microscope. 60 min after copulation young zygotes are observed in which two nuclei and two chloroplasts of parental strains are present (Fig. 2a). The plastids are easily differentiated by their morphology: they contain well-developed pyrenoids and stigmae.

2-3 h after zygote formation the onset of nuclei fusion can be observed. After the fusion of nuclei, 0.5-1 h later, the first stages of fusion of chloroplasts are revealed (Fig. 3a, b). The process of fusion begins in both cases with the formation of a 'membrane bridge', formed by the contact of outer membranes of nuclear and chloroplast membranes. The fusion gradually increases (Fig. 4a).

4-5 h after gamete copulation completely fused nuclei can be observed. In some cases two nucleoli are discerned. The fusion band of chloroplasts involves the whole width of the plastid. Two zones are clearly distinguished (due to the differences of the membrane organization) in a single structural unit which is the zygote chloroplast formed by the gamete plastids (Fig. 4b). In a completely form-

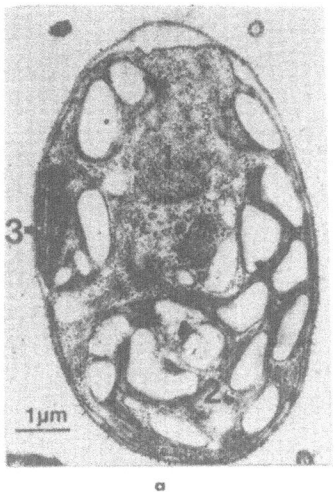
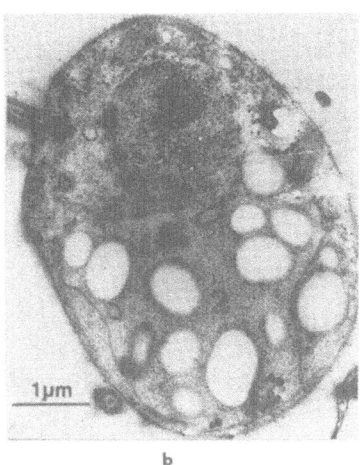

Fig. 1. Structural organization of gametes of *Chlamydomonas reinhardi*: (a) initial wild strain (W.T.) and (b) mutant Y-4: 1 – nucleus, 2 – chloroplast, 3 – stigma.

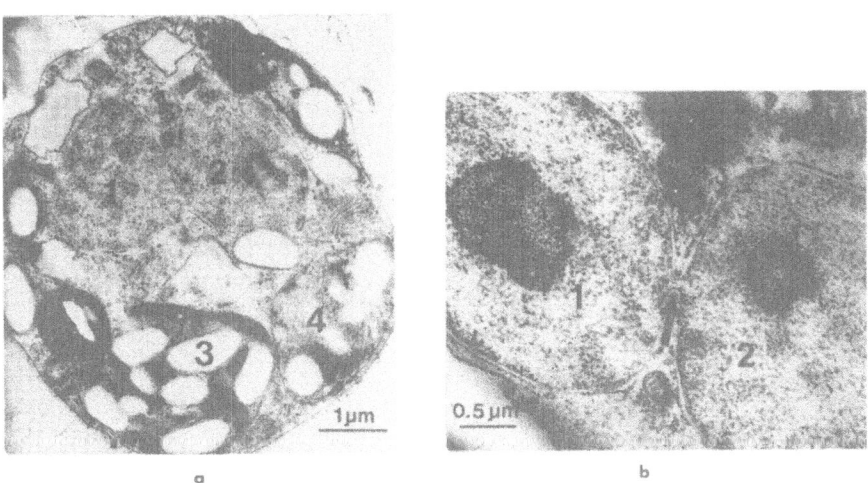

Fig. 2. (a) Zygote with two nuclei (1 and 2) and two chloroplasts of the W.T. (3) and Y-4 (4) gametes. (b) Onset of fusion of gamete nuclei (1 and 2). The arrow indicates the fusion region of the two nuclei.

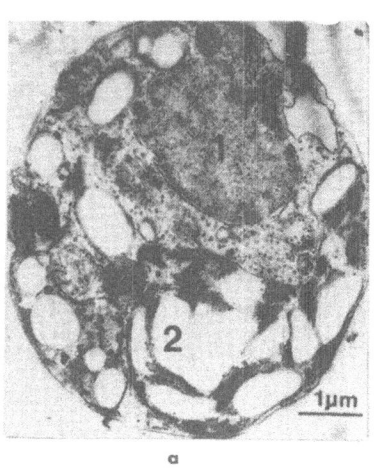 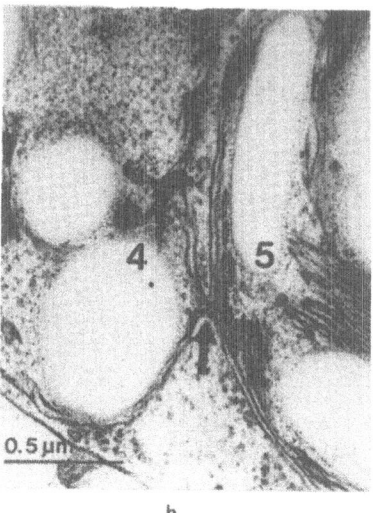

Fig. 3. (a) Zygote in which the fusion of nuclei (1) is almost completed and the fusion of gamete chloroplasts of W.T. (2) and Y-4 (3) just beginning. (b) Zygote fragment in the region where plastids fusion begins (4 and 5). The arrow indicates the fusion region of two chloroplasts.

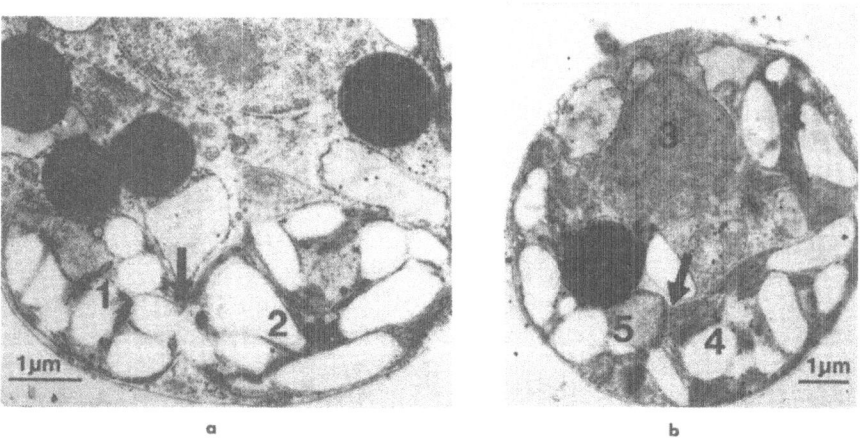

Fig. 4. (a) Zygote fragment in the fusion region of two chloroplasts (1 and 2). (b) Zygote containing one diploid nucleus (3) and one chloroplast formed by the fusion of chloroplasts of the W.T. (4) and Y-4 (5). The arrow in both cases indicates the fusion region of two chloroplasts.

ed chloroplast of the zygote (in 7 h-old zygotes) a single structure is found. Its lamellar organization is similar to that of the W.T. chloroplasts with a well-developed system of thylakoids. Consequently, one chloroplast of the zygote is formed by the fusion of two plastids of the gametes.

Discussion

A genetic analysis of *Chlamydomonas* pigment mutants shows as a rule that the character of the changed pigmentation of colonies of mutants crossed with the W.T. is inherited according to the Mendelian type (STOLBOVA 1971; LADYGIN 1972). Due to the haploidy of vegetative cells of this alga the zygote produces always half of the progeny of the mutant type, another half of the W.T. Such a splitting in the first generation agrees well with the distribution of nuclear genes. The diploid zygote nucleus contains both the mutant gene and its allele of the W.T. When zoospores are formed, during meiosis, the chromosome with the mutant gene is distributed into half the zoospores, while the homologous chromosome containing its allele of the W.T. goes to another half. All the zoospore plastids are formed from one chloroplast of the zygote, and they seem not to differ from each other at first. Only after the plastid of each zoospore comes under the control of its genome is an exact control over the level of pigment accumulation established. Consequently, the lamellar system develops normally only in a half of the progeny while in another half the chloroplast structure remains reduced.

The above consideration is correct only if it is possible to prove that the chloroplast plastome of the zygote is not distributed like the genome. Existence of such a mechanism would require the fulfilment of these conditions: (1) preservation and fusion of gamete chloroplasts in the course of zygote chloroplast formation, (2) preservation and unification of plastomes of parental gamete chloroplasts when the zygote plastom is formed, and (3) existence of the reduction mechanism allowing precise distribution of plastomes into daughter cells as it takes place in the nucleus during meiosis.

As has been shown in the present work and elsewhere on cells of the W.T. of *Chlamydomonas* (BASTIA et al. 1969; CAVALIER-SMITH 1970), the first of these obligatory conditions is present. The data concerning possible fusion of plastomes are still contradictory. Some investigators think that DNA of the chloroplasts of both gametes is preserved in the zygote chloroplast (GILLHAM 1969; DICKINSON & HESLOP-HARRISON 1970), others indicate that DNA of the paternal cell is lost (SAGER & LANE 1969; SAGER & RAMANIS 1970). The division of DNA located in the chloroplast region in the course of vegetative propagation, the mechanism of which may be similar to mitosis (SPREY 1968; BISALPUTRA & BISALPUTRA 1970; DICKINSON 1970), indicates that complex mechanisms are possibly involved in distributing chloroplast genetic information. These mechanisms act not only during vegetative but also during sexual reproduction of zoospore chloroplasts from chloroplasts of the zygote (BISALPUTRA et al. 1971).

Conclusions

The results of our studies on plastids behaviour from the moment of gamete fusion up to the formation of the zygote chloroplast enable a conclusion that both chloroplasts of the gametes in *Chlamydomonas* are preserved during copulation and while fusing they form one chloroplast of the zygote. The membrane organization of gamete plastids in the zygote remains almost unchanged up to the complete fusion of chloroplasts. Fusion of gamete nuclei in the zygote begins 0.5-1 h before the onset of chloroplast fusion is revealed.

REFERENCES

BASTIA, D., CHIANG, K.-S. & SWIFT, H., *J. Cell Biol.* 43: 11a, 1969.
BISALPUTRA, T. & BISALPUTRA, A. A., *J. Ultrastruct. Res.* 32: 417, 1970.
BISALPUTRA, T., SHIELDS, C. M. & MARKHAM, J. W., *J. Microscop.* 10: 83, 1971.
BRATEN, T., *J. Cell. Sci.* 9: 621, 1971.
CAVALIER-SMITH, T., *Nature* 228: 333, 1970.
DICKINSON, H. G. & HESLOP-HARRISON, J., *Cytobios* 2 (6): 103, 1970.
FJELD, A., *Exp. Cell. Res.* 69: 449, 1971.
GILLHAM, N. W., *Amer. Natur.* 103: 365, 1969.
GRANICK, S., *Amer. Natur.* 99: 193, 1965.
HASTINGS, P. J., LEVINE, E. E., COSBEY, E., HUDOCK, M. O., GILLHAM, N. W., SURZYCKI, S. J., LOPPES, R. & LEVINE, R. P., *Microbial genet. Bull.* 23: 17, 1965.
LADYGIN, V. G., *Genetika* 6, 42, 1970.
LADYGIN, V. G., in: FRANK, G. M. (ed.): Biofizika Zhivoï Kletki, Vol. 3, p. 93, Pushchino 1972.
LEVINE, R. P., *Science* 162: 768, 1968.
SAGER, R. & LANE, D., *Fed. Proc.* 28: 347, 1969.
SAGER, R. & RAMANIS, Z., *Proc. nat. Acad. Sci. U.S.A.* 65: 593, 1970.
SPREY, B., *Planta* 78 115, 1968.
STOLBOVA, A. V., *Genetika* 7 (11): 124, 1971.
SURZYCKI, S. J., GOODENOUGH, U. W., LEVINE, R. P. & ARMSTRONG, J. J., *Symp. Soc. exp. Biol.* 24 13 1970

GENETIC CONTROL OF BIOSYNTHESIS IN CHLOROPLASTS

GENETIC AND EVOLUTIONARY RELATIONSHIPS BETWEEN PLASTIDS AND THE NUCLEAR-CYTOPLASMIC SYSTEM

L. BOGORAD

Harvard University, The Biological Laboratories, Cambridge, Massachusetts 02138, U.S.A.

Each green plant cell has at least three separate genetic systems. How these interact is a conspicuous genetic aspect of photosynthesis. Each of the genetic systems, is, furthermore, part of a potentially semi-independent metabolic system whose activity could affect the expression of genes in the other genomes. What are the potential mechanisms of integration among the chloroplasts, the mitochondria, and the nucleo-cytoplasm? Which of these mechanisms is used? We have been exploring some specific aspects of these general problems.

A Comparison of DNA-dependent RNA Polymerases of *Zea mays*

We have been investigating the possibility of intracellular integration through selective gene transcription by examining and comparing the DNA-dependent RNA polymerases of the chloroplasts and of the nucleo-cytoplasmic system of *Zea mays* (BOTTOMLEY et al. 1971; STRAIN et al. 1971; BOGORAD et al. 1972; MULLINIX et al. 1973).

Some properties of the two nuclear enzymes and of a solubilized chloroplast

Table 1. RNA polymerases of *Zea mays*.

	Nuclear			Chloroplast
	I	IIa	IIb	
Elution from DEAE: $(NH_4)_2SO_4$ [M]	0.08	0.18	0.22	0.21 M
Inhibition by 10 μg ml^{-1} α-amanitin [%]	0	90		0
DNA preference	N^D	N^S	N^D	C^S
$\dfrac{25\ mM\ Mg^{++}}{8\ mM\ Mn^{++}}$	2.5	2.5		5

N^D = double-stranded nuclear DNA
N^S = single-stranded nuclear DNA
C^S = single-stranded chloroplast DNA

enzyme are shown in Table 1. The enzyme which is most easily eluted from DEAE (enzyme I) is not inhibited by α-amanitin. Enzyme II consists of two fractions which elute differently from DEAE and differ markedly in their template preference. Enzyme IIb prefers double-stranded maize nuclear DNA while enzyme IIa is more active with single-stranded maize nuclear DNA. All of the nuclear enzymes have different properties than the chloroplast enzyme. The latter elutes from DEAE at about the same position as enzyme IIa and b but is not inhibited by α-amanitin, it prefers single-stranded chloroplast DNA, and its relative activity in the presence of magnesium and manganese distinguishes it sharply from the nuclear enzymes (BOTTOMLEY et al. 1971; STRAIN et al. 1971).

Maize nuclear RNA polymerase IIa is composed of polypeptide subunits with molecular masses of 180 000; 160 000; 35 000; 25 000; 20 000 and 17 000 dalton (MULLINIX et al. 1973). The maize chloroplast enzyme has 180 000; 140 000 dalton subunits and polypeptides of approximately 100 000; 95 000; 85 000 and 40 000 dalton are frequently associated with it. The 180 000 dalton subunits of the two enzymes cannot be distinguished from one another on high resolution sodium dodecyl-containing polyacrylamide gels (SMITH & BOGORAD 1974). Thus nuclear enzyme IIa and the plastid enzyme are distinctly different but one subunit may be common and other subunits of one enzyme could be modified forms of polypeptides present in the other polymerase. Sharing of structural elements could be an integrative mechanism; differences may tune the enzymes to distinctive signals.

Genes Specifying Chloroplast Components

Especially because there are distinctive transcription systems in the nucleus and plastids, the location of a gene could affect its level of expression. A useful system for locating genes for organelle components should have these qualities: an identifiable gene product such as a functionally alterable protein or RNA which is present in large amounts and can be purified so that it can be characterized; and, a system in which genetics can be done. If the gene product is something which can also affect a fundamental activity of the organelle like protein synthesis, all the better.

We have chosen to try to locate genes for chloroplast ribosomal proteins in *Chlamydomonas*. At the time the plan for this work was being considered it was known (a) that *Chlamydomonas reinhardi* has chloroplast and nuclear genetic systems (SAGER & RAMANIS 1970), (b) that *Chlamydomonas* is sensitive to the antibiotic erythromycin (SAGER & RAMANIS 1970); (c) that erythromycin-resistant mutants of *Chlamydomonas* can be obtained (SAGER & RAMANIS 1970); (d) that erythromycin blocks bacterial protein synthesis; and (e) that erythromycin-resistant mutants of *Escherichia coli* have altered ribosomes (OTAKA et al. 1970) and thus at least one kind of mutation to resistance against this antibiotic involves a change in some property of the large subunit of the susceptible ribosome. Within this framework the first step in trying to locate genes for chloroplast ribosome components in *Chlamydomonas* was to determine whether erythromycin binds to specific *Chlamydomonas* ribosomes or to particular ribosomal subunits. We found that ^{14}C-erythromycin binds to 52S subunits of chloroplast ribosomes of *Chlamydomonas* and not to any of the other ribosomal subunits (METS & BOGORAD 1971).

The next step was to isolate erythromycin-resistant mutants and to determine whether they are resistant because cells have become impermeable to erythromycin, or because they have mutated to be able to metabolize erythromycin into some harmless form, or because the 52S ribosomal subunits have changed.

Mutants of *Chlamydomonas* were generated by treating cells with ethylmethane sulfonate. Erythromycin-resistant mutants were selected by plating cells on medium containing erythromycin. Of a large number of isolates, we have examined only nine in detail.

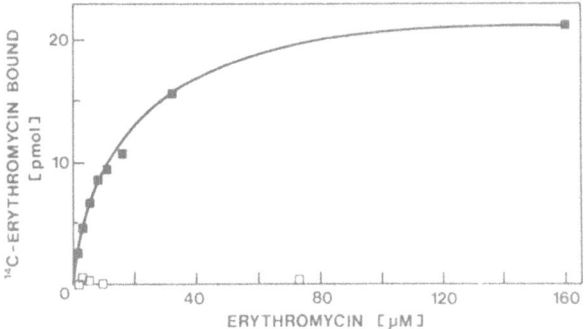

Fig. 1. The binding of ^{14}C-erythromycin to crude ribosomes. The binding to ribosomes from wild-type *Chlamydomonas reinhardi* shown by filled squares; binding to ribosomes from mutant ery-Mlc is shown by open squares. Reaction mixtures of 0.1 ml containing buffer ^{14}C-erythromycin, and 0.27 mg of nucleic acid in a crude ribosome preparation were incubated for 30 min at 25 °C, diluted with 3 ml buffer, and poured over Millipore HA filters. The filters were washed 6 times with 3 ml portions of buffer, dried and counted in toluene-based scintillation fluid. The ^{14}C-erythromycin, 24 counts min^{-1} pmol^{-1}, was the generous gift of Dr. R.H. Williams (Lilly Research Laboratories). (From METS & BOGORAD 1971.)

The solid line in Fig. 1 shows the curve for the saturation of wild-type *Chlamydomonas* ribosomes with erythromycin. The open squares show the behavior typical of each of the nine mutants we have examined. Under the conditions for optimal binding of ^{14}C-erythromycin by wild-type ribosomes no antibiotic was bound by the ribosomes of any of the mutants we have examined. This shows that the ribosomes themselves have been altered in each of these mutant strains.

Having established that some properties of ribosomal subunits had indeed been altered it was time to move to the next step in this investigation — to determine by genetic analyses the site of the gene for each of the mutations we had isolated.

It is shown in Table 2 that we can subdivide the nine mutant strains into three classes which we have designated as ery-M1, ery-M2 and ery-U1. The class ery-M1 includes four mutant strains designated a, b, c and d, etc. When any of the mutants in classes ery-M1 or M2 is mated with wild-type cells half of the offspring are resistant and half, like the wild-type, are sensitive to erythromycin. This is the characteristic Mendelian distribution of the traits. (The zygote is the only 2N stage in the normal life cycle of *Chlamydomonas*.)

Table 2. Erythromycin resistant mutations in *Chlamydomonas reinhardi*.

Locus:	ery-M1	Strains:	a. 1L5	b. 11S7	c. 13L1	d. 15S4
	ery-M2		a. 1S3	b. 1S5	c. 2L2	d. 12S3
	ery-U1		a. 2L1			

Crosses:
1. ery-M1 or -M2 × wild-type = 2 resistant, 2 sensitive.
2. ery-M1 × ery-M1 = all resistant.
3. ery-M2 × ery-M2 = all resistant.
4. ery-M1 × ery-M2 = some resistant (parental) types, some sensitive (non-parental) types, some zoospores not viable.
5. ery-U1 mt$^+$ × wild-type mt$^-$ = all resistant.
6. wild-type mt$^+$ × ery – U1 mt$^-$ = all sensitive.

The basis for stating that there are two classes of Mendelian or nuclear mutations among our strains is the following: if any ery-M1 strain is mated with any other, all 4 offspring are resistant. The same is true when any two ery-M2 strains are crossed. On the other hand, if a member of one of these Mendelian classes is mated with any member of the other class, zygotes form and three types of tetrads have been found: (a) four resistant products, (b) two resistant, one sensitive and one lethal product; and (c) two sensitive and two lethal products. Type '(a)' is recognized as yielding the parental types; i.e. the trait segregated without combining. The lethal products are judged to have arisen by combination of this M1 and M2 mutations while sensitive products are complementary. The latter taken to be recombinants carry wild-type genes at loci M1 and M2.

J. DAVIDSON in our laboratory has found that ery-M1 is linked to the character 'paralyzed flagella-2' and is thus in the Mendelian linkage group XI. All of the ery-M1 mutations are allelic.

The nineth mutant is designated ery-U1 and is represented by a single strain, 2L1. When mating type '+' cells of this strain are crossed with wildtype strains of mating type '–' all of the offspring are resistant. In the reciprocal cross, in which the wild-type cells are the mating type '+', all four offspring are of the wild type and are susceptable to erythromycin. Thus, the character of this mutant is transmitted in a uniparental, i.e. non-Mendelian, fashion. Dr. Ruth SAGER has kindly undertaken to map this mutation and has found that it is linked relatively closely to SM (streptomycin marker) in the uniparental genome of *Chlamydomonas reinhardi*.

Thus, we have found at least two nuclear genes and one chloroplast gene which in some way specify a character of a chloroplast 52S ribosomal subunit. The precise mapping work is being continued.

The next step in this investigation was to determine the biochemical basis of the altered erythromycin binding in each of these strains. This first requires that pure ribosomal subunits be isolated in large quantity.

Conditions have been worked out in which all the subunits of the two classes of ribosomes present in *Chlamydomonas* dissociate from one another. It has been possible to separate each of the types of ribosomal subunits on sucrose gradients in the zonal centrifuge. Fig. 2 shows the distribution of such subunits in the gradient. The identity and purity of the subunits in each of the peaks has been established by recentrifugation of samples from these bands and by gel electrophoresis of the ribosomal RNAs of these bands (VASCONCELOS & BOGORAD 1971; BOGORAD et al. 1972).

As is well known (e.g. GESTELAND & STAEHLIN 1967, VASCONCELOS & BOGORAD 1971), the proteins and RNAs of ribosomal subunits can be easily

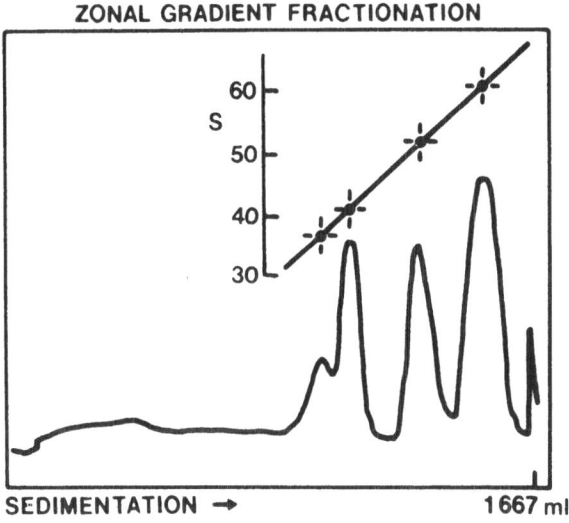

Fig. 2. Separation of *Chlamydomonas reinhardi* ribosomal subunits by sucrose density gradient centrifugation. Wild-type cells were frozen, broken, thawed in 25 mM $MgCl_2$; 25 mM KCl; 25 mM Tris-Cl pH 7.5 buffer, and centrifuged to remove membranes and other insoluble components. Ribosomes were then purified by centrifugation through a layer of buffered 1 M sucrose. The pellet was resuspended in 2.5 mM $MgCl_2$; 200 mM KCl; 25 mM Tris-Cl pH 7.5 buffer and centrifuged through a 1 517 ml 7.25–40% (w/v) sucrose gradient in the same buffer in which the ribosomes had been suspended. After centrifugation at 33 000 rpm for 23 h the rotor was slowed down and the contents were pumped through a UV monitor. The 254 nm absorbance profile is shown here together with the approximate S values for the material at each peak position (BOGORAD et al. 1972; METS & BOGORAD 1972).

dissociated by lithium chloride and urea and the proteins can be examined on polyacrylamide gel electrophoresis in urea. By this method we have found that a ribosomal protein of molecular weight about 6 000 of the 52S ribosomal subunits of mutant 12S3, i.e. ery-M2d, has a slightly slower electrophoretic mobility than the comparable protein of the wild type. That is, a protein which is present in the wild type is absent from ery-M2d but is substituted for by another protein which moves slightly slower in electrophoresis in urea gels (METS & BOGORAD 1972). The precise nature of the alteration which results in a change of electrophoretic mobility is still under investigation.

One dimensional disc gel electrophoresis in urea resolves proteins primarily on the basis of their net charge. Many properties of a protein which might alter its ability to cooperate in the binding of erythromycin might not be revealed in this manner. We have consequently undertaken to develop a two-dimensional system in which ribosomal proteins are first separated electrophoretically in loose (4%) urea

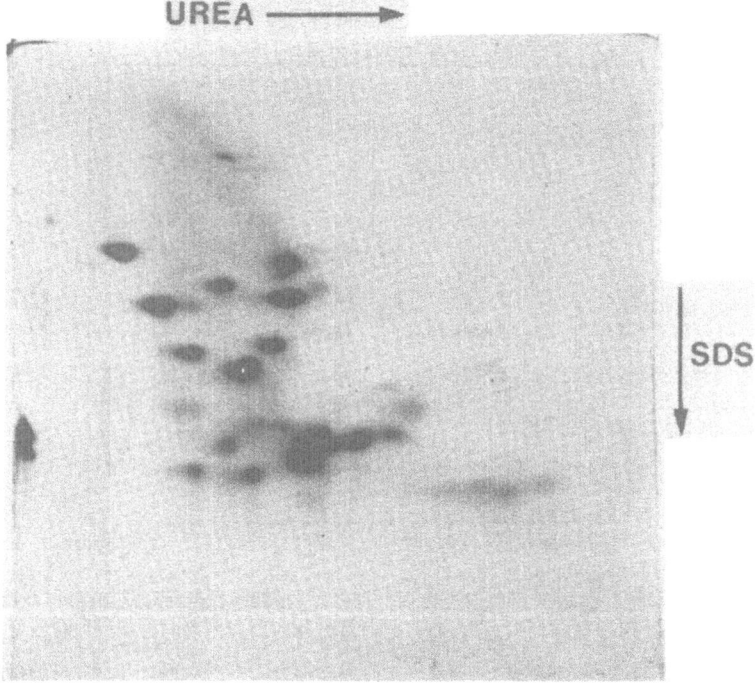

Fig. 3. The proteins of 52S chloroplast ribosomal subunits prepared from wild-type *Chlamydomonas reinhardi* and separated by electrophoresis in urea-containing polyacrylamide gel cylinders at pH 5.0 (left to right) and then by electrophoresis in an SDS-containing polyacrylamide slab (top toward bottom) (BOGORAD et al. 1972; METS & BOGORAD 1972; METS & BOGORAD 1974).

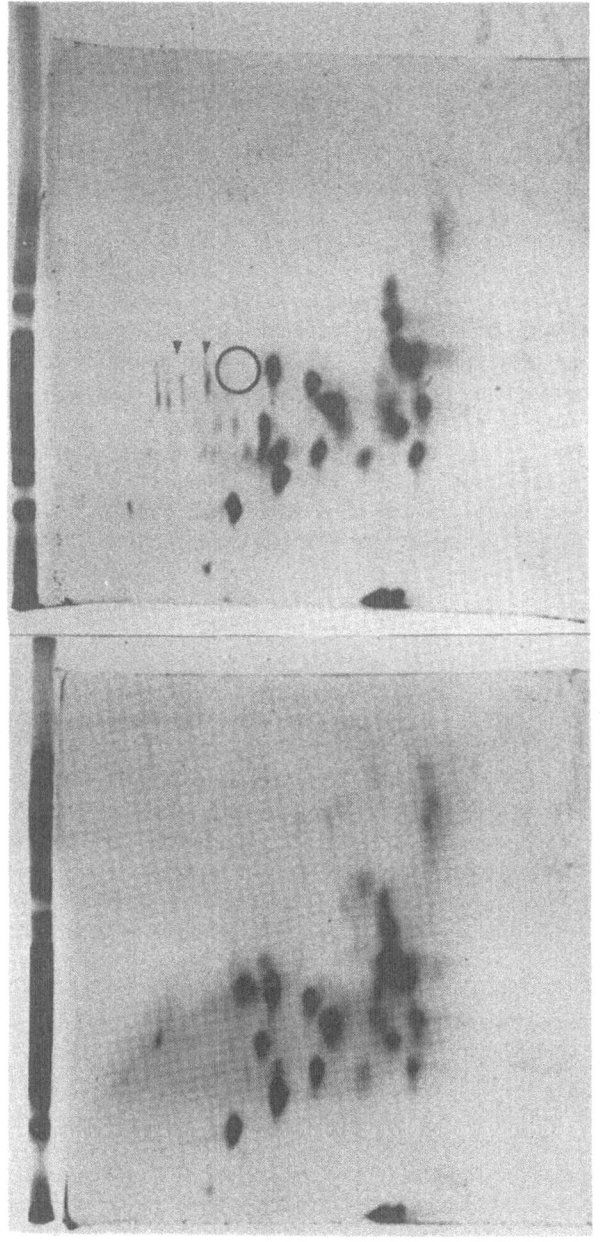

Fig. 4. A composite of two-dimensional gels (left) of proteins from the 52S ribosomal subunits of wild-type cells (right) and from mutant strain ery-U1a (BOGORAD et al. 1972; METS & BOGORAD 1972; METS & BOGORAD 1974). The position of the protein missing from ery-U1a is marked with a circle. The four new components are present as bands just above the circle; the positions of two of them are indicated by the arrow heads. Duplicate 'first-dimension' gels were run of each sample. One was used and the other was stained. The stained gels are shown here above each of the 'second dimension' gel slabs to illustrate the relationships between the two gels.

gels at pH 5 and are then separated in a second dimension in SDS (METS & BOGORAD (1974). In this two-dimensional gel system, separation in one direction should be almost entirely dependent on charge and hardly affected by the size of the protein. The second dimension separation should be entirely on the basis of molecular size. A photograph of a two-dimensional gel of the proteins of the 52S ribosomal subunit of wild-type *Chlamydomonas* is shown in Fig. 3. Fig. 4 is a composite of a pair of two-dimensional gels; one of the proteins of the 52S ribosomal subunits from wild-type cells and one of those from the mutant strain ery-U1a. It is clear that this uniparentally transmitted mutation results in the absence of a protein with a molecular weight of about 33 000 which is normally present in wild-type cells. New polypeptides with about the same electrophoretic mobility as that missing from the wild type, but which differ in their migration on SDS gels, are present instead. It seems likely that the alteration which affects binding of erythromycin also affects the aggregating properties of these polypeptides. We are now attempting to gain additional information about the precise nature of these modifications.

The data presented here show that two genomes of *C. reinhardi* contain genes which specify plastid ribosomal proteins (METS & BOGORAD 1972). We do not know yet whether expression of the nuclear genes can regulate the level of plastid ribosomes.

Analysis of Theories of the Origin and Evolution of Organelles

It is also interesting to consider theories of the origin and evolution of organelles in view of the evidence that both nuclear and chloroplast genes may in some ways specify proteins of chloroplast ribosomes. We do not yet know for certain that 'specify' means alterations in the primary amino acid sequences although it seems very likely with ery-M2d. It is clear, however, that proteins of chloroplast ribosomes are altered in some way by changes in the genes at the ery-M1 and -M2 loci.

There are two logical views of how modern eucaryotic cells could have arisen. One is that they formed by addition. That is, the addition of preexisting procaryotic cells, such as blue-green algae or photosynthetic bacteria or their ancestors, to nucleated cells which had already evolved to have different sizes and types of ribosomes, types of chromosomes, etc. The other possible view is that, rather than by addition, the modern eucaryotic cells arose by subdivision of the contents of a procaryotic cell. The first of these hypotheses has been termed 'the endosymbiont hypothesis' and the second we will call here the 'cluster-clone' hypothesis (Fig. 5). The steps in evolution according to the latter view would have had to be (a) the clustering of genes; (b) the formation of membranes around these clusters (genomes) to separate them from the rest of the cell; and finally (c) the replication of each of the types of membrane-enclosed clusters to form clones of chloroplasts, clones of mitochondria, clones of nuclei, etc.

We must also assume that it is most likely that ribosomes arose only once in evolution. According to the endosymbiont hypothesis, the ribosome of the pro-

Cluster – Clone :

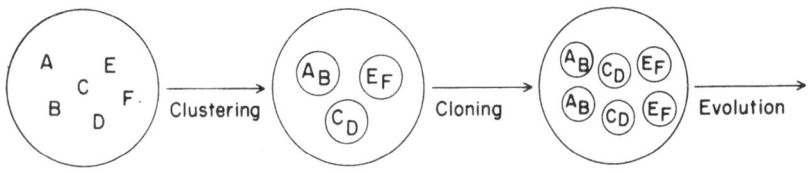

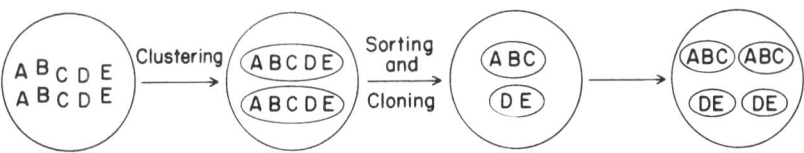

Fig. 5. The cluster-clone hypothesis of the origin of eucaryotic systems. A, B, C, D, E, and F represent genes.
Upper figures: The 'loose' genes group to form A + B, C + D and E + F clusters. A membrane forms around each of these clusters. At the next stage each cluster type replicates in the cloning step. Reorganization and recombination of genes in the various clones may occur during subsequent evolution.
Lower figures: Each 'loose' gene is present in two copies. Each initial clone contains one copy of each gene. By subsequent reduction and selection different types of clones could arise. Although it is not shown in this particular example, one gene could persist in more than one type of clone.

caryotic cell entered the host together with all of the genes for its proteins and RNAs. How have these genes become separated? Let us consider two possibilities: (a) nuclear transformation by gene transfer (Fig. 6a) and (b) protein and gene substitution (Fig. 6b).

We do not know how the genetic systems within the cell interact but it does not require very much straining to imagine that a copy of a gene from a symbiont might be incorporated into the host genome just as in bacterial transformation or in viral transformation of a host cell. The original gene in the symbiont might or might not subsequently be lost. In this way — very directly — the gene for a character of the symbiont could come to occur in the host's genome.

'Protein and gene substitution' is a more indirect way for a gene of an organellar constituent to appear in the nuclear genome. Suppose that a mutation occurred in a gene coding for a chloroplast ribosomal protein and the mutational change was so great that the product was no longer useful. Suppose that simultaneously there had been in the soluble phase throughout the cell — and in the organelle — a protein present for an entirely different purpose but which could substitute for the original chloroplast ribosomal protein. A protein for a cytoplas-

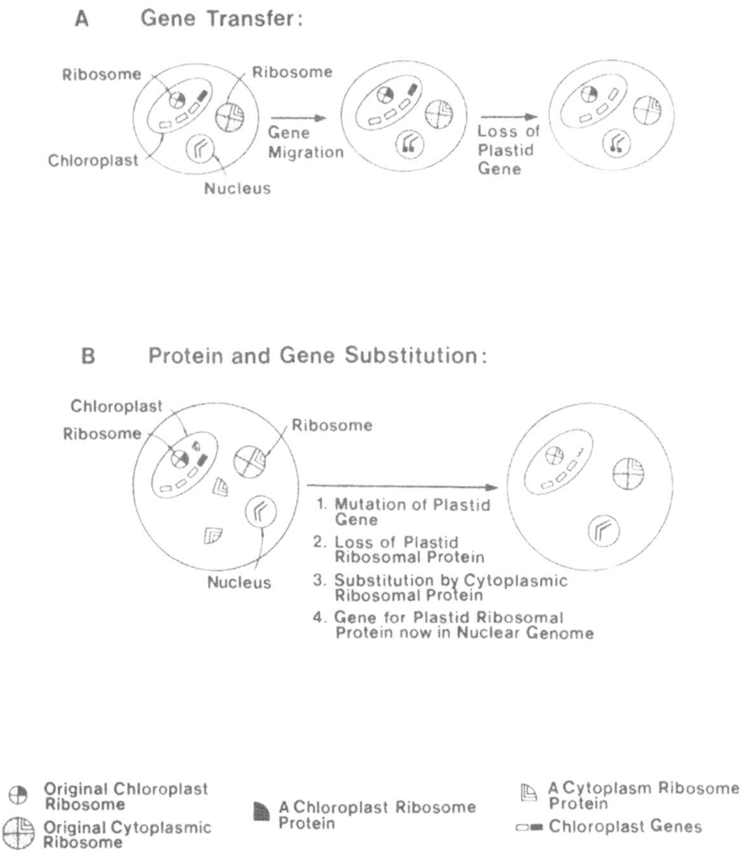

A   Gene Transfer:

Fig. 6. Hypotheses to explain the presence of genes specifying chloroplast ribosomal proteins in the nuclear genome assuming the endosymbiotic origin of eucaryotic cells. A. Nuclear transformation by gene transfer. A copy of the chloroplast gene first appears in the nucleus and is then subsequently lost from the plastid genome. B. Protein and gene substitution. The sequence of events leading to the substitution is noted in the figure.

mic ribosome is an obvious candidate for the substituent. If the gene for the 'new protein' is in the nuclear genome, after protein substitution the gene for this chloroplast ribosomal component is in the nuclear genome.

Obviously, some selective forces would work and only those cells in which such a substitution could occur would be able to survive. (In fact, one might expect that at an earlier point in the adaptation of the symbiont to the cell some substitution might have occurred. Then the pressure for the retention of a particular gene might favor the nuclear one if that protein is in fact being used twice –

once as a substitute on a chloroplast ribosome and once for its original function.)

These, then, are two ways in which the gene for a component which entered with the symbiont could now occur in the nuclear genome.

In the cluster-clone hypothesis the progenitor is a single procaryotic cell. If each gene is represented in the procaryotic progenitor only once, each gene can be present in only a single cluster. At the other extreme, every gene could be present in every cluster as in the case of modern multinucleate cells. One of the many intermediate possibilities could be that some genes are represented in more than one compartment and other genes are represented in only one of the various compartments of the cell.

The procaryote would only have a single type of ribosome. Suppose we start with the possibility that ribosomal genes are present in at least two clusters – the cluster which will give rise to the chloroplast clone and the cluster which will give rise to the nuclear clone. Then evolution of ribosomes, of chromosomes, and of certain other features which distinguish organelles of the nucleo-cytoplasmic system would have had to occur along two separate lines. Evolution along two separate lines is presumed to have occurred in different groups of organisms before the postulated recombination took place according to the endosymbiont view. On the other hand, the cluster-clone view requires that these evolutionary events occurred within organisms after the clustering of genes and the establishment of separate plastid, mitochondrial and nuclear clones.

Evolution within the cell after the establishment of the initial organelle pattern has generally not been considered. One important aspect of this evolution is the possible transfer of genes. Another is that elemination of some functions present in more than one compartment might not have gone exactly the same way in every species. In fact, this would seem very unlikely for, as in the case of social insects, the unit of selection is the entire community. In the present case, we are discussing the community of the nucleo-cytoplasm and the organelles. Information storage and labor could be divided among the members of the community in many different ways to give equally successful and viable organisms. It would be surprising to find that the relationships between the organelles and the nucleo-cytoplasmic system are the same in every individual which will be studied.

Summary

The interactions of organellar and nuclear genetic systems are important in the development and maintenance of photosynthetic activity of chloroplasts. Investigations of two aspects of intracellular integration are described.

Three different DNA-dependent RNA polymerases have been identified in cells of maize leaves. Two of these are located in the nucleus and one in the chloroplast. One of the nuclear enzymes has been purified and its polypeptide subunit composition established. The chloroplast enzyme has been brought to a similar state of purity. Each of these enzymes contains two very large polypeptide chains (molecular weight exceeding 130 000) and three or four smaller polypeptide chains. Each enzyme contains a 180 000 dalton subunit. It is not known whether these are identical nor whether any subunit of one enzyme is derived from a polypeptide of the other.

At least two nuclear genes and one chloroplast gene of *Chlamydomonas* specify proteins of the 52S subunit of chloroplast ribosomes.

The endosymbiont hypothesis of the origin of eucaryotic cells is examined and the cluster-clone hypothesis is presented. Mechanisms by which genes for plastid ribosome components might have become segregated in different genomes are presented and discussed.

Acknowledgements

This work was supported in part by the National Institutes of Health of the United States Public Health Service, by the National Science Foundation (GB-34205), and by the Maria Moors Cabot Foundation of Harvard University.

REFERENCES

BOGORAD, L., METS, L. J., MULLINIX, K. P., SMITH, H. J. & STRAIN, G. C. in: GOODWIN, T. W. (ed.): Proc. of IUBS – Biochem. Soc. Symp., Leeds, England 1972.
BOTTOMLEY, W., SMITH, H. J. & BOGORAD, L., *Proc. nat. Acad. Sci. U.S.A.* 68: 2412, 1971.
GESTELAND, R. F. & STAEHLIN, T., *J. mol. Biol.* 24: 149, 1967.
METS, L. J. & BOGORAD, L., *Science* 174: 707, 1971.
METS, L. & BOGORAD, L., *Proc. nat. Acad. Sci. U.S.A.* 69: 3779, 1972.
METS, L. J. & BOGORAD, L., *Anal. Biochem.* 57: 200, 1974.
MULLINIX, K. P., STRAIN, G. C. & BOGORAD, L., *Proc. nat. Acad. Sci. U.S.A.* 70: 2386, 1973.
OTAKA, E., TERAOKA, H., TAMAKI, M., TANAKA, K. & OSAWA, S., *J. mol. Biol.* 48: 499, 1970.
SAGER, R. & RAMANIS, Z., *Proc. nat. Acad. Sci. U.S.A.* 65: 593, 1970.
SMITH, H. J. & BOGORAD, L., *Proc. nat. Acad. Sci. U.S.A.* 71: 4839, 1974.
STRAIN, G. C., MULLINIX, K. P. & BOGORAD, L., *Proc. nat. Acad. Sci. U.S.A.* 68: 2647, 1971.
VASCONCELOS, A. C. L. & BOGORAD, L., *Biochim. biophys. Acta* 228: 492, 1971.

INTERACTIONS AMONG CELLULAR COMPARTMENTS IN EUGLENA DURING CHLOROPLAST DEVELOPMENT

J. A. SCHIFF

Department of Biology, Brandeis University, Waltham, Mass. 02154 U.S.A.

Introduction

Euglena gracilis KLEBS var. *bacillaris* PRINGSHEIM offers many inducements to students of organelle development and inheritance. SEYMOUR HUTNER, LUIGI PROVASOLI and their numerous co-workers at Haskins Laboratories persuaded the organism to grow luxuriently in axenic culture on a variety of completely defined media over the extraordinarily wide pH range of three to eight. They also showed that this strain of *Euglena* could be made to lose visible plastids and pigmentation reversibly by growth in the dark and irreversibly through growth on streptomycin (PROVASOLI et al. 1948). CRAMER & MYERS (1952) studied its photosynthetic properties while PRINGSHEIM & PRINGSHEIM (1952) who had originally isolated this strain, also showed that growth above 32 °C would bring about irreversible loss of plastids. Out of these observations has grown a body of knowledge and techniques for exploiting these useful attributes.

Fig. 1 summarizes some of the techniques we employ in our laboratory to study chloroplast development and inheritance in this organism. Since *Euglena* is an efficient organotroph and since many carbon sources do not impair the formation and function of chloroplasts, plastid development and even plastid elimination can be studied under conditions where photosynthesis is gratuitous. By employing media containing carbon sources, photosynthesis and plastids can be impaired or eliminated without appreciably affecting the growth or viability of the cells.

In the light the growing organism contains about ten chloroplasts. This number is kept fairly constant as the cells divide, through concomitant chloroplast division (Fig. 2) (GOJDICS 1934). Chloroplast division has been observed in algae for some time, the earliest observations going back more than 50 years (see BOLD 1951). These observations provided some of the earliest evidence that these organelles might possess some degree of autonomy.

When grown in the dark, *Euglena gracilis* var. *bacillaris* contains precursor bodies of the chloroplasts, the proplastids which are smaller in size and possess far less internal structure (SCHIFF 1970; KLEIN et al. 1972; SALVADOR et al. 1971). Since the cells always retain the ability to be induced by light to form chloroplasts when grown indefinitely in the dark, the proplastids must also replicate themselves at each division. This process of plastid replication can be studied

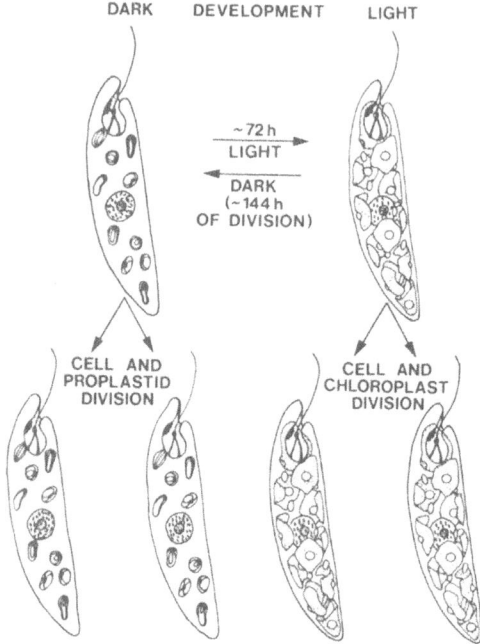

Fig. 1. Plastid replication and development in *Euglena gracilis* (light-grown cell after LEE-DALE 1967).

in either darkness or light in dividing cells and, carried back far enough in time, prompts questions about plastid origins.

The dark-grown proplastid-containing cells can be placed upon a medium lacking essential nutrients which prevents the cells from dividing (STERN et al. 1964b). Under these conditions, when exposed to light, the proplastids will develop into chloroplasts unencumbered, from the experimenter's point of view, with variables due to cell and plastid division (Fig. 1.). This process we call plastid development to distinguish it from plastid replication in dividing cells.

With the flourishing of experimental genetics at the beginning of the century sparked by the rediscovery of MENDEL's works, came the great scientific generalizations which extended and confirmed his principles but along with the confirmation important exceptions to these principles were found as well. Beginning at least with CORRENS, it became apparent that there were genetic determinants in higher plants, particularly those connected with plastid phenotypes, which did not strictly obey Mendelian principles. The behaviour of these characteristics was not identical in reciprocal crosses since the progeny always exhibited the phenotype of the female parent. Since the plastids were known to have some degree of autonomy and since the bulk of the cytoplasm of the zygote came from the egg

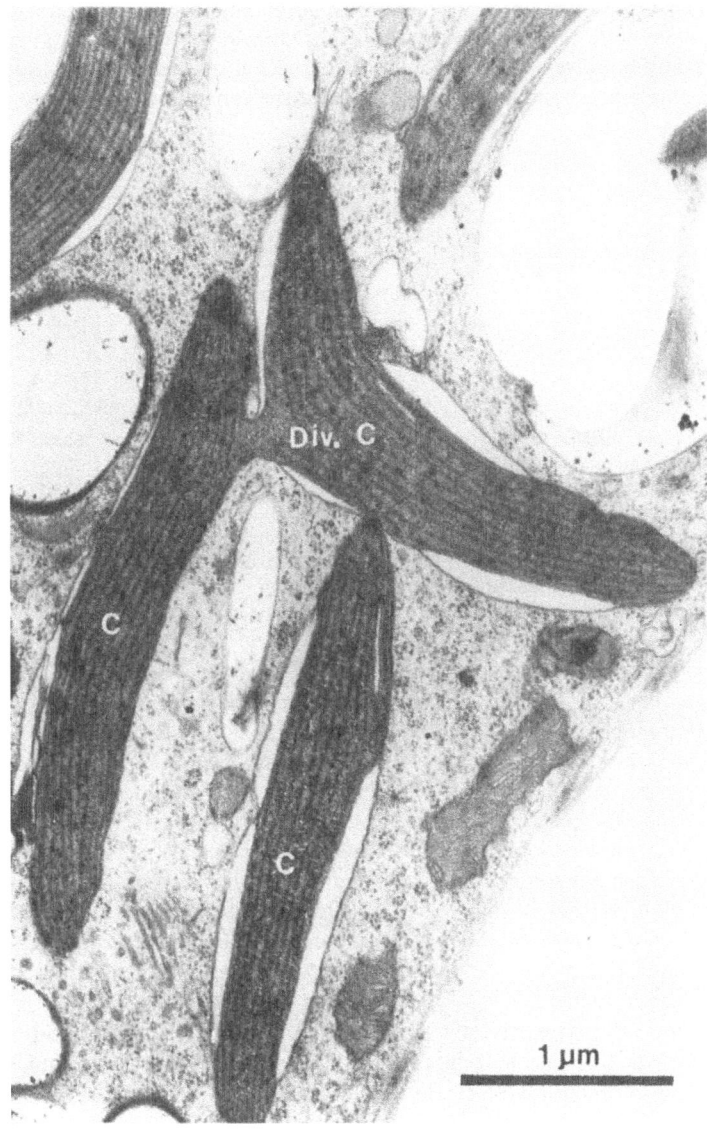

Fig. 2. A dividing plastid (div. c) in *Euglena gracilis* var. *bacillaris*. C-chloroplast.

rather than the pollen, the notion was advanced that this matroclinal inheritance was due to genetic factors which were localized in the cytoplasm and, perhaps, in the plastids themselves. We now know the situation to be more complicated than this, but this picture is a fair representation of the general outlines of the problem (RHOADES 1946).

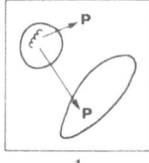 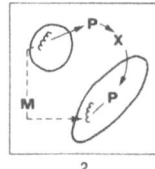 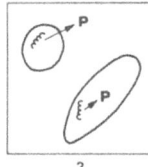

Fig. 3. Simplified hypothetical schemes for possible genetic interactions between organelles. In all cases, the circular structure represents the nucleus and the ellipse depicts the chloroplast. In alternative *1*, an informational unit in the nucleus codes for the proteins of the entire cell including the chloroplast. Alternative *3* represents the other extreme in which the nucleus and chloroplast have independent informational units, the nuclear unit codes for generalized cell protein, and the chloroplast unit determines chloroplast proteins. Alternative *2* is necessitated by genetic studies with higher plants and shows two additional modes of interaction. The nucleus may code for a protein(s) which manufactures nutrients (X) required by the chloroplast. Alternatively, the nucleus may manufacture a mutagen (M) which irreversibly mutates the chloroplast informational unit (SCHIFF & EPSTEIN 1965, 1966, 1968).

On the basis of available information from many organisms, it is possible to make some general models of chloroplast inheritance. Fig. 3 (SCHIFF & EPSTEIN 1965, 1966, 1968) shows three extremes for the purposes of discussion. In the first alternative information for the construction of a plastid resides in the nucleus; this codes for production of everything else in the cell as well. This model predicts that the chloroplast (or proplastid, or any other organelle) is constructed *de novo* in each generation from information supplied by the nucleus. The large numbers of mutations in maize, barley, and other organisms which affect chloroplast phenotypes and which behave in a perfectly Mendelian fashion can be used in support of this model. As we will see later, however, other interpretations of these data are possible.

The third alternative visualizes independent informational units in the nucleus and in the plastid. The unit in the nucleus codes for proteins produced in the nucleus and cytoplasm exclusive of the organelle in question. The organelle itself contains an informational unit which codes for its own proteins. When the cell replicates, the two informational units are replicated independently leading to the possibility of autonomous chloroplast division.

In between these two extremes of interpretation are the possibilities for genetic interaction between the nucleus and the organelle shown in the second alternative of Fig. 3. These possibilities have been suggested by experiments with higher plants. RHOADES (1946) discovered a mutant affecting chloroplast phenotype in maize which he called *iojap*, the *iojap* gene is chromosomal and behaves in a mendelian manner. Plants homozygous for the mutant gene produce abnormal chloroplasts. These abnormalities persist and are perpetuated, however, when the females with abnormal chloroplasts are crossed with males having a normal chromosomal genetic constitution. RHOADES suggested, therefore, that the nucleus and plastid might have different genomes but that a mutant gene in the nucleus could produce a mutagen which irreversibly mutated the genome of the plastids which then continue to replicate the abnormality even after the nuclear constitution was returned to normal. Similar and even more complex interactions have been found in other plants, particularly *Oenothera* (CLELAND 1962).

Nuclear mutations affecting chloroplast phenotypes but which are entirely normal in Mendelian behavior (cited above in connection with alternative 1) can also be reinterpreted here. It is possible that the nucleus and plastid have independent genomes but that during the course of evolution the plastid has become nutritionally dependent on the rest of the cell for one or more metabolites (represented by 'X' in alternative 2). A nuclear mutation which prevented the formation of these nutrients would lead to chloroplast abnormalities even though there was no direct informational dependence of the chloroplast on the nucleus.

As we will see the evidence favors some version of model 2. We have already indicated that the plastid is capable of division and that genetic studies in higher plants implicate matroclinal inheritance indicating that there is a plastid genome distinct from, and in addition to, the nuclear genome. Further along we will supply evidence that the plastid seems to be nutritionally and informationally dependent on the rest of the cell.

The Development of the Proplastid into the Chloroplast in Euglena

Dark-grown cells of *Euglena* contain proplastids which are 1-2 μm in size and are roughly spheroidal (Fig. 4 and 5) (SCHIFF 1970; KLEIN et al. 1972; SALVADOR et al. 1971). The organelle is bounded by a membrane at least two layers in thickness as found for the fully mature chloroplast (GIBBS 1960). Ribosome-like bodies are present in the proplastids which are smaller than those in the surrounding cytoplasm consistent with the finding that the plastid ribosomes are 70S and those of the cytoplasm are 87S (RAWSON & STUTZ 1968, 1969; SCOTT & SMILLIE 1969; HEIZMANN 1970; AVADHANI & BUETOW 1972; LEDOIGT, COHEN & SCHIFF, unpublished). Antibiotics which selectively interfere with 70S ribosomes selectively affect chloroplast development and inheritance in *Euglena* without affecting cell viability and growth. This undoubtedly arises from the lack of sensitivity of the 87S ribosomes to these inhibitors together with a mitochondrial resistance to their action. It has not been possible so far to demonstrate ribosomes in electron micrographs of *Euglena* mitochondria (Fig. 5) and should they be completely absent or very different this could explain the high selectivity of 70S inhibitors for plastids in *Euglena*. There have been two recent reports of ribosomes and RNAs from the mitochondria of *Euglena*, however (KRAWIEC &

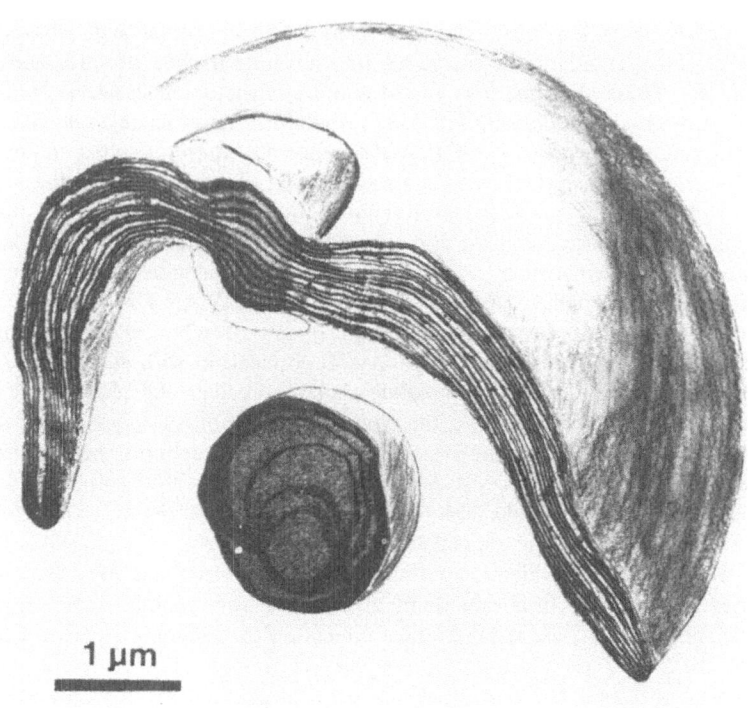

Fig. 4. Three dimensional visualization of the proplastid and chloroplast from thin sections.

EISENSTADT 1970; AVADHANI & BUETOW 1972) and if these components perform indispensable functions it may be that the mitochondrion is impermeable to these antibiotics.

The proplastids also contain prolamellar bodies which are composed of tubular elements but these are much less highly organized than those in higher plants. Also, in *Euglena*, the prolamellar bodies are smaller than in higher plants and the primary membranes or thylakoids extending from them are more extensive (KLEIN et al. 1972).

When dark-grown resting cells of *Euglena* are placed in the light chloroplast development begins. Chlorophyll synthesis for the first 12-14 h is continuous at a low rate (STERN et al. 1964b). During this lag period, there is little or no change in the amount of membrane material in the plastids (KLEIN et al. 1972). Photosynthesis makes its appearance at about four hours of development (STERN et al. 1964b; SCHIFF 1963).

From 14 hours onward chlorophyll synthesis is rapid and photosynthesis grows concommittantly as shown in Fig. 6 (STERN et al. 1964b). During this period, until the completion of development at 72-96 hours of development, the plastids

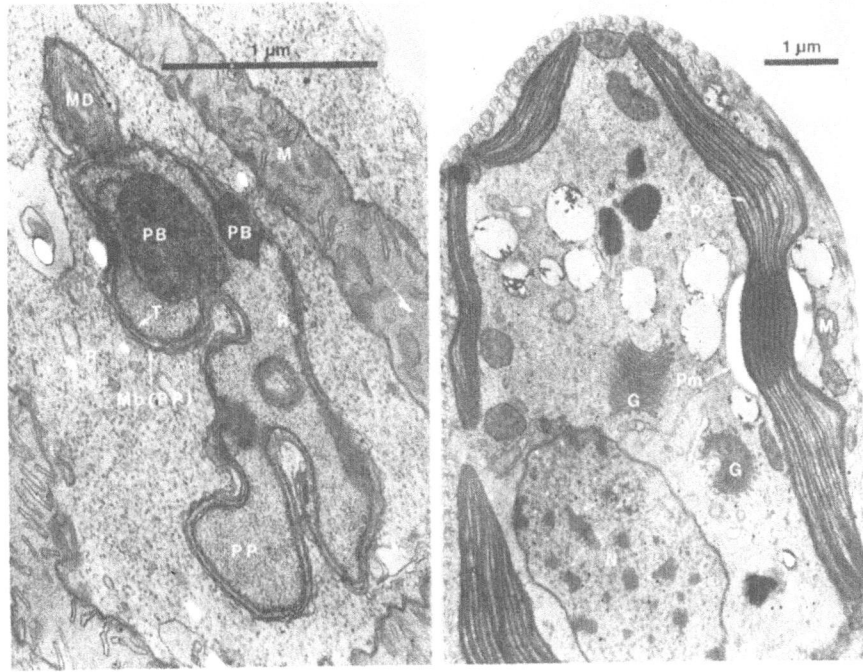

Fig. 5. Sections of *Euglena gracilis* var. *bacillaris* showing a proplastid (dark-grown cell) and a chloroplast after light-induced development (SCHIFF 1970). Abbreviations: C, chloroplast; G, Golgi; L, lamella; M, mitochondrion; Mb, membrane(s); MD, membrane depot; N, nucleus; PB, prolamellar body; Pm, paramylum; Po, polyphosphate; PP, proplastid; R, ribosome-like particles; T, thylakoids.

increase in size and reorganize into a discoid shape (BEN-SHAUL et al. 1964; KLEIN et al. 1972). At the same time thylakoids fuse with each other to form stacks or bands composed of 2-4 thylakoids closely appressed to each other, and much new membrane material is formed and incorporated into thylakoids and bands. The pyrenoid differentiates between 24 and 48 h of development. By the end of development, the proplastid has increased about 60 fold in volume in becoming the chloroplast (Fig. 4, 5, and 7).

The period of chloroplast development in *Euglena* is a period of rapid synthesis of many constituents. Molecular synthesis as evidenced by the appearance of chloroplast-specific antigens, enzymes, lipids, carotenoids and t-RNAs (see SCHIFF & EPSTEIN 1965, 1968 for references, with the addition of: BARNETT et al. 1969, CARELL et al. 1970) is extensive and is associated with an increase in the rate of oxygen uptake by the cells. In many ways, the situation is reminiscent of adaptive enzyme formation. Presented with lactose in the medium as a potential substrate *E. coli* cells produce the necessary enzymes to utilize this substrate.

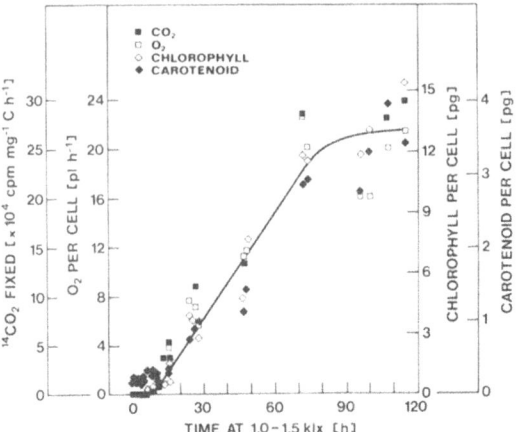

Fig. 6. Kinetics of the appearance of chlorophyll, carotenoids, photosynthetic oxygen evolution, and photosynthetic carbon dioxide fixation during chloroplast development in *Euglena*. Zero time represents measurements on dark-grown cells; time is measured from inception of light-induced chloroplast development (STERN et al. 1964b).

Presented with light, a potential substrate, dark-grown cells of *Euglena* produce the necessary machinery to use this light, but unlike lactose utilization which requires the formation of a few activities, photosynthesis requires the induction of many activities and much structure, all properly coordinated. This sequential induction and coordination of synthesis we call chloroplast development.

At least three interesting questions are prompted by the material presented so far: (1) Where do the building blocks for plastid development originate? (2) Where does the genetic information necessary to build the chloroplast originate? (3) How did the plastid originate?

Where do the Building Blocks for Plastid Development Originate?

Several lines of evidence indicated that the developing plastid might not provide the energy and intermediates for its own synthesis (SCHIFF & ZELDIN 1968). Photosynthesis does not make its appearance about 4 h of chloroplast development and the rate is quite low before 12-14 h. Also, the optimal illuminance for chloroplast development (1.5 klx) is much lower than the optimum for photosynthesis at all stages of chloroplast development (22 klx— STERN et al. 1964b) suggesting that plastid development is not very sensitive to the rate of photosynthesis. Further, the onset of chloroplast development is associated with an increase in the rate of cellular respiration (SCHIFF 1963) suggesting that the cytoplasm is called upon to supply energy for chloroplast development and this is in agreement with the finding that the proplastids of *Euglena* have a large surface in

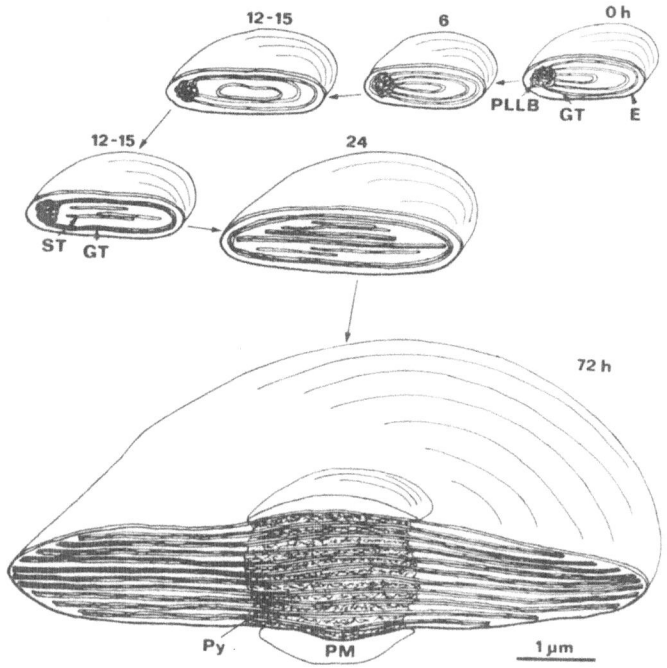

Fig. 7. Diagrammatic representation of the important structural features of light-induced plastid development in *Euglena* roughly to scale. Abbreviations: E, envelope; Gt, girdle thylakoid; PLLB, prolamellar-like body; PM, paramylum; Py, pyrenoid; ST, straight thylakoid. For details see text. (From KLEIN et al. 1972.)

contact with the cytoplasm through extensive invaginations which frequently contain mitochondria and other cellular organelles (SCHIFF 1970) (Fig. 8). Dinitrophenol, an uncoupler of mitochondrial oxidative phosphorylation, inhibits chloroplast development (EVANS 1971).

We decided to check whether photosynthesis was necessary for plastid development by carrying out the developmental process in the presence of DCMU, a specific and powerful inhibitor of photosynthesis which acts by preventing the emergence of electrons from system II (SCHIFF et al. 1967). This DCMU treatment should deny the developing plastid ATP produced during non-cyclic photophosphorylation and reducing power in the form of TPNH. Concentrations of DCMU which block photosynthetic carbon dioxide fixation completely do not impair chlorophyll formation during chloroplast development to any great extent (Fig. 9.). If the DCMU is washed away at the end of plastid development (72 h) full photosynthetic activity appears indicating that all constituents necessary for normal development of the plastids have been formed in the presence of DCMU

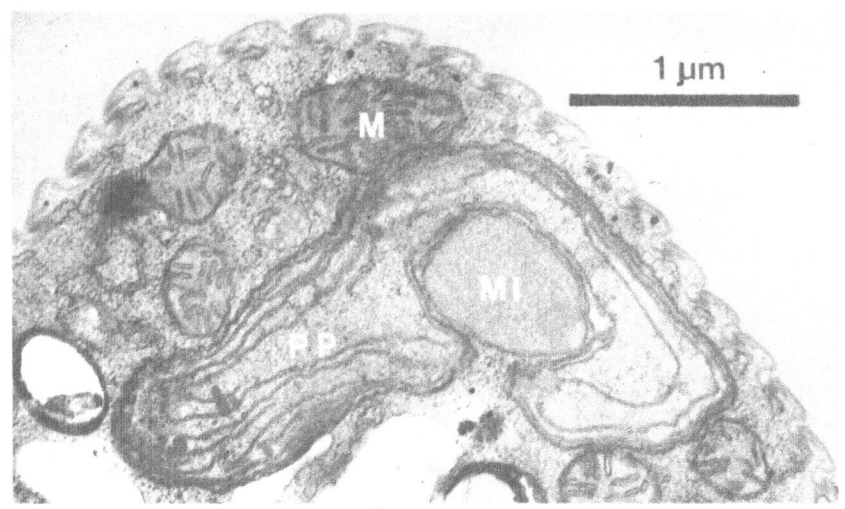

Fig. 8. Section through a dark-grown resting *Euglena* cell which has undergone light-induced chloroplast development for 24 h showing a close association of mitochondria (M) and proplastid (PP) as well as a microbody (MI) in an invagination of the proplastid.

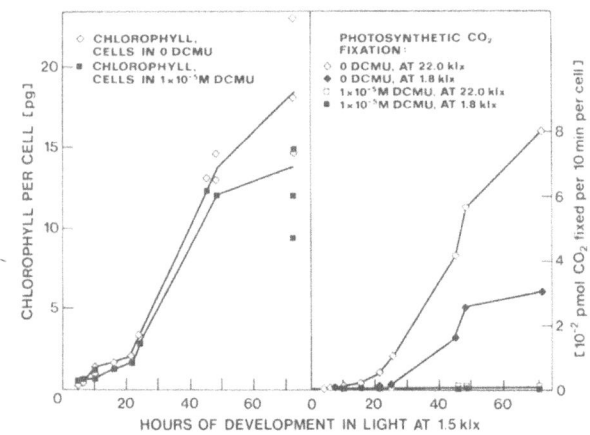

Fig. 9. Chlorophyll synthesis and photosynthetic carbon dioxide fixation by dark-grown resting cells of *Euglena* exposed to light in the presence or absence of DCMU (SCHIFF et al. 1967).

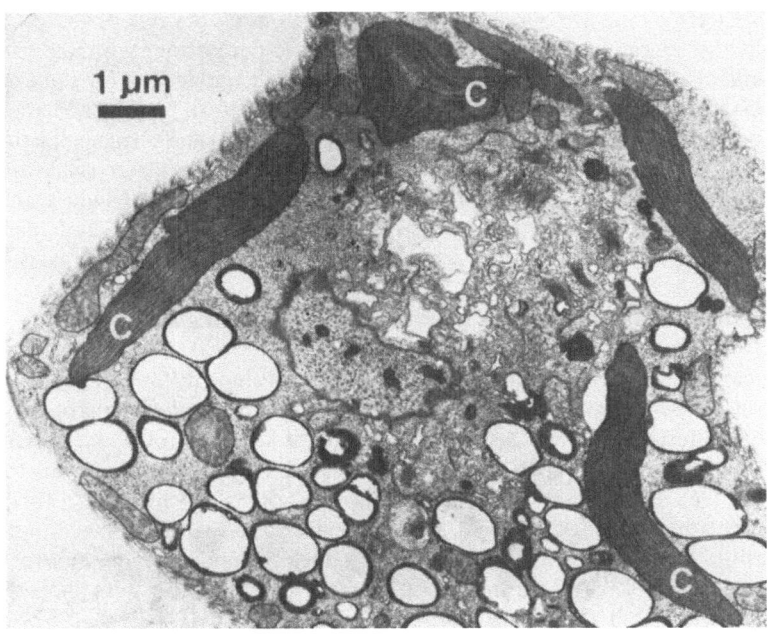

Fig. 10. Dark-grown resting *Euglena* cell after 72 h of development at 1.5 klx in the presence of 10 μM DCMU (SCHIFF et al. 1967). C, chloroplast.

(SCHIFF et al. 1967). Fig. 10 shows that plastids of normal size are formed with extensive membrane structures. An interesting peculiarity of these plastids is that they usually lack pyrenoids. Since the pyrenoid has classically been thought to be the place where the products of photosynthesis are organized into polysaccharide, it is possible that the products of photosynthesis normally induce the pyrenoid; in the presence of DCMU there are no photosynthetic products, hence no pyrenoids. (Pyrenoid differentiation will be discussed further along in this paper.)

These data indicate that the developing plastid need not rely upon its own photosynthesis for the energy and building blocks for its own development and strongly suggest that the developing plastid calls upon the rest of the cell for many constituents. This would explain why those many mutants of maize and other higher plants which affect chloroplast phenotypes are normal Mendelian mutants which map to nuclear chromosomal loci. These are probably mutations which deny the developing plastid a source of certain constituents ordinarily supplied by the rest of the cell.

After we had completed this work, we realized that nature had already performed this experiment for us. Many organisms such as *Chlorella*, *Chlamydomonas*, and pine cotyledons form normal chloroplasts and chlorophyll in the dark

(BOGORAD 1950; SAGER 1958). In these organisms the rest of the cell must supply the energy and building blocks for plastid development. Similarly, those mutants of *Chlamydomonas* and *Scenedesmus* which are blocked for some step in photosynthesis but make otherwise normal chloroplasts can be enlisted to support this model (LEVINE & VOLKMANN 1961; BISHOP 1966). Higher plants and *Chlorella protothecoidies* also make some plastid structures in the presence of CMU (KLEIN & NEUMANN 1966; MATSUKA & HASE 1966). It would seem that chloroplast nutrition requires extensive contributions of energy and materials from the rest of the cell, and this seems to be true of many different organisms.

Where Does the Information for Building a Chloroplast Originate?

The reason this question is not trivial is that at least three different DNA genomes exist in the *Euglena* cell (Tab. 1). The existence of the *Euglena* chloroplast DNA was first inferred from data provided by studies of ultraviolet (UV) inactivation of plastid inheritance (LYMAN et al. 1959, 1961). UV quantitatively eliminated the ability of *Euglena* cells to transmit the plastid-forming potentiality to their progeny. This inactivation displayed a multiplicity of 30 indicating that there were 30 entities concerned and the action spectrum for the phenomenon indicated that a nucleoprotein was the chromophore involved. The UV effect was photoreactivable, suggesting that a DNA nucleoprotein was involved (SCHIFF et al. 1961). Since the number of entities was large, and since no loss of cell viability occurred at UV doses which produced 100% loss of plastid-forming ability, the entities were assumed to be cytoplasmic (LYMAN et al. 1961; SCHIFF et al. 1961). This was later confirmed with UV microbeam experiments (GIBOR & GRANICK 1962a, b).

It then became possible to demonstrate a unique species of DNA which was associated with the ability of the cells to produce plastids (LEFF et al. 1963). This DNA was shown to be chloroplast-localized (BRAWERMAN & EISENSTADT 1964a, b; EDELMAN et al. 1964; RAY & HANAWALT 1964) and was found to be absent in mutant strains cloned from dividing cells which had been treated with streptomycin, UV or temperatures above 34 °C (EDELMAN et al. 1965; RAY & HANAWALT 1965). It soon became apparent that there was yet another species of DNA with a base composition different from either chloroplast or nuclear DNA which could be localized to the mitochondria (EDELMAN et al. 1966). The properties of these DNAs are shown in Table 1.

Also shown in Table 1 are the amounts of DNA per organelle. The value for the amount per mitochondrion is very approximate since the exact number of mitochondria in a *Euglena* cell has not been determined with precision – the value shown here is probably somewhat too high. However, it is clear that the plastid contains about 4.5×10^8 dalton of DNA and if we assume the thirty UV sensitive entities are equally divided among the ten plastids, then each plastid contains three DNA genomes of about 1.5×10^8 dalton each. Considering the analytic uncertainty upon which this value is based (it is calculated from early values of

Table 1. DNA of *Euglena gracilis*.

Source and method	Main band		Chloroplast satellite		Mitochondrial satellite		References
	AT	GC*	AT	GC	AT	GC	
E. g. bacillaris							
Density	52	48	74	26	69	31	LEFF et al. (1963); EDELMAN et al. (1964, 1965, 1966)
Thermal denaturation	45	55	70	30	–	–	EDELMAN et al. (1964)
Analysis	47	53	76	24	–	–	RAY & HANAWALT (1964)
Z strain							
Density	51	49	76	24	67	33	BRAWERMAN & EISENSTADT (1964a, b)
Analysis	49	51	75	25	–	–	BRAWERMAN & EISENSTADT (1964a, b)
Thermal denaturation**	47-52	48-53	74-79	21-26	–	–	BRAWERMAN & EISENSTADT (1964a, b)
Molecular weight, as isolated	$20\text{-}40 \times 10^6$		8.3×10^7 (circular)		$2.9\text{-}3.2 \times 10^6$		RAY & HANAWALT (1964, 1965); SCHORI et al. (1970); MANNING et al. (1971)
Denaturation studies	Double-stranded		Double-stranded		Double-stranded		See above references
Density [g cm^{-3}]	1.707		1.686		1.691		
Total DNA [%]	97.0		1.5		1.5		
pg per cell**	0.485		0.0075		0.0075		EDELMAN et al. (1964, 1966)
pg per organelle**	0.485		0.00075		0.000027		
Dalton organelle **	2.9×10^{11}		4.5×10^8		$\sim 16 \times 10^6$		

* Includes approximately 2.3% methyl cytosine (BRAWERMAN et al. 1962; RAY & HANAWALT 1964). Main band is assumed to represent the nuclear DNA.
** Calculated from data given in reference.

the amount of chloroplast DNA in whole cells) it is not very different from the value of 8.3×10^7 found recently for each circular DNA molecule from the chloroplasts of *Euglena* (MANNING et al. 1971). Thus each plastid genome contains a substantial amount of potential genetic information, enough to code for about 100-200 proteins of 50 000 molecular weight, using the older figures, or about 50-100 using the newer ones. The mitochondrion, on the other hand, is not so fortunate. It contains only enough information to code for about 20 such proteins (being generous). Thus as has been found for many other systems (see MILLER 1970 for recent reviews), the mitochondrial genome of *Euglena* is rather small while the plastid has a rather considerable genome although somewhat smaller than those found in bacterial cells.

When the plastid DNA was first found, we had the simplistic notion that it might supply all of the information for plastid synthesis (alternative 3 in Fig. 3). Early experiments with $^{32}PO_4^{3-}$, where the bulk RNAs of the cell became labeled during chloroplast development yielded results which indicated that this notion was too simple (ZELDIN & SCHIFF 1967). It was found that the RNAs of both the chloroplast compartment and of the non-chloroplast compartments of the cell became labeled, although the chloroplast RNA had about threefold higher specific activity. This has since been found to be true for plastid development in radish cotyledons (INGLE 1968) and *Ochromonas danica* (GIBBS 1969) a flagellated ameboid member of the yellow algae. These results are consistent with an activation of RNA synthesis both within and without the plastid brought about by light activation of plastid development. It is also consistent with the view already advanced that the synthetic machinery of the cytoplasm provides intermediates for plastid formation.

Some relevant findings bearing on the question of the origin of information for chloroplast development have been obtained from studies of streptomycin inhibition of development. As mentioned early in this paper, streptomycin causes plastid elimination and a loss of plastid DNA in dividing cells without impairing either cell viability or cell division, suggesting that streptomycin acts on the chloroplast in a highly selective manner. In bacterial systems, this inhibitor has been shown to exert a major effect by blocking translation on ribosomes during protein synthesis and is selective for the 70S bacterial ribosomes thought to be closely related to the chloroplast ribosomes (MODOLELL & DAVIS 1968; WEISBLUM & DAVIES 1968). We surmised that streptomycin might block plastid replication in *Euglena* by selectively inhibiting translation on plastid ribosomes leading to the non-production of enzymes (such as the plastid DNA polymerase) necessary for plastid DNA replication, without affecting ribosomes external to the chloroplast. In order to determine whether streptomycin does indeed block protein synthesis in plastids, we turned to a study of the effects of this inhibitor on chloroplast development and the formation of plastid enzymes in non-dividing cells (BOVARNICK et al. 1968). Unexpectedly, streptomycin proved to exhibit at least two patterns of action on the formation of certain chloroplast enzymes (BOVARNICK et al. 1970; J. BOVARNICK and J. SCHIFF, unpublished): the synthesis of pig-

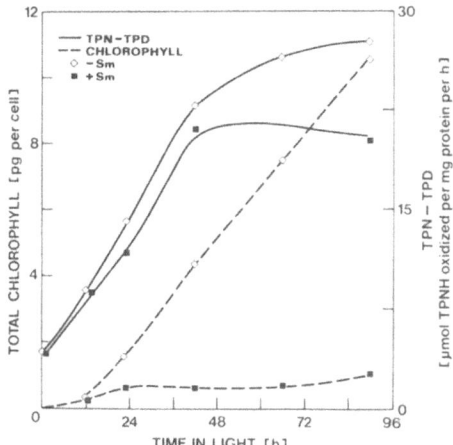

Fig. 11. Chlorophyll and TPN triose phosphate dehydrogenase (TPN-TPD) appearance during light-induced chloroplast development in dark-grown resting cells of *Euglena* in the presence and absence of 0.05% streptomycin (Sm). (BOVARNICK *et al.* 1970; BOVARNICK & SCHIFF, unpublished).

ments and of ribulose diphosphate carboxylase (an enzyme of the photosynthetic carbon dioxide fixation cycle) and of cytochrome 552, a carrier in photosynthetic electron transport, were inhibited from 15 h onward in development resulting in a 90% inhibition by 72 h (Fig. 11, Table 2). Another enzyme of the photosynthetic carbon cycle, the TPN-linked triose phosphate dehydrogenase, behaved in an entirely different manner. Its appearance during development (Fig. 11) commences without a 12 hour lag and the final levels achieved demonstrate that its synthesis is not strongly inhibited by streptomycin. To be sure that this pattern was consistent, chloramphenicol, another specific inhibitor of translation on 70S ribosomes, was employed with comparable results (Table 3): all parameters of plastid development measured were inhibited by chloramphenicol except for the TPN-triose phosphate dehydrogenase whose formation was virtually unaffected (BOVARNICK et al. 1970). As one would expect, the synthesis of the TPN-triose phosphate dehydrogenase is strongly inhibited by cycloheximide, an inhibitor of translation on the 87S ribosomes of the cytoplasm (EGAN & SCHIFF, unpublished).

Which DNA genomes are concerned with the synthesis of these two groups of enzymes? Comparing the activities in a mutant of *Euglena* (W_3BUL) in which chloroplast DNA is undetectable (Table 2) we see that those enzymes which are thought to be synthesized on chloroplast ribosomes are not detectable, suggesting that chloroplast DNA controls their synthesis. The TPN triose phosphate dehydrogenase and the DPN triose phosphate dehydrogenase are both present at the same levels as in the dark-grown cells suggesting that these enzymes are coded by

Table 2. Enzyme formation during chloroplast development in non-dividing Euglena (wild type and W_3 BUL) in the presence of streptomycin (from BOVARNICK et al. 1970, BOVARNICK & SCHIFF, unpublished).

	Wild type			W_3 BUL
	Dark-grown	Dark-grown + light, 72-96 h	Dark-grown + light, 72-96 h. + streptomycin	Light-grown
Structure	Proplastid	Chloroplast	Chloroplast rudiment	Largely absent
Plastid DNA	+	+	+	0
Total chlorophyll [pg per cell]	0	10.50±0.86	0.88±0.22	0
Total carotenoid [pg per cell]	0.26±0.07	2.62±0.19	0.85±0.12	0.13
PS CO_2 fixation [× 10 pmol per cell per h]	0.08±0.02	28	0.44	0
Cytochrome 552 [× 10 μmol per cell]	0	2.79±0.39	0.37±0.60	0
Ribulose diphosphate carboxylase [μmol CO_2 fixed per mg protein per h]	0.14±0.02	4.11±0.34	0.16±0.06	0
G-3-P dehydrogenase				
TPN } [μmol PNH oxidized per mg	4.10±0.70	27.00±6.83	16.00±6.25	4.01
DPN } protein per h]	43.00±14.00	27.00±5.42	32.00±4.91	40.10

0 = below limit of detection; + = present.

Table 3. Enzyme formation during chloroplast development in *Euglena* in the presence of chloramphenicol (from BOVARNICK et al. 1970; BOVARNICK & SCHIFF, unpublished).

	Dark-grown 0 hours	Dark-grown + light, 72 h	Dark-grown + light, 72 h + chloramphenicol [2 mg per ml]
Chlorophyll [pg per cell]	0	10.07±1.59	4.95±0.83
Photosynthesis [pmol CO_2 fixed per cell per 10 min]	0	0.265±0.058	0.07±0.02
RuDP carboxylase [μmol CO_2 fixed per mg protein per h]	0.053 0.046	2.11 9.68 3.32	0.807 1.43 0.59
TPN triose-P dehydrogenase [μmol TPN oxidized per mg protein per h]	6.17 4.26	31.40 40.37 33.75	18.60 37.34 28.46

nuclear DNA. We cannot yet say that the structural genes for the first group are chloroplastic but further studies may well show this to be the case. Nuclear DNA, then, can provide information for the synthesis of a chloroplast protein; elegant studies in yeast have shown that the structural gene for the synthesis of a mitochondrial cytochrome is in the nucleus (SHERMAN et al. 1968). Further evidence from *Euglena* is also at hand since there are now reports that certain chloroplast-specific amino acyl t-RNA synthetases (REGER et al. 1970) and a plastid DNAase (EGAN & CARELL 1972) may be synthesized outside the chloroplast.

These studies with *Euglena* suggest that the ribulose diphosphate carboxylase and the cytochrome 552 are both coded in the chloroplast DNA and are synthesized on chloroplast ribosomes. (I am indebted to Dr. K. BOARDMAN for pointing out that if the chloroplast DNA codes for chloroplast ribosomes, as indicated by the hybridization experiments of STUTZ (1971), the absence of these ribosomes in the mutant lacking chloroplast DNA could prevent the synthesis of various chloroplast proteins whether or not they are coded in the chloroplast DNA. Thus absence of a chloroplast enzyme from the mutant does not conclusively prove that it is coded or controlled in the chloroplast DNA.) The TPN triose phosphate dehydrogenase of the chloroplast and the DPN triose phosphate dehydrogenase of the cytoplasm are probably coded in the nucleus and synthesized on non-chloroplast ribosomes. Several models can be offered to account for the relationship between the DPN and the TPN-triose phosphate dehydrogenases. The first model assumes that both enzymes are coded in the nucleus in separate cistrons. Thus each enzyme could be regulated separately at the gene level in the nucleus. Indeed, the two enzymes in other species have proven to be so similar that it has been impossible to separate them by conventional fractionation procedures (SCHULMAN & GIBBS 1968; ANDERSON 1969). Since in certain photosynthetic bacteria the DPN, but not the TPN enzyme is found, while the photosynthetic eucaryotic cell has both, it is possible that the TPN gene represents a duplication, during the course of evolution, of the DPN gene with minor modifications. A reciprocal relation between the levels of the activities of the two enzymes in the same cell has been suggested (HUDOCK & FULLER 1965) prompted by data similar to those of Table 2, where the DPN enzyme seems to fall somewhat as the TPN enzyme increases. This has been interpreted to indicate a possible interconversion between the two enzymes in certain organisms (HUDOCK & FULLER 1965). Having both enzymes coded in the nucleus would certainly facilitate such regulatory interactions. A second picture which can be made consistent with the *Euglena* data would have only the DPN enzyme coded in the nucleus. The DPN enzyme might then be modified in a minor way (*e.g.* through the addition or deletion of one or a few amino acids or the degree of oxidation of SH groups (see HUDOCK & FULLER 1965) in the cytoplasm or chloroplast to yield the TPN enzyme. The conversion of the DPN enzyme to the TPN enzyme in the presence of high TPN concentrations has recently been proposed (MÜLLER et al. 1969). Whatever the mechanism, it is clear that nuclear-chloroplast and nuclear-mitochondrial interactions are to be expected, and perhaps in the opposite direction as well, in eucaryotic cells. All possible permutations and combinations of informational crosstalk among the three organelles might eventually be found.

The level of the TPN triose phosphate dehydrogenase in the mutant is at the same level as in dark-grown wild-type cells (Table 2). This indicates that the synthesis of the TPN enzyme is repressed to the same extent in both types of cells, if we assume that the specificity of the DPN enzyme for DPN is absolute so that the TPN activity represents the level of the TPN triose phosphate dehydrogenase. To date non-chloroplast DPN triose phosphate dehydrogenases seem to possess this high selectivity for the DPN coenzyme (ARNON et al. 1954; PUPILLO 1972; R. McGOWAN, personal communication). If the control is at the level of a nuclear gene, normal conditions of derepression by light involve either a signal from the proplastids to the nucleus, or a separate non-chloroplast photoreceptor as suggested previously in connection with non-chloroplast RNA synthesis (ZELDIN & SCHIFF 1968).

The Temporal Sequence of Nutritional and Informational Interactions During Chloroplast Development

It would seem from the foregoing that the developing chloroplast can obtain, and probably must obtain, much of the energy and building blocks for development from the rest of the cell. In addition, it is becoming clear that the information necessary to construct an organelle does not come entirely from the organelle's genome nor entirely from the nuclear genome, but rather from both. We deal, therefore, with organelles which receive at least part of their small and large molecules and, probably, most of their energy during development, from outside the organelle.

Chloroplast development is characterized by a lag period during which chlorophyll synthesis is slow followed by a period of rapid synthesis (STERN et al. 1964b). In *Euglena* this lag period is about twelve hours in length and it is striking that not much seems to happen within the developing plastid during this period. Chloroplast enzymes such as the ribulose diphosphate carboxylase are already at significant levels in the dark-grown cell and do not change much during the first 12 h of development. The internal structure of the developing plastid changes little during this period and there does not appear to be any significant membrane synthesis during this time (KLEIN et al. 1972). In agreement with this, inhibitors of translation on 70S ribosomes have very little effect during this period. In short, the 12 h lag seems to be a period of minimal activity within the chloroplast. It would seem that this period is devoted mainly to filling available membrane sites with chlorophyll, although at least one photosynthetic component whose synthesis is streptomycin sensitive seems to be made during this period (BOVARNICK et al. 1968).

In contrast, non-chloroplast activities seem to be very prominent during this lag period. The TPN triose phosphate dehydrogenase which seems to be synthesized on the 87S ribosomes of the cytoplasm from information provided by the nucleus is synthesized without a 12 h lag during the first hours of development. The overall respiration of the cell increases at the inception of chloroplast development and probably represents the energy demands of the many cytoplasmic activities called forth to support chloroplast development. In agreement, the first twelve hours of chloroplast development is very sensitive to cycloheximide, an inhibitor of translation on the 87S cytoplasmic ribosomes (SCHWARTZBACH & SCHIFF 1971).

Using this information we have constructed the following picture of the sequence of events in chloroplast development (Fig. 12). When light strikes the dark-grown cell, there is an activation of cytoplasmic systems to provide the necessities for chloroplast development. These include energy, small molecules (undoubtedly including early precursors of chlorophyll) and large molecules such as the TPN triose phosphate dehydrogenase and the enzymes necessary to mobilize energy and make the small molecules. Within the chloroplast, however, very little is known to occur other than a low rate of chlorophyll synthesis to fill sites on existing membranes and at least one other component associated with photosynthesis. This makes sense since the dark-grown plastids seem to have significant levels of many enzymes, membranes, *etc.* which seem to be adequate for the first twelve hours of development. We would think, then, that the first twelve hours of plastid development are mainly devoted to a feeding of the plastid by the rest of the cell to raise the levels of those constituents supplied from outside the plastid to levels of other chloroplast constituents already within the plastid from the beginning of development. At the point where these externally-supplied constituents become adequate within the plastid, the lag period ends and the period of rapid synthesis inside the plastid commences leading to the great expansion in volume of the plastid between 14 and 72 h and the large accumulations of membranes, enzymes and other constituents.

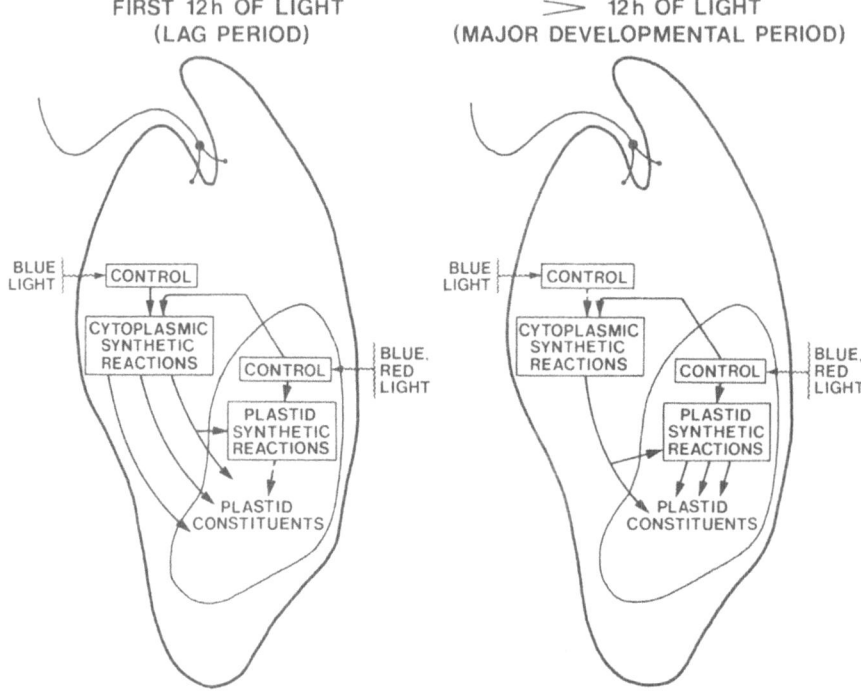

Fig. 12. Model for events of lag period and period of active development. See text for details (SCHWARTZBACH & SCHIFF 1971).

The Control of Chlorophyll Synthesis

A potentially useful approach for exploring the induction and early stages of chloroplast development has been provided by the finding that a brief preillumination of the dark-grown cells followed by a dark period results in the elimination of the usual lag in chlorophyll formation when the cells are subsequently returned to continuous light, a phenomenon we call potentiation (Fig. 13) (HOLOWINSKY & SCHIFF 1968, 1970; KLEIN et al. 1972). Since the optimal length of the dark period for lag elimination is about 12 h it seems reasonable to assume that a brief exposure to light triggers some events which ordinarily take place in the lag period of continuous illumination (also 12 h in length) and permits them to occur in the dark period following pre-illumination. This should allow a separation of the inductive phase from the later consequences of induction since the latter can be studied during the dark period.

Preillumination and the subsequent dark period provide an increased potential for chlorophyll synthesis when the cells are returned to continuous light. The

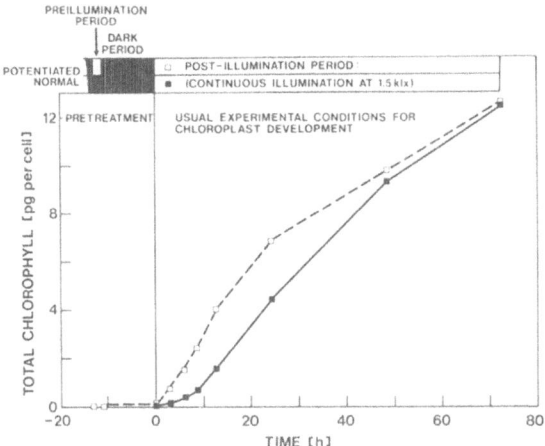

Fig. 13. Time course of chlorophyll accumulation in control and potentiated *Euglena* cells. Zero time is taken as the beginning of the post-illumination period. Cells from a 3-day dark-grown resting culture were exposed to 90 min of preillumination (white light, 1.5 klx) starting at 12 h and to a dark period (potentiated cells) before exposure to continuous illumination with white light (post-illumination). Control cells experienced an uninterrupted dark period until they were exposed to continuous illumination at 0 h (HOLOWINSKY & SCHIFF 1968, 1970).

events of the preillumination period (in terms of subsequent chlorophyll synthesis) require oxygen (KLEIN et al. 1972) and are sensitive to cycloheximide and chloramphenicol (SCHWARTZBACH & SCHIFF 1971). The events of the subsequent dark period do not require oxygen and are cycloheximide and chloramphenicol sensitive in terms of their effect on subsequent chlorophyll synthesis when the cells are returned to continuous light. The oxygen dependence of the light period would suggest that this is a time of mitochondrial activation to provide energy and intermediates, recalling that one of the earliest events of light induction of plastid development is an increase in the rate of cellular respiration (SCHIFF 1963) and that plastid development in *Euglena* is blocked by dinitrophenol, an uncoupler of mitochondrial oxidative phosphorylation (EVANS 1971). In addition we might also remember that mitochondria and other organelles such as microbodies are frequently seen in close association with the developing plastids (Fig. 8). Both the light period and dark period seem to require protein synthesis in both the cytoplasm and the plastid to provide the potential for rapid chlorophyll synthesis when the cells are returned to continuous light. As far as we know, preillumination triggers the accumulation of a potential for increased chlorophyll synthesis but does not seem to induce the formation of other plastid proteins. No consistent increases in the levels of the TPN-triose phosphate dehydrogenase, a plastid-associated DNAase, or of ribulose diphosphate carboxylase can be detected in the dark period following preillumination (EGAN & SCHIFF, unpublished; BOVARNICK & SCHIFF, unpublished). This is consistent with the observation that under normal developmental conditions the formation of these enzymes is strictly light dependent rather than light-triggered; if the cells are darkened during chloroplast development, synthesis of the enzymes stops abruptly and resumes when the cells are returned to light (BRAWERMAN & KONIGSBERG 1960). Studies of membrane formation (KLEIN et al. 1972) show that there is little increase in the number or amount of membranes during preillumination or the dark period following, and on return of the preilluminated cells to continuous light the events of development are only slightly accelerated. In effect, since there is little increase in membrane formation during the first twelve hours after the return of the preilluminated cells to continuous light, the chlorophyll which is formed at an accelerated rate in these cells must be used to fill sites on existing membranes. Since the appearance of photosynthetic carbon dioxide fixation is also accelerated during this time, much of the photosynthetic machinery must already be present, and indeed we know that the levels of ribulose diphosphate carboxylase in the dark-grown cells do not change during the first twelve hours of normal development and seem to be adequate for the rates of photosynthesis achieved during the lag period. In other words, it seems that preillumination triggers the formation of an increased potential for chlorophyll synthesis and little else; the synthesis of other components seems to require continuous illumination. (An exception is an increase in photoreactivating enzyme activity which is triggered by a brief illumination and which increases subsequently in the dark — DIAMOND & SCHIFF, unpublished.)

The action spectrum for induction during the preillumination period showed evidence of two pigment systems (HOLOWINSKY & SCHIFF 1970). One is a red-blue system related to the protochlorophyll(ide) — chlorophyll(ide) transformation. the other resembles the high energy blue response (see ROLLIN 1966). Evidence for two pigment systems in development comes also from studies of the intensity dependence of chloroplast development in *Euglena* (STERN et al. 1964a). Since a light stimulation of RNA labeling can be observed in mutants of *Euglena* in which plastid DNA is undetectable (ZELDIN & SCHIFF 1968) and which lack the protochlorophyll(ide)-related system, it seems reasonable to think that the non-chloroplast or cytoplasmic phase of plastid development is, to some extent, under the control of the blue system — although not uniquely, since reasonably complete plastid development can occur in red light (BOVARNICK et al. 1969). A picture consistent with the available evidence, slender as it is, suggests a synergistic effect of the protochlorophyll-related system (or red-blue system) with the blue system in controlling the cytoplasm in plastid-containing cells (see Fig. 10). Presumably the red-blue system controls chlorophyll formation and chloroplast-centered synthetic activities from within the plastid. It is conceivable that phytochrome, which enters the chloroplast development picture in the flowering plants, exerts control over the non-chloroplast cytoplasmic aspects of chloroplast development in higher plants.

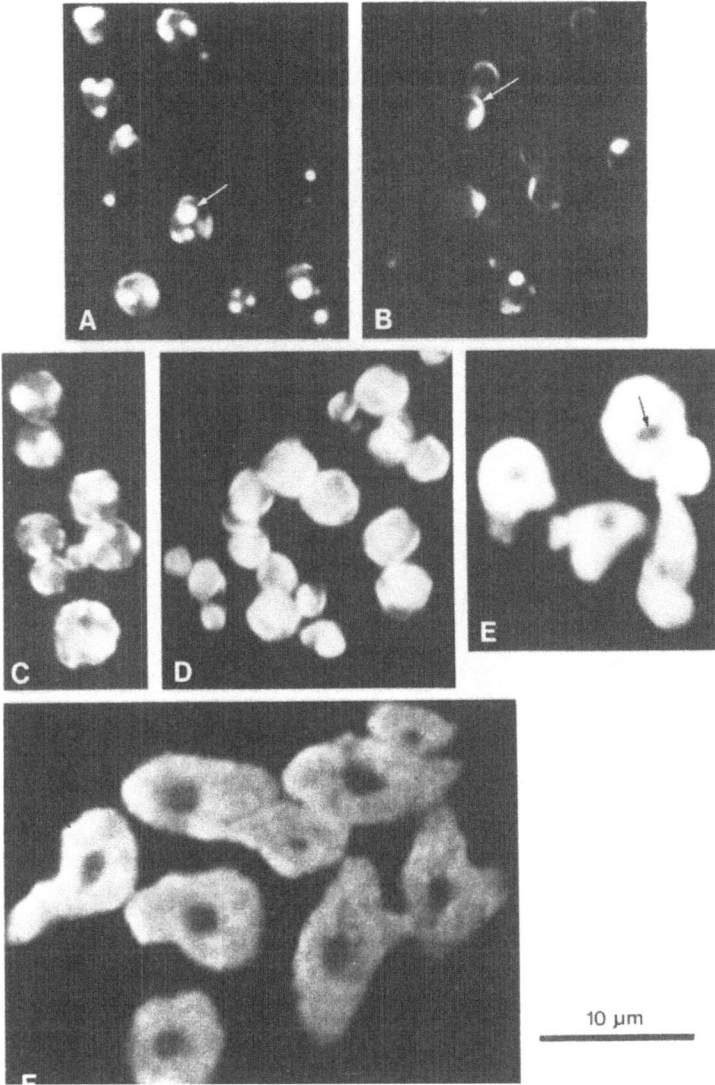

Fig. 14. Fluorescence micrographs of plastids of cells of dark-grown resting *Euglena gracilis* var. *bacillaris* illuminated for varying lengths of time. *(A)* 1.5 h of continuous illumination. Bright spots (indicated by arrow) represent fluorescent centers equivalent to prolamellar-like bodies observed by electron microscopy. *(B)* Same as *(A)*. Note possible girdle thylakoids (arrow) continuous with the prolamellar bodies. *(C)* 12 h of continuous illumination. *(D)* 24 h of continuous illumination. *(E)* 48 h of continuous illumination. Dark spot (arrow) is the pyrenoid. *(F)* 72 h of continuous illumination.

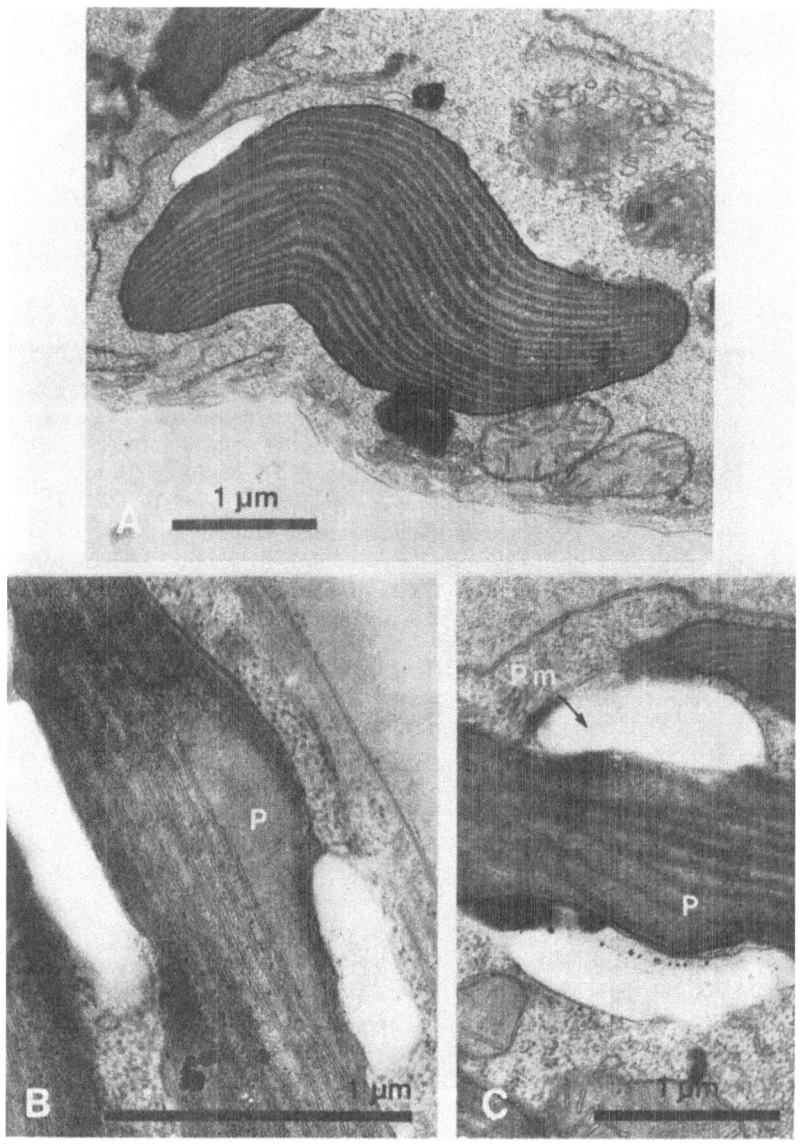

Fig. 15. (A) Sections through dark-grown resting cells of *Euglena* which have completed chloroplast (P) development for 72 h in light in the presence of DCMU. *(B)* Same, 15 min following removal of DCMU. *(C)* Same, 4 h following removal of DCMU. Pm-paramylum.

The Differentiation of the Pyrenoid

During normal plastid development in *Euglena* the pyrenoid differentiates between 24 and 48 h of illumination (BEN-SHAUL et al. 1964; KLEIN et al. 1972) (Fig. 14). As noted above, if the cells are prevented from carrying out photosynthesis by inclusion of the inhibitor DCMU during plastid development, plastids capable of carrying out photosynthesis are formed but these lack pyrenoids. We suggested that the formation of a pyrenoid may be a developmental induction where the products of photosynthesis normally induce the formation of a pyrenoid to process them. Since we had shown that the removal of DCMU at the end of development resulted in an immediate restoration of photosynthesis (SCHIFF et al. 1967) it was of interest to determine whether the pyrenoid also reappeared. As may be seen from Fig. 15 the pyrenoid begins to differentiate rapidly after the removal of DCMU, within minutes in fact (WEATHERBEE & SCHIFF 1972). Since HOLDSWORTH had shown that pyrenoids from *Eremosphaera viridis* contain ribulose diphosphate carboxylase (RuDPCase) or fraction I protein we measured this activity during the return of the pyrenoid. As may be seen from Fig. 16,

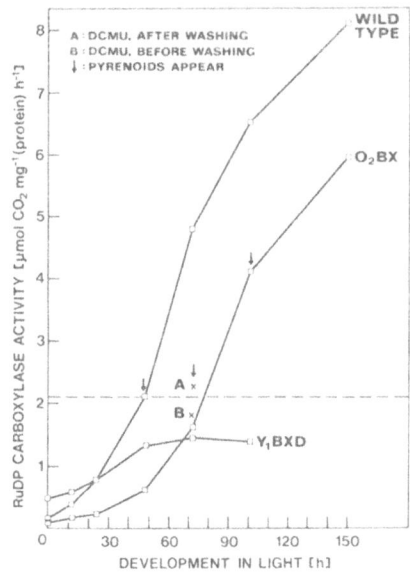

Fig. 16. Ribulose diphosphate carboxylase levels during light-induced chloroplast development in dark-grown resting wild type and mutant cells of *Euglena*. The horizontal dotted line shows the level of ribulose diphosphate carboxylase achieved by wild type cells at the point of pyrenoid differentiation (arrow). Other arrows show the point of pyrenoid appearance in mutant O_2 and levels of enzyme (x) before *(B)* and after *(A)* removal of DCMU at 72 h, in normal developing wild type cells (WEATHERBEE & SCHIFF 1972).

RuDPCase achieves a level of about 2 μmol CO_2 fixed per mg chlorophyll per hour at the time the pyrenoid normally differentiates, in cells without DCMU. In the DCMU-treated cells, at 72 h, the normal conclusion of development, there are no pyrenoids and the level of RuDPCase is below this level of 2. After washing out the DCMU, the level of RuDPCase rises above 2 as the pyrenoid differentiates. In a mutant, Y_1 which never makes pyrenoids, the level of RuDPCase achieved is below 2, while in a mutant O_2 which has a delayed development, the pyrenoid differentiates late and the level of RuDPCase achieves or surpasses 2 at this later time (WEATHERBEE & SCHIFF 1972). Thus it appears that the appearance of the pyrenoid is strongly correlated with a level of RuDPCase exceeding 2 μmol per mg protein per hour. Since fraction I protein is large, it seems reasonable to assume that this protein provides the granular matrix of the pyrenoid between the pyrenoid thylakoids. Since the reappearance of the pyrenoid and the increase in RuDPCase is so rapid, it seemed likely that this was an organizational process that did not require protein synthesis. Consistent with this, the increase in RuDPCase activity is not sensitive to streptomycin or cycloheximide (WEATHERBEE & SCHIFF 1972). Other laboratories working with higher plants have suggested that one of the types of subunits of the RuDPCase are coded in the nucleus while the other type arises in the plastid (KAWASHIMA & WILDMAN 1972). Perhaps the pyrenoid arises from the proper organization of these subunits into structures concomitant with increased RuDPCase activity.

Conclusions

An old but attractive notion which is currently popular supposes that the organelles such as the chloroplast arose from endosymbiotic invaders resembling, perhaps, blue-green algae which established themselves in the host cells and whose division and other aspects of development became subject to control by the host (EDELMAN et al. 1967). The results obtained to date are not inconsistent with this view. The plastid in particular resembles a procaryotic cell within the envelope of a nucleated cell. This procaryotic organelle has its own DNA, and 70S ribosomes and divides like its free-living counterparts.

It is quite possible that the nutritional, informational and regulatory interactions we have detailed in this paper for the *Euglena* chloroplast represent accommodations between invader and host which have grown up during the evolution together of the associates. Thus the invader has adopted certain constituents provided by the host to replace or augment its own synthesis and has thereby become subject to regulation by the host. Still remaining, however, are many basic processes which the plastid carries out for itself using its own information and synthetic machinery.

In answer to the questions posed in this paper then, organelles rely both on the capacities of the cell they are in and on their own machinery for synthesis and for information. Regulatory interactions also exist and perhaps the most interesting one, for which we have practically no information, is how the division of the host and the organelles are coordinated.

Three recent symposia collect a great deal of our current knowledge of the development and inheritance of chloroplasts and mitochondria and the reader is referred to them for many excellent studies of mitochondrial and plastid development and inheritance (MILLER 1970; BOARDMAN et al. 1971; FORTI et al. 1972). Updated references for several sections of this paper will be found in SCHIFF (1973, 1974).

Acknowledgment

The technical assistance of Miss NANCY O'DONOGHUE in preparing the electron micrographs is greatly appreciated. The research reported here and the preparation of the manuscript was supported by GM14595 from NIH.

REFERENCES

ANDERSON, L., in: Proc. XIth Int. Bot. Congr., p. 3, Seattle, Washington 1969.
ARNON, D. I., ROSENBERG, L. L. & WHATLEY, F. R., *Nature* 173: 1132, 1954.
AVADHANI, N. G., BUETOW, D. E., *Biochem.J.* 128: 353, 1972.
BARNETT, W. E., PENNINGTON, C. J. Jr. & FAIRFIELD, S. A., *Proc.nat.Acad.Sci. U.S.A.* 63: 1261, 1969.
BEN-SHAUL, Y., SCHIFF, J. A. & EPSTEIN, H. T., *Plant Physiol.* 39: 231, 1964.
BISHOP, N. I., *Annu.Rev.Plant Physiol.* 17: 185, 1966.
BOARDMAN, N. K., LINNANE, A. W. & SMILLIE, R. M. (ed.): Autonomy and Biogenesis of Mitochondria and Chloroplasts, North-Holland Publ. Co., Amsterdam – London 1971.
BOGORAD, L., *Bot.Gaz.* 111: 221, 1950.
BOLD, H. C., in: SMITH, G. M. (ed.): Manual of Phycology, Chronica Botanica, Waltham, Mass. 1951.
BOVARNICK, J. G., CHANG, S. W. & SCHIFF, J. A., *Plant Physiol.* 43: S-6, 1968.
BOVARNICK, J. G., FREEDMAN, Z. & SCHIFF, J. A., *Plant Physiol.* 46: S-21, 1970.
BOVARNICK, J. G., ZELDIN, M. H. & SCHIFF, J. A., *Developm. Biol.* 19: 321, 1969.
BRAWERMAN, G. & EISENSTADT, J. M., *J.mol.Biol.* 10: 403, 1964a.
BRAWERMAN, G. & EISENSTADT, J. M., *Biochim. biophys. Acta* 91: 477, 1964b.
BRAWERMAN, G., HUFNAGEL, D. A. & CHARGAFF, E., *Biochim. biophys.Acta* 61: 340, 1962.
BRAWERMAN, G. & KONIGSBERG, N., *Biochim. biophys. Acta* 43: 374, 1960.
CARELL, E. F., EGAN, J. M. & PRATT, E. A., *Arch.Biochem. Biophys.* 138: 26, 1970.
CLELAND, R. E., *Advance. Genet.* 11: 147, 1962.
CRAMER, M. & MYERS, J., *Arch. Mikrobiol.* 17: 384, 1952.
EDELMAN, M., COWAN, C. A., EPSTEIN, H. T. & SCHIFF, J. A., *Proc.nat.Acad.Sci. U.S.A.* 52: 1214, 1964.
EDELMAN, M., EPSTEIN, H. T. & SCHIFF, J. A., *J.mol.Biol.* 17: 463, 1966.
EDELMAN, M., SCHIFF, J. A. & EPSTEIN, H. T., *J.mol.Biol.* 11: 769, 1965.
EDELMAN, M., SWINTON, D., SCHIFF, J. A., EPSTEIN, H. T. & ZELDIN, B., *Bact.Rev.* 31: 315, 1967.
EGAN, J. M., Jr. & CARELL, E. F., *Plant Physiol.* 50: 391, 1972.
EVANS, W. R., *J.biol.Chem.* 246: 6144, 1971.

FORTI, G., AVRON, M. & MELANDRI, A. (ed.): Photosynthesis, Two Centuries after its Discovery by Joseph Priestley. Vol. 1, 2, 3. Junk, The Hague 1972.
GIBBS, S. P., *J. Ultrastruct.Res.* 4: 127, 1960.
GIBBS, S. P., in: Proc. XIth Int.Bot.Congr., p. 69, Seattle, Washington 1969.
GIBOR, A. & GRANICK, S., *J. Protozool.* 9: 327, 1962a.
GIBOR, A. & GRANICK, S., *J. Cell Biol.* 15: 599, 1962b.
GOJDICS, M., *Trans.amer.microscop.Soc.* 53: 299, 1934.
HEIZMANN, P., *Biochim. biophys. Acta* 224: 144, 1970.
HOLOWINSKY, A. & SCHIFF, J. A., *Plant Physiol.* 45: 339, 1970.
HOLOWINSKY, A. W. & SCHIFF, J. A., *Plant Physiol.* 43: S-7, 1968.
HUDOCK, G. A. & FULLER, R. C., *Plant Physiol.* 40: 1205, 1965.
INGLE, J., *Plant Physiol.* 43: 1850, 1968.
KAWASHIMA, N. & WILDMAN, S. G., *Biochim. biophys. Acta* 262: 42, 1972.
KLEIN, S. & NEUMAN, J., *Plant Cell Physiol.* 7: 115, 1966.
KLEIN, S. & SCHIFF, J. A., *Plant Physiol.* 49: 619, 1972.
KLEIN, S. & SCHIFF, J. A., HOLOWINSKY, A. W., *Developm. Biol.* 28: 253, 1972.
KRAWIEC, S. & EISENSTADT, J. M., *Biochim. biophys. Acta* 217: 120, 1970.
LEEDALE, G. F., Euglenoid Flagellates, Prentice Hall, Englewood Cliffs 1967.
LEFF, J., MANDEL, M., EPSTEIN, H. T. & SCHIFF, J. A., *Biochem. biophys. Res. Commun.* 13: 126, 1963.
LEVINE, R. P. & VOLKMANN, D., *Biochem. biophys. Res. Commun.* 6: 264, 1961.
LYMAN, H., EPSTEIN, H. T. & SCHIFF, J. A., *J. Protozool.* 6: 264, 1959.
LYMAN, H., EPSTEIN, H. T. & SCHIFF, J. A., *Biochim. biophys. Acta* 50: 301, 1961.
MANNING, J. E., WOLSTENHOLME, D. R., RYAN, R. S., HUNTER, J. A. & RICHARDS, O. C., *Proc.nat.Acad.Sci. U.S.A.* 68: 1169, 1971.
MATSUKA, M. & HASE, E., *Plant Cell Physiol.* 7: 149, 1966.
MILLER, P. L. (ed.): Control of Organelle Development, Soc.exp.Biol. Symp.24, Cambridge Univ. Press 1970.
MODOLELL, J. & DAVIS, B. D., *Proc.nat.Acad.Sci. U.S.A.* 61: 1279, 1968.
MÜLLER, B., ZIEGLER, I. & ZIEGLER, H., *Europe. J. Biochem.* 9: 101, 1969.
PRINGSHEIM, F. & PRINGSHEIM, O., *New Phytol.* 51: 65, 1952.
PROVASOLI, L., HUTNER, S. H. & SCHATZ, A., *Proc.Soc.exp.Biol.Med.* 69: 279, 1948.
PUPILLO, P., *Phytochemistry* 11: 153, 1972.
RAWSON, J. R. & STUTZ, E., *Plant Physiol.* 43: S-18, 1968.
RAWSON, J. R. & STUTZ, E., *Biochim. biophys. Acta* 190: 368, 1969.
RAY, D. S. & HANAWALT, P. C., *J. mol. Biol.* 9: 812, 1964.
RAY, D. S. & HANAWALT, P. C., *J. mol. Biol.* 11: 760, 1965.
REGER, B. J., FAIRFIELD, S. A., EPLER, J. L. & BARNETT, W. E., *Proc.nat.Acad.Sci. U.S.A.* 67: 1207, 1970.
RHOADES, M. M., Cold Spring HarborSymp. quant. Biol. 11: 202, 1946.
ROLLIN, P. (ed.): Symposium on Photomorphogenesis, *Photochem. Photobiol.* 5: 347, 1966.
SAGER, R., *Brookhaven Symp. Biol.* 11: 101, 1958.
SALVADOR, G., LEFORT-TRAN, M., NIGON, V. & JOURDAN, F., *Exp. Cell Res.* 64: 457, 1971.
SCHIFF, J. A., Carnegie Inst. Wash. Yearbook 62: 375, 1963.
SCHIFF, J. A., in: MILLER, P. L. (ed.): Control of Organelle Development, Soc.exp.Biol. Symp. 24, p. 277, Cambridge Univ. Press 1970.
SCHIFF, J. A., *Adv. Morphogen.* 10: 265, 1973.
SCHIFF, J. A., in: AVRON, M. (ed.): Proc. Third Int. Congress Photosynthesis, Elsevier, Amsterdam 1974.
SCHIFF, J. A. & EPSTEIN, H. T., in: LOCKE, M. (ed.): Reproduction: Molecular, Sub-cellular and Cellular, 24th Annu. Symp. Soc.Developm. Growth, p. 131, Academic Press, New York – London 1965.

SCHIFF, J. A. & EPSTEIN, H. T., in: GOODWIN, T. W. (ed.): Biochemistry of Chloroplasts, Vol. I, p. 341, Academic Press, New York – London 1966.
SCHIFF, J. A. & EPSTEIN, H. T., in: BUETOW, D. E. (ed.): The Biology of Euglena, Vol. 2, p. 286, Academic Press, New York – London 1968.
SCHIFF, J. A., LYMAN, H. & EPSTEIN, H. T., *Biochim. biophys.Acta* 50: 310, 1961.
SCHIFF, J. A. & ZELDIN, M. H., *J.Cell Physiol.* 72: 103, 1968.
SCHIFF, J. A. & ZELDIN, M. H., RUBMAN, J., *Plant Physiol.* 42: 1716, 1967.
SCHORI, L., BEN-SHAUL, Y. & EDELMAN, M., *Israel J. Chem.* 8: 117p, 1970.
SCHULMAN, M. D. & GIBBS, M., *Plant Physiol.* 43: 1805, 1968.
SCHWARTZBACH, S. D. & SCHIFF, J. A., *Plant Physiol.* 47: 45-S, 1971.
SCOTT, N. S. & SMILLIE, R. M., *Currents mod. Biol.* 2: 339, 1969.
SHERMAN, F., STEWART, J. W., PARKER, J. H., INHABER, E., SHIPMAN, N. A., PUTTERMAN, G. J., GARDISKY, R. L. & MARGOLIASH, E., *J. biol. Chem.* 243: 5446, 1968.
STERN, A. I., EPSTEIN, H. T. & SCHIFF. J. A., *Plant Physiol.* 39: 226, 1964a.
STERN, A. I., SCHIFF, J. A. & EPSTEIN, H. T., *Plant Physiol.* 39: 220, 1964b.
STUTZ, E., in: BOARDMAN, N. K., LINNANE, A. W. & SMILLIE, R. M. (ed.): Autonomy and Biogenesis of Mitochondria and Chloroplasts, p. 277, North-Holland Publ. Co., Amsterdam – London 1971.
WEATHERBEE, J. A. & SCHIFF, J. A., *Plant Physiol.* 49: 28-S, 1972.
WEISBLUM, B. & DAVIES, J., *Bact.Rev.* 32: 493, 1968.
ZELDIN, M. H. & SCHIFF, J. A., *Plant Physiol.* 42: 922, 1967.
ZELDIN, M. H. & SCHIFF, J. A., *Planta* 81: 1, 1968.

CHLOROPLAST MESSENGER RNA FROM ETIOLATED PEA SEEDLINGS

K. A. ALIEV, SAÏDA MUZAFAROVA, H. RADZHABOV, MAVYUDA KHOLMATOVA, GALINA ULUGBEKOVA & YU. S. NASYROV

Institute of Plant Physiology and Biophysics, Tajik Academy of Sciences, Dushanbe, U.S.S.R.

Recent evidence shows that chloroplasts contain their own DNA with physicochemical properties different from those of the nuclear DNA (KIRK & TILNEY-BASSETT 1967; ODINTSOVA et al. 1969) as well as an entire specific protein-synthesizing system (ribosomes, tRNA and aminoacyl-tRNA ligase — ALIEV & FILIPPOVICH 1968; FILIPPOVICH et al. 1969, 1970).

However, a number of laboratories have recently reported that only some of the chloroplast proteins and enzymes are synthesized in the chloroplast (NASYROV & ALIEV, 1970, 1972; GOODENOUGH et al. 1971; NASYROV et al. 1971). Experiments with some higher and lower plants have shown that chloramphenicol, a specific inhibitor of protein synthesis on 70S ribosomes, exerts an inhibiting effect on the synthesis of cytochromes a, b_{559}, f, phosphoribulokinase and ribulose 1,5-diphosphate carboxylase, i.e. proteins and enzymes involved in electron transport during photosynthesis and some enzymes for the photosynthetic assimilation of carbon. Formation of other proteins such as ferredoxin, phosphoglucomutase and glyceraldehyde 3-P dehydrogenase is blocked by cycloheximide — a specific inhibitor on 80S ribosomes — but is not affected by chloramphenicol which indicates that their synthesis is located outside the chloroplast (SMILLIE et al. 1967; HOOBER et al. 1969; GOODENOUGH & LEVINE 1970; STERN & FRIEDMAN 1970; BRÄNDLE & ZETSCHE 1971; ELLIS & HARTLEY 1971; GOODENOUGH et al. 1971; MARGULIES 1971; MUNNS et al. 1972).

Hence the synthesis of lamellar proteins and components of the electron-transport chain of photosynthesis is controlled by the cooperative interaction of genetic factors of the nucleus and the chloroplasts. Proteins of cytoplasmic origin are assumed to be essential for the formation of Photosystem 1, whereas proteins of plastid origin seem to be involved in the formation of Photosystem 2.

It has been supposed that synthesis of lamellar proteins in the photosynthetic membranes occurs at the translational level. Discovery of certain mRNA forms in chloroplasts during morphogenesis might serve as a valid experimental proof of this hypothesis.

It is known that chloroplast DNA has cistrons of the ribosomal RNA. For *Euglena gracilis* the 16S and 23S ribosomal RNA's were found to be coded by different sites on the chloroplast DNA. According to TEWARI & WILDMAN (1968), these sites constitute 4% of the chloroplast DNA. These data also indicate the dual nuclear-plastid control of the chloroplast transcription-translation systems. It is probable not only that formation of the same component takes place in different parts of the cell under the control of different genomes, but also that their synthesis and formation de novo undergo great changes in the course of onthogenetic development of these structures (NASYROV & ALIEV 1970; NASYROV et al. 1971).

We have established that antibiotics inhibiting the synthesis of proteins and enzymes on 70S and 80S ribosomes have different effects on the incorporation of labelled amino acids into chloroplast proteins during their formation (NASYROV & ALIYEV 1972; NASYROV et al. 1971). After 5 h illumination cycloheximide strongly inhibits the incorporation of labelled amino acids into chloroplast proteins, whereas chloramphenicol displays no action upon this process. After 20 h illumination the reverse situation has been observed, i.e. the inhibiting effect of cycloheximide sharply decreases while chloramphenicol rapidly inhibits the incorporation of ^{14}C-amino acids into chloroplast proteins. Hence two protein-synthesizing systems are involved in chloroplast morphogenesis: the cytoplasmic protein-synthesizing system, which is inhibited by cycloheximide, and the plastid protein-synthesizing system, which is sensitive to chloramphenicol.

Table 1. Effects of actinomycin D upon 8-^{14}C-adenine incorporation into chloroplast RNA of the etiolated pea seedlings (from the unpublished experiments of S. MUZAFAROVA, K. ALIEV & Yu.S. NASYROV). The apical parts of 8-10 d old etiolated pea seedlings previously illuminated for 5, 10 and 35 h were incubated for 3 h with actinomycin D (10-20 μg ml^{-1}) in darkness, then re-incubation with 8-^{14}C adenine (50 μg ml^{-1}) for 40 min in the light.

Illumination [h]	Addition of actinomycin D	Specific activity [counts min^{-1} mg^{-1} RNA]	Inhibition [% of control]
5	–	234	0
	+	218	7.3
20	–	906	0
	+	776	14.4
35	–	1072	0
	+	563	47.4

Experiments on the inhibition of transcription by actinomycin D revealed that the sensitivity of RNA synthesis to this antibiotic increased in the course of chloroplast development (Table 1). Up to 5 h of illumination the incorporation of ^{14}C-adenine into chloroplast RNA is rather insignificant. Further illumination led to increasing ^{14}C-adenine incorporation into RNA. Actinomycin induced a strong inhibition of RNA synthesis after long illumination periods – 50% inhibition after 35 h. These data indicate that during the early stages of etioplast development the plastid transcription system is not yet completely formed.

Experiments on protein biosynthesis after 40 min incubation in the presence

Table 2. Effects of actinomycin D upon ^{14}C-lysine incorporation into chloroplast proteins of etiolated pea seedlings. The apical parts of 8-10 d old etiolated pea seedlings illuminated for 0, 5, 10, 20, 35, 50 h were incubated for 3 h in actinomycin D (10-20 μg ml^{-1}) in darkness, then re-incubated with ^{14}C-lysine (50μCi ml^{-1}) during 40 min in the light.

Illumination [h]	Addition of actinomycin D	Specific activity [counts min^{-1} mg^{-1} protein]	% of control
0	−	172	100
	+	186	100
5	−	235	100
	+	227	97.1
10	−	295	100
	+	273	92.2
20	−	570	100
	+	356	62.4
35	−	726	100
	+	412	56.7
50	−	1217	100
	+	547	45.1

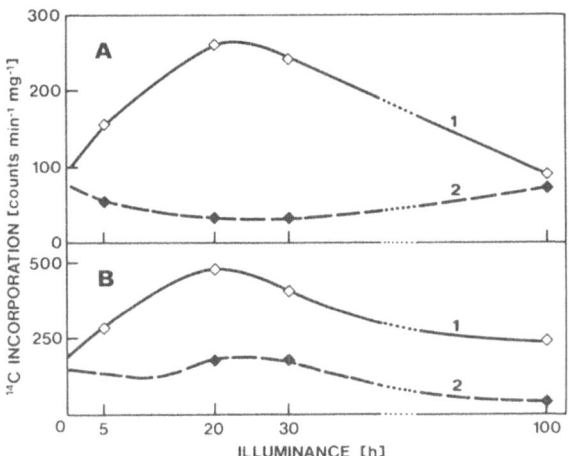

Fig. 1. The effect of rifampicin on ^{14}C-uridine incorporation into RNA (A) and ^{14}C-leucine into chloroplast proteins (B) of etiolated pea seedlings. 1 − control; 2 − with rifampicin.

of actinomycin D revealed a high rate of ^{14}C-lysine incorporation into chloroplast proteins in the early phases of plastid morphogenesis (Table 2). The actinomycin-resistant synthesis of protein amounts to more than 90% of the control. After 20 h and, more particularly, 35 h of illumination the actinomycin-sensitive system of protein synthesis in chloroplasts rapidly accumulates. This indicates that during the early stages of chloroplast development (up to 5-10 h illumination) an available stable form of mRNA is functioning.

Experiments with a specific inhibitor of the synthesis of RNA-polymerase, rifampicin, provided information on the inhibition of ^{14}C-uridine incorporation into RNA in the course of chloroplast development (Fig. 1 A). The inhibitory effect of rifampicin on protein synthesis was also studied; this inhibition gradually increased with time, particularly after 20 h illumination of the etiolated seedlings (Fig. 1 B). These data (confirmed by the formation of polyribosomes) constitute additional evidence for the existence and functioning of available mRNA during the early stages of chloroplast development.

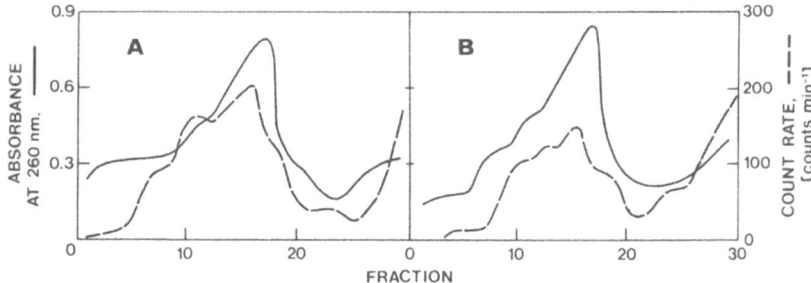

Fig. 2. The effect of 10 µg actinomycin D (B) upon ^{14}C-leucine incorporation into polyribosomes of etioplasts of pea seedlings. A – control.

Centrifugation of the ribosomal fraction in sucrose gradients revealed the formation of polysomes from the components earlier synthesized in etioplasts (up to 5 h of illumination). The pattern of polysomes distribution (Fig. 2) is not altered by the presence of actinomycin D (20 µg ml^{-1}); the rate of protein synthesis in etioplasts is similarly unaffected. Hence the mRNA stored in the etioplasts is probably responsible for the synthesis of proteins in the early stages of plastid morphogenesis, and this form of mRNA may be required to trigger the plastid protein-synthesizing system. In our experiments only ribosomal RNA was synthesized in the early stages of chloroplast development. Incorporation of ^{14}C-uridine has been observed in the region of 23S and 16S RNA (Fig. 3).

The formation of mRNA and tRNA takes place during later plastid development. We have carried out a series of experiments aimed at analyzing the action of different concentrations of actinomycin D and rifampicin upon synthesis of all

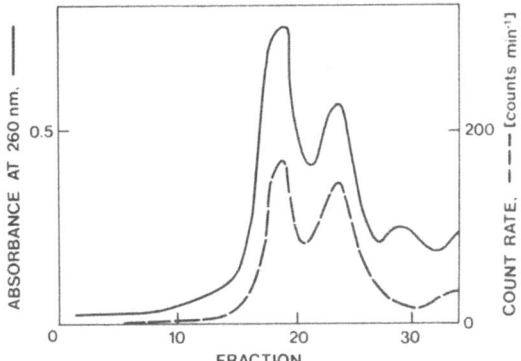

Fig. 3. Distribution of ^{14}C-RNA of etiolated pea seedlings in a sucrose concentration gradient. Etiolated seedlings were incubated with ^{14}C-uridine for 12 h and then RNA was extracted by the phenol-detergent method and purified on Sephadex G-100. RNA was centrifuged for 12 h in a sucrose density gradient (15-30%) at 24 000 rev. per min.

types of RNA. Since actinomycin D and rifampicin do not disturb the structure of the protein-synthesizing apparatus and its function, they were employed in a study of the function of mRNA in the cell in vivo.

Thus the apical parts of etiolated pea seedlings, after 5 h illumination, were treated for 3 h with different concentrations of actinomycin D or rifampicin and then incubated for 2 h with ^{14}C-uridine. The chloroplast RNA's obtained were centrifuged in a sucrose concentration gradient for 14 h at -1 to -2 °C.

In the control sample all types of chloroplast RNA were labelled (i.e. soluble RNA, 23S and 16S peaks of rRNA, and the heavy fraction of heterogeneous RNA with a coefficient of sedimentation about 28-30S). In the sample treated with a low concentration of actinomycin D most radioactivity was localized in the region of the heavy fraction of heterogeneous RNA (Fig. 4, bottom). This heterogeneous fraction of RNA was detected in etioplasts. The increase of radioactivity was observed in this fraction for up to 20 h of illumination; the label disappeared after 30-40 h illumination of etiolated pea seedlings.

The quantitative estimation of all types of chloroplast RNA isolated in another experiment after 10 h illumination and treatment with different concentrations of actinomycin D is summarized in Table 3.

Low concentrations of the antibiotic (10-20 μg ml^{-1}) had no significant effect on the formation of the heterogeneous fraction of RNA or tRNA. On the other hand, the inhibition of 16S and 23S rRNA constituted 70-80% of the control. Hence, low concentrations of actinomycin D slowly inhibit ^{14}C-uridine incorporation into mRNA, but almost completely block formation of ribosomal RNA.

The ability of the labelled heterogeneous fraction of RNA to form, in vitro, polyribosomes when added to free chloroplast ribosomes purified on a sucrose gradient was tested in another experiment (Fig. 5). The increase in absorbance and

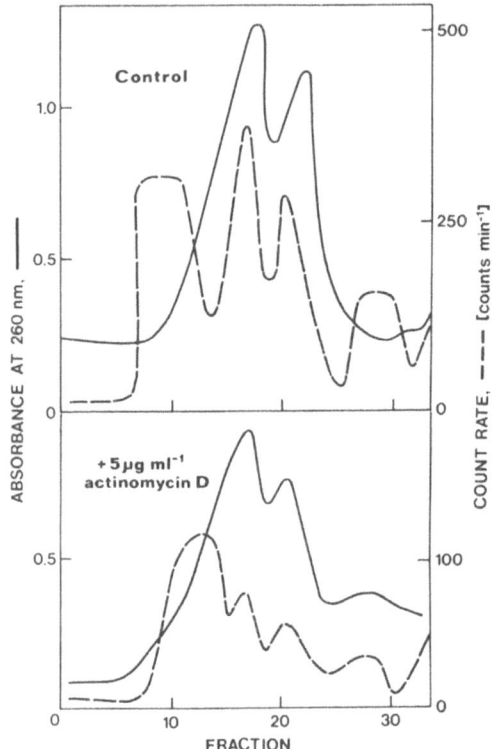

Fig. 4. The effect of actinomycin D on sedimentation distribution of ^{14}C-RNA in pea seedling chloroplasts. Etiolated pea seedlings were illuminated with 6 klx for 5 h, after which the upper part of the seedlings were incubated for 5 h with actinomycin D and reincubated with ^{14}C-uridine for 2 h. Chloroplast RNA was extracted by a phenol-detergent method and purified on Sephadex G-100 with following centrifugation in a sucrose density gradient (15-30%) for 12 h at 24 000 rev. per min at + 2 °C in the preparative ultracentrifuge VAC-60.

radioactivity in the region of polysomes led to the conclusion that the 30S heterogeneous fraction of RNA present after 10 h illuminance is capable of forming polyribosomes in vivo.

In the presence of actinomycin D in etioplasts during 10 h almost no inhibition of amino acids incorporation into proteins is observed. This may be explained by the presence of relatively stable forms of mRNA which perform synthesis of the functional and lamellar proteins in the initial stages of chloroplast development. After 20 h illumination the inhibition of protein synthesis rapidly increases during further treatment with actinomycin D (Table 4). The blocking of different newly formed mRNA's by actinomycin D is clearly expressed.

Table 3. Percent inhibition of synthesis of different fractions of RNA by actinomycin D. The apical parts of 8-10 d old etiolated pea seedlings, after 5 h illumination, were treated with different concentrations of actinomycin D (10, 15, 20, 30, 50, 100 µg ml^{-1}) during 3 h, then re-incubated with ^{14}C-uridine for 2 h. Chloroplast RNA's were obtained by the phenol-detergent method and were centrifuged in the sucrose gradient 5-20% during 14 h at 2 °C.

Actinomycin D [µg]	rRNA		Heterogeneous RNA	tRNA
	23S	16S		
10	27	18	9	2
15	74	82	14	18
20	72	78	27	48
30	69	81	34	67
50	82	77	57	66
100	85	79	87	74

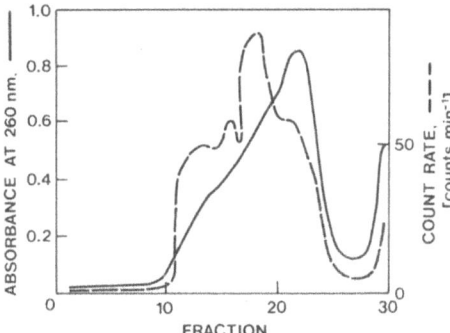

Fig. 5. Sedimentation distribution of ^{14}C-RNA in the presence of chloroplast ribosomes. Purified 28S ^{14}C-RNA from pea seedling chloroplasts was incubated with 70S chloroplast ribosomes purified in sucrose density gradient for 30 min, followed by centrifugation in sucrose density gradient (15-30%) for 3 h at 24 000 rev. per min.

Formation of the heterogeneous fraction of mRNA proved to be sensitive to rifampicin, which blocked the formation of all types of chloroplast RNA (Fig. 6). 16S rRNA was found to be more sensitive to rifampicin than 23S rRNA. This indicates that chloroplast rRNA, tRNA and 30S mRNA are transcribed on DNA of plastids. The experiments with RNase confirmed that the isolated fraction is really mRNA (Fig. 7).

The above data on the synthesis of different types of RNA and proteins during chloroplast development constitutes good evidence that the stable mRNA form with the coefficient of sedimentation 30S accounts for the initiation of transla-

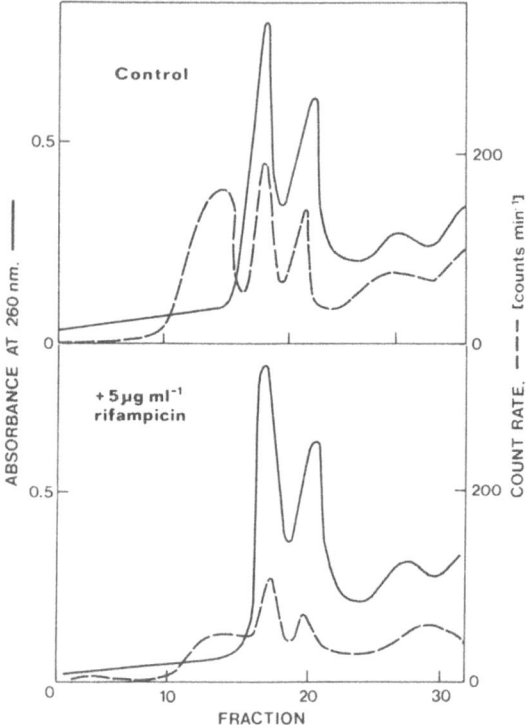

Fig. 6. The effect of rifampicin on the sedimentation distribution of ^{14}C-RNA in pea seedling chloroplasts. Etiolated pea seedlings were illuminated for 5 h at 6 klx, then incubated with 5 μg ml^{-1} rifampicin and reincubated with ^{14}C-uridine for 2 h. The RNA's obtained were centrifuged in sucrose density gradient (15-30%) for 12 h at 24 000 rev. per min.

tion. This mRNA is probably involved in the synthesis of functional (enzymatic) and lamellar proteins of chloroplasts. This RNA form is not detected in the chloroplast during later stages of development.

The appearance of the heterogeneous component of mRNA in chloroplasts coincides with the formation of the protein-synthesizing system. Along with the development of the chloroplast lamellar system the synthesis of all RNA types intensifies and the role of the plastid protein-synthesizing system simultaneously increases. These facts agree with the concept of the existence in chloroplasts of a protein-synthesizing system closely associated with the lamellar system, and indicate that the development of this system takes place at the beginning of chloroplast morphogenesis (OPARIN et al. 1972).

A short appearance of new 30S RNA coincided with the beginning of intensive ^{14}C-amino acid incorporation into proteins; this RNA type is probably involved

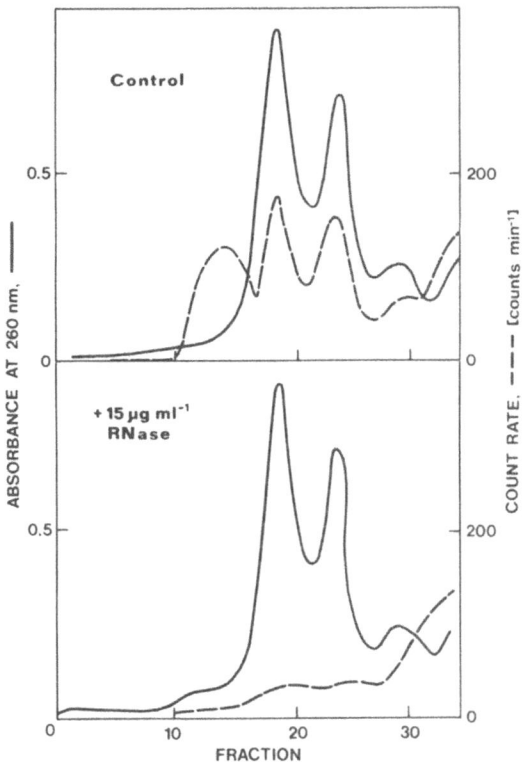

Fig. 7. The effect of RNase upon ^{14}C-uridine incorporation into chloroplast RNA. Proplastids of pea seedlings were incubated with RNase for 30 min at 25 °C and centrifuged in sucrose density gradient (15-30%) for 12 h at 24 000 rev. per min.

in the 'trigger' mechanism of the plastid protein-synthesizing system.

It may be expected, of course, that mRNA with a coefficient of sedimentation of 28-30S is present only in the early development phase of the chloroplast, whereas fully developed chloroplasts contain only components with low constants of sedimentation. However, we should be careful of such conclusions: it is possible that, in the later stages of chloroplast development, the quantity and activity of RNase is higher than in the first hours of plastid development and during the extraction of RNA in this period of development the degradation of 30S RNA to smaller components may take place.

Some authors have claimed to isolate mRNA with a coefficient of sedimentation of 30S and even 40S from animal cytoplasm. ATTARDI et al. (1966) reported that the main component of the rapidly-labelled polysome RNA in *HeLa* cells exists as a 6 to 25S fraction as well as a larger 30S fraction. According to MILLER

Table 4. Incorporation of ^{14}C-amino acids into chloroplast proteins in presence of actinomycin D. The apical parts of 8-10 d old pea seedlings were incubated for 3 h in actinomycin D (10-20 µg ml^{-1}), then re-incubated in ^{14}C-hydrolysate of protein (50 µCi ml^{-1}). The reaction was stopped by addition of 20% TCA. The samples were taken at the actinomycin block after 15, 30 and 45 min and 1, 2, 5 and 10 h.

Illumination [h]	Time after treatment by actinomycin D	Specific activity [counts min^{-1} mg^{-1} protein]	% inhibition
0	control	320	—
	15 min	306	5
	30 min	311	3
	45 min	297	7
	1 h	332	0
	2 h	365	0
	5 h	307	8
	10 h	295	0
20	control	2320	—
	15 min	1725	26
	30 min	1859	20
	45 min	1978	19
	1 h	2015	20
	2 h	1982	19
	5 h	1715	27
	10 h	837	64
	20 h	282	88

et al. (1968), mRNA's from polysomes are distributed in sucrose gradients in the region of 28S RNA.

Hence in the first hours of chloroplast development we observed new formation of all types of RNA. Their formation is a nonsynchronous process. Light-induced synthesis of 16S and 23S ribosomal RNA commences in the first hours of illumination and gradually intensifies in the course of chloroplast development. After 5 h of illumination of etiolated seedlings the short appearance of the new 30S RNA was observed in etioplasts, coinciding with the beginning of activity of the plastid protein-synthesizing system. This 30S RNA probably participates in the biogenesis of the photosynthetic membranes.

Summary

The appearance of a heavy component (30S) of the radioactive RNA in certain stages of the early morphogenesis of pea chloroplasts was ascertained. It formed polyribosomes, and disintegrated after RNase treatment. Actinomycin D inhibited synthesis of ribosomal RNA, but showed no effect on the synthesis of the heavy component. The 30S RNA probably contributes to the 'trigger' mechanism of the protein-synthesizing system of plastids.

REFERENCES

ALIEV, K. A. & FILIPPOVICH, I. I., *Mol. Biol. (Moskva)* 2: 364, 1968.
ATTARDI, G., PARNAS, H., HWANG, M-I. H. & ATTARDI, B., *J. mol. Biol.* 20: 145, 1966.
BRÄNDLE, E. & ZETSCHE, K., *Planta* 99: 46, 1971.
ELLIS, R. J. & HARTLEY, M. R., *Nature-new Biol.* 233: 193, 1971.
FILIPPOVICH, I. I., SVETAÏLO, E. N. & ALIEV, K. A., in: Khloroplasty i Mitokhondrii, p. 248, Nauka, Moskva 1969.
FILIPPOVICH, I. I., SVETAÏLO, E. N. & ALIEV, K. A., in: Funktional'naya Biokhimiya Kletochnykh Struktur, p. 132, Nauka, Moskva 1970.
GOODENOUGH, U. W. & LEVINE, R. P., *J. Cell Biol.* 44: 547, 1970.
GOODENOUGH, U. W., TOGASAKI, R. K., PASZEWSKI, A., LEVINE, R. P., in BOARDMAN, N. K., LINNANE, A. W. & SMILLIE, R. M. (ed.): Autonomy and Biogenesis of Mitochondria and Chloroplasts, p. 224, North-Holland Publ. Co., Amsterdam-London 1971.
HOOBER, J. K., SIEKEVITZ, P. & PALADE, G. E., *J. biol. Chem.* 244: 2621, 1969.
KIRK, J. T. O. & TILNEY-BASSETT, R. A. E., The Plastids, Their Chemistry, Structure, Growth and Inheritance, W. H. Freeman and Co., London-San Francisco 1967.
MARGULIES, M. M., *Biochem. biophys. Res. Commun.* 44: 539, 1971.
MILLER, A. O. A., DHONT, E., DE VREESE, A. M. & VAN NIMMEN, L., *Biochem. biophys. Res. Commun.* 33: 702, 1968.
MUNNS, R., SCOTT, N. S. & SMILLIE, R. M., *Phytochemistry* 11: 45, 1972.
NASYROV, Yu. S. & ALIEV, K. A., *Dokl. Akad. Nauk Tadzh. SSR* 13: 50, 1970.
NASYROV, Yu. S. & ALIYEV, K., in: FORTI, G., AVRON, M. & MELANDRI, A. (ed.): Photosynthesis, Two Centuries after Its Discovery by Joseph Priestley, p. 2545, Junk, The Hague 1972.
NASYROV, Yu. S., ALIEV, K. A., ABDULLAEV, Kh. A. & MUZAFFAROVA, S., in: NASYROV, Yu. S. (ed.): Geneticheskie Aspekty Fotosinteza, p. 5, Donish, Dushanbe 1971.
ODINTSOVA, M. S., MIKELSKA, E. I. & TURISHCHEVA, M. S., in: Funktsional'naya Biokhimiya Kletochnykh Struktur, p. 148, Nauka, Moskva 1969.
OPARIN, A. I., FILIPPOVICH, I. I. & BEZSMERTNAYA, I. N., *Fiziol. Rast.* 19: 995, 1972.
SMILLIE, R. M., GRAHAM, D., DWYER, M. R., GRIEVE, A. & TOBIN, N. F., *Biochem. biophys. Res. Commun.* 28: 604, 1967.
STERN, R. & FRIEDMAN, R. M., *Nature* 226: 612, 1970.
TEWARI, K. & WILDMAN, S. G., *Biochem. biophys. Res. Commun.* 35: 502, 1968.

RELATIONSHIP BETWEEN THE PROTEIN-SYNTHESIZING SYSTEM AND THE STRUCTURE OF CHLOROPLASTS

IRINA I. FILIPPOVICH, BOTOGOZ A. ALINA, IRINA N. BEZSMERTNAYA, ALISA M. TONGUR & A. I. OPARIN

A.N. Bakh Institute of Biochemistry, Academy of Sciences of the U.S.S.R., Moscow, U.S.S.R.

It is well known that chloroplasts contain DNA and major components of the transcription and translation system each of which differs from the corresponding component of the rest of the cytoplasm or the nucleus. This is particularly characteristic of chloroplast ribosomes which are significantly dissimilar to cytoplasmic ribosomes of eukaryotes and are similar to ribosomes of prokaryotes. The components of the pre-ribosomal stage of protein synthesis (tRNA and aminoacyl-tRNA-synthetase) and the mechanism of the initiation of the protein synthesis are not identical to cytoplasmic components either.

It was shown recently that, in spite of the fact that chloroplasts have transcription and translation systems of their own, many of the proteins and enzymes occurring in the chloroplasts are not synthesized within them. Most are synthesized outside chloroplasts with participation of cytoplasmic ribosomes and under the control of the nuclear genome. The synthesis of some enzymes of chloroplasts is under dual nuclear chloroplast control. The role of DNA and the translation system of chloroplasts has been demonstrated only for certain proteins and enzymes localized in chloroplasts.

All these data regarding the peculiar features of the molecular organization of the main components of the protein-synthesizing system and its capacity to synthesize a limited number of proteins and enzymes undoubtedly extend our knowledge of the translation system of chloroplasts. However, in order to reach a good understanding of its properties, function in chloroplasts, role in the formation of organoids and the mechanism of its regulation, it is important to ascertain its localization and spatial organization in the fine structure of chloroplasts.

This line of research was preceded by the discovery of the capacity of fragments of disrupted chloroplasts to incorporate intensively ^{14}C-amino acids in vitro (SISAKYAN et al. 1962; PARTHIER & WOLLGIEHN 1963) and by the detection of RNA strongly bound with the lamellar system (GNANAM & KAHN 1967). It was also shown that the removal of an appreciable portion of free ribosomes from chloroplasts produced practically no effect on their capacity to synthesize protein. Later, a fraction of particles that was extremely active in protein synthesis was isolated from disrupted chloroplasts treated with Na-deoxycholate (FILIPPOVICH et al. 1967).

In view of the above facts it is beyond doubt that in chloroplasts a proportion of ribosomes is bound with the structure of lamellae and that the synthesis of protein in these organoids is realized mainly by the ribosomes bound with the lamellar system.

Thus, it has been established that chloroplasts contain two types of ribosomes that differ in their topography: free ribosomes and ribosomes bound with lamellae. Free ribosomes are located in the matrix of chloroplasts and can be liberated from them after osmotic shock, and bound ribosomes can be freed by treatment of chloroplasts fragments (pelleted at 40 000 x g from the suspension of chloroplasts disrupted by osmotic shock) with detergents (0.5% Na-deoxycholate and 0.4% Triton X-100).

The ultracentrifugation of both fractions of ribosomes in a sucrose linear gradient (5-20%) revealed substantial differences in their structural composition. The fraction of free ribosomes exhibited one component identified as 70S ribosomes whereas the fraction of bound ribosomes displayed five components, four of which were located in the area that is normally occupied by polyribosomes (FILIPPOVICH et al. 1970). In response to mild RNase treatment (1×10^{-9} g ml^{-1}, at 2 °C) only one of these components (fraction 5) dissociated into ribosome monomers, providing evidence that this fraction contains polyribosomes. The latter was confirmed by electron microscopic studies. Other components were insensitive to RNase during mild treatment. The electron microscopic examination of fractions from the sucrose linear gradient showed that the RNase sensitive fraction actually contained polyribosomes, and other fractions included fragments of lamellae and thylakoids that were degraded to different extents by the Na-deoxycholate treatment. The study of these structures made it possible to follow the step-by-step disassembly of thylakoids and grana under the influence of the detergent.

The present communication will discuss only the main stages of this process.

Following the treatment of chloroplast fragments with 0.5% Na-deoxycholate and centrifugation in a sucrose linear gradient (5-20%), thylakoids in the grana were displaced relative to each other and, consequently, split off (Fig. 1a). Then lipids of the central part of the thylakoids underwent solubilization resulting in the fall-out and subsequent detachment of the 'core' (Fig. 1b, c). Due to this, rims remained from the thylakoids (Fig. 1d). This shows that thylakoids are heterogeneous. Unlike their central portion, their peripheral part remained stable under the influence of Na-deoxycholate.

The subsequent treatment of the bound ribosome fraction with 0.4% Triton X-100 altered the distribution pattern of fraction components in the sucrose gradient (PHILIPPOVICH et al. 1970).

The material of fraction 3 which contained chloroplast fragments (thylakoids, grana) disappeared entirely and the absorbance of the gradient samples containing rims of thylakoids decreased. At the same time the absorbance of fractions 5 and 8 increased markedly; these fractions contained ring-shaped structures composed of the particles identified as 70S ribosomes (Fig. 1e) with respect to their sedi-

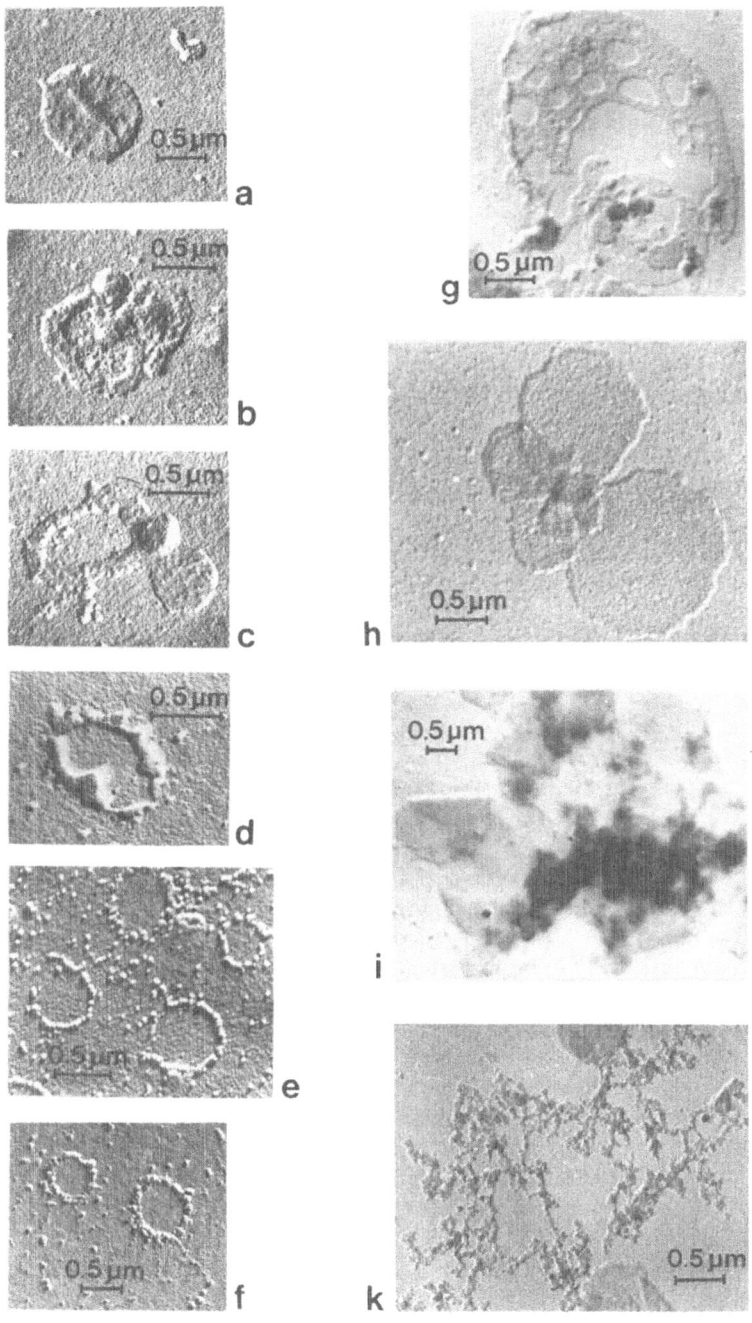

Fig. 1. The substructures isolated from chloroplasts. For explanation see the text.

mentation properties, size and function. Fraction 5 contained mainly polyribosomes in the form of monolayered circles and fraction 8 contained multilayered circles (Fig. 1f).

The size and configuration of isolated polyribosomes is related to the age of plants. From chloroplasts of 14-day old pea seedlings cyclic polyribosomes can be obtained, while from 8-day old pea seedlings polyribosomes built of several interconnected cyclic polyribosomes and networks of gigantic polyribosomes (PHILIPPOVICH et al. 1973) can be isolated.

The differences between polyribosomes obtained from young and mature chloroplasts seem to be associated with the fact that the disassembly of grana of chloroplasts from 8-day old pea seedlings is not accompanied by the dissociation of grana into separate thylakoids; due to this, the grana liberate from the lipoprotein moiety as a whole. This can be clearly seen in small areas of grana.

The detection of cyclic polyribosomes is not an essentially new fact. Polyribosomes having the form of circles and rosettes associated with the membranes of the endoplasmic reticulum have been found in cells of animals and plants. Helix-shaped polyribosomes were also detected in the matrix of chloroplasts by BARTELS & WEIER (1967). When studying ultrathin sections of kidney bean leaves FALK (1969) described cyclic polyribosomes with configurations similar to those of polyribosomes of the endoplasmic reticulum, and which were located on the surface of thylakoids. We, on the contrary, believe that cyclic polyribosomes are situated not on the surface but within thylakoids. This is suggested by the fact that Triton X-100 treatment of bound ribosomes changes substantially the sucrose linear gradient profile: the absorbance of the fractions containing mono- and multilayered cyclic polyribosomes increases and that of fractions containing thylakoids and their circular fragments decreases. The occurrence of cyclic polyribosomes within thylakoids is well supported by the presence of intermediate forms that result from the stripping of thylakoidal circles which contain packed polyribosomes as well as by the similarity of the diameters of cyclic polyribosomes and thylakoids. The diameter of cyclic polyribosomes is close to that of thylakoidal rims, and gigantic networks of polyribosomes are easily inscribed within multi-thylakoidal grana.

The concept that polyribosomes occur within thylakoids is in agreement with the finding that, following the treatment of thylakoids with formaldehyde, polyribosomes are released onto their surface (PHILIPPOVICH et al. 1970). This can be attributed to the fact that the formaldehyde treatment of thylakoids compresses them strongly and hence the polyribosomes are squeezed out from the thylakoids onto their surface through the most sensitive central part.

The above data seems to convince us that polyribosomes are localized within thylakoids. At the same time the question arises whether polyribosomes occur only in grana thylakoids or in stroma lamellae as well.

In search of the solution to this problem, fragments of chloroplasts of 10-day old etiolated pea seedlings illuminated for 19 h were separated by centrifugation in a sucrose discontinuous gradient (1, 1.5, 2 M) into fractions one of which was rich in stroma lamellae and the other in grana (PHILIPPOVICH et al. 1973).

The centrifugation of the suspension of chloroplasts disrupted by osmotic shock in the sucrose discontinuous gradient results in the separation of chloroplast substructures into three layers (Fig. 2). The middle layer (fraction II) contains separate and interconnected lamellae of the stroma (Fig. 1g, h). Many of these are damaged in the central part (Fig. 1g). These lamellae with the damaged central part form large networks.

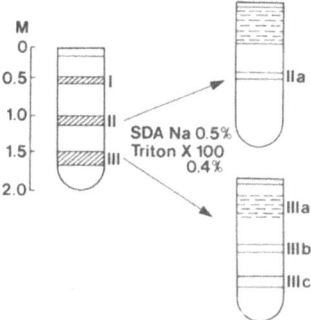

Fig. 2. Scheme of chloroplast membrane fractionation in a discontinuous sucrose gradient. For explanation of fractions see the text.

The lower dark-green layer of the discontinuous gradient (fraction III) contains lamellae to which separate thylakoids are adjacent as well as a great number of interconnected thylakoids and grana containing large fragments of chloroplasts. Thylakoids and grana are always connected with lamellae of the stroma or with strands diverging from it (Fig. 1i).

Thus, the middle layer displays mainly lamellae whereas the lower layer exhibits predominantly large fragments of chloroplasts and lamellae with which thylakoids and grana are connected.

Fractions II and III differ significantly in the ratio of their content of protein, RNA and DNA (OPARIN et al. 1972). Fraction III contains twice as much RNA and DNA (related to protein) than fraction II (Table 1). Therefore, there is a

Table 1. The ratio of the content of protein, RNA and DNA in fractions of chloroplasts.

Fractions (see Fig. 2)	Protein	RNA	DNA	RNA
	[% of their total sum]			[$\mu g\ mg^{-1}$ protein]
I	92.0	8.0	0	87.3±8.6
II	90.9	8.6	0.5	95.3±8.2
III	83.3	15.6	1.1	188.8±9.1

distinct correlation in the distribution of grana and nucleic acids among the fractions. Fraction III is rich in grana and is characterized by a high content of RNA and DNA.

After the treatment of fraction II with detergents (0.5% Na-deoxycholate for 10 min and 0.4% Triton X-100 for 15 min at 2 °C) and centrifugation in the sucrose discontinuous gradient, solubilized lipids and chlorophyll are localized in the upper part of the tube whereas the structural components devoid of chlorophyll occur in the same area of the discontinuous gradient as the initial fraction (Fig. 2). The main structural component of this fraction is large networks consisting of rims of lamellae of the stroma. Unlike the initial fraction this fraction contains no intact undamaged lamellae.

As a result of the detergent treatment and centrifugation in the sucrose discontinuous gradient structures of fraction III disassociate into two main layers: the upper layer (fraction IIIb) and the lower layer (fraction IIIc). Fraction IIIb contains fragments of the grana having the form of chains which are often connected with separate lamellae.

Fraction IIIc contains long branched chains (Fig. 1k, ca. one fifth of the intact network). In certain points they are also bound with lamellae. Their general structure looks very much like networks of gigantic polyribosomes isolated previously from chloroplasts of 8-day old pea seedlings. This resemblance suggests that the structures found in these fractions of the sucrose discontinuous gradient are polyribosomes.

Table 2. Incorporation of ^{14}C-amino acids into proteins by chloroplast fractions.

Fractions (see Fig. 2)	Treatment	Rate of ^{14}C incorporation [count min^{-1} mg^{-1} (protein)]	
		1	2
II		43± 2	27± 2
III		346± 22	434±14
IIIb	0.5% sodium deoxycholate	272± 57	
IIIc	and 0.4% Triton X-100	1759±112	

The concept of the polyribosomal nature of the above structures also finds support in the high biological activity of the fractions containing these structures. After incubation of chloroplast fragments with ^{14}C-amino acids and subsequent centrifugation in the sucrose discontinuous gradient radioactivity was mainly found in proteins of the grana containing fraction III (Table 2). This intensive incorporation is realized by the protein-synthesizing system localized in the same fraction since during incubation of each fraction with ^{14}C-amino acids proteins of fraction III are characterized by a specific activity which is several times higher than that of proteins of fraction II. It is important to note that the specific activity of proteins after the detergent treatment of fraction III increases noticeably. Thus, the de novo synthesized proteins remain strongly bound with the structures of fraction IIIc. As the electron microscopic polyribosomes-like structures can be discerned in this very fraction, it can be suggested that the de novo synthesized polypeptides remain nascent in these structures.

This suggestion is consistent with the earlier finding (CHEN & WILDMAN 1970) that 30-40% of amino acids incorporated into protein are retained by the fraction of membrane-bound ribosomes of tobacco chloroplasts.

Thus, the fraction rich in thylakoids and grana differs from the fraction of the stroma lamellae in having a higher content of RNA and DNA as related to protein and in the capacity to incorporate intensively ^{14}C-amino acids into proteins. After the treatment of this fraction with 0.5% Na-deoxycholate and 0.4% Triton X-100 and repeated centrifugation in the sucrose discontinuous gradient it dissociates further into two fractions. Under the electron microscope one of these fractions shows partially disrupted grana and the other displays large networks of polyribosomes which are still incompletely liberated from the protein, including the de novo synthesized protein.

The data presented favour the concept of the occurrence of polyribosomes in the grana of chloroplasts. The existence of polyribosomes bound with lamellae of the stroma seems to be doubtful because similar treatment of the fraction rich in stroma lamellae (fraction II) with detergents does not result in the liberation of polyribosomes. The capacity of this fraction to synthesize protein is also very weak.

The findings discussed in the present communication may be regarded as an addition to the literature data giving evidence for the relation between the transcription and translation systems and the structure of chloroplasts. It is well known that there is a close relationship between lamellae of chloroplasts and DNA (GREEN & BURTON 1968; WERZ & KELLNER 1968; WOODCOCK & FERNÁNDEZ-MORÁN 1968; BISALPUTRA & BURTON 1969; PHILIPPOVICH et al. 1970; MACHE & WAYGOOD 1970), DNA-polymerase (SPENCER & WHITFELD 1969), DNA-dependent RNA-polymerase (TEWARI & WILDMAN 1969; BOTTOMLEY 1970), ribosomes (FILIPPOVICH et al. 1967), and polyribosomes (FALK 1969; CHEN & WILDMAN 1970).

All this gives grounds to believe that in the near future not only components and systems responsible for the process of photosynthesis but also the systems responsible for the formation and function of the photosynthetic membrane will be taken into consideration when developing models of the molecular organization of thylakoidal membranes. At present this is hindered by a lack of data concerning the localization and structural organization of many components of these systems as well as by the inadequate information about the mechanisms of their function in close contact with the lamellar structure. Today it seems impossible, for instance, to adjust the protein-synthesizing system to the structure of chloroplasts at the molecular level, taking into account all delicate features of the complex organization of both this system and the photosynthetic membrane. At the same time, on the basis of our findings we have attempted to develop a

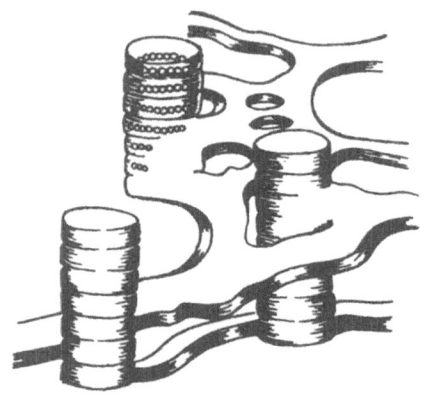

Fig. 3. Scheme of localization of polyribosomes in ultrastructure of chloroplasts.

preliminary scheme of the spatial organization of polyribosomes in the ultrastructure of chloroplasts (Fig. 3). This scheme illustrates our concept, according to which the grana contain gigantic polyribosomes consisting of many interconnected cycles each of which is enclosed within a thylakoid.

Summary

Peculiar features of the protein synthetizing system of chloroplasts manifest both in the molecular organization of its components and in the polyribosome arrangement within the chloroplast structure. The latter was investigated by means of a detergent-induced gradual degradation of grana and thylakoids of chloroplasts from pea seedlings and subsequent separation of the resultant structures in sucrose discontinuous and linear gradients. Electron microscopy of linear gradient fractions revealed structures that were in different stages of disassembly of grana and thylakoids, e.g. polyribosomes having the shape of rings and large nets composed of interconnected circular polyribosomes. The first formations were well encompassed by thylakoids and the second by grana. It seems probable that these polyribosomes included in grana are responsible for protein synthesis in chloroplasts.

REFERENCES

BARTELS, P. G. & WEIER, T. E., *J. Cell Biol.* 33: 243, 1967.
BISALPUTRA, T. & BURTON, H., *J. Ultrastruct. Res.* 29: 224, 1969.

BOTTOMLEY, W., *Plant Physiol.* 46: 437, 1970.
CHEN, J. L. & WILDMAN, S. G., *Biochim. biophys. Acta* 209: 207, 1970.
FALK, H., *J. Cell Biol.* 42: 582, 1969.
FILIPPOVICH, I. I., ALINA, B. A., TONGUR, A. M. & OPARIN, A. I., in: Geneticheskie Funktsii Organoidov Citoplazmy, p. 52, Nauka, Leningrad 1974.
FILIPPOVICH, I. I., SPANDAR'YAN, O. A., SVETAÏLO, E. N. & SISAKYAN, N. M., *Dokl. Akad. Nauk SSSR* 172: 1214, 1967.
FILIPPOVICH, I. I., TONGUR, A. M., ALINA, B. A. & OPARIN, A. I., *Biokhimiya* 35: 247, 1970.
GNANAM, A. & KAHN, J. S., *Biochim. biophys. Acta* 142: 486, 1967.
GREEN, B. R. & BURTON, H., *Science* 168: 981, 1968.
MACHE, R. & WAYGOOD, E. R., *Can. J. Bot.* 48: 173, 1970.
OPARIN, A. I., FILIPPOVICH, I. I. & BEZSMERTNAYA, I. N., *Fiziol. Rast.* 19: 995, 1972.
PARTHIER, B. & WOLLGIEHN, R., *Naturwissenschaften* 50: 598, 1963.
PHILIPPOVICH, I. I., BEZSMERTNAYA, I. N. & OPARIN, A. I., *Exp. Cell Res.* 79: 159, 1973.
PHILIPPOVICH, I. I., TONGUR, A. M., ALINA, B. A. & OPARIN, A. I., *Exp. Cell Res.* 62: 399, 1970.
SISAKYAN, N. M., FILIPPOVICH, I. I. & SVETAÏLO, E. N., *Dokl. Akad. Nauk SSSR* 147: 488, 1962.
SPENCER, D. & WHITFELD, P. R., *Arch. Biochem. Biophys.* 132: 477, 1969.
TEWARI, K. K. & WILDMAN, S. G., *Biochim. biophys. Acta* 186: 358, 1969.
WERZ, G. & KELLNER, G., *J. Ultrastruct. Res.* 24: 109, 1968.
WOODCOCK, C. L. F. & FERNÁNDEZ-MORÁN, H., *J. mol. Biol.* 31: 627, 1968.

THE USE OF PLASTID AND GENE MUTANTS OF HIGHER PLANTS IN STUDYING THE GENETIC CONTROL OF PLASTID FUNCTIONS

R. HAGEMANN, F. HERRMANN & Th. BÖRNER

Department of Genetics, Section of Biosciences, The University, DDR-402 Halle/S., G.D.R.

Plastid (= plastome) mutants are suitable and important objects for studies on the functions of the genetic information in the plastids. The genetic blocks in plastome mutants are caused by mutations in the plastid DNA itself. Therefore it seems possible to define plastome-controlled metabolic reactions and to obtain knowledge about the molecular functions of the plastid DNA by studying plastome mutants and by characterizing their structural, functional and biochemical changes (HAGEMANN 1971).

Recently we investigated plastome mutants of *Antirrhinum majus* and *Pelargonium zonale*, and compared them with nuclear gene mutants of similar phenotypes.

The plastids of the plastome mutant *en:alba-1* (*extra-nuclear: alba-1*) of *A. majus* — usually white — become pale green under dim light and contain a level of chlorophyll up to 38% of the control. However, they are not capable of photosynthesis. They have an impaired photosystem 1, whereas photosystem 2 is intact, and the plastid enzymes specific for photosynthesis are also active in the mutant plastids (HERRMANN 1971a). Analysis of SDS-solubilized lamellar proteins of the mutant plastids in polyacrylamide gels demonstrated an absence of the chlorophyll-protein-complexes I and Ia (and the corresponding protein bands 2 and 1) which are associated with photosystem 1. The chlorophyll-protein-complexes and protein bands associated with photosystem 2 are present (Fig. 1). Thus the lack of a functional photosystem 1 in *en:alba-1* plastids is due to the absence of the protein components of the pigment complexes I and Ia (HERRMANN 1971b).

The plastids of the plastome mutant *en:viridis-1* of *A. majus* are pale green. Their chlorophyll content amounts to 37% of the control. But the mutant plastids cannot perform photosynthesis. The plastids are active in photosystem 1, but there is a defect in photosystem 2. The electrophoretic pattern of SDS-solubilized chlorophyll-protein-complexes and lamellar protein shows changes in protein bands associated with photosystem 2; protein band 7 is not detectable and bands 8-12 have decreased intensities in comparison to the control. The main chlorophyll-protein-complex II (= protein band 13) of photosystem 2 is present. Therefore the lamellar protein components 7-12 are also necessary for photosystem 2 associated activities (HERRMANN 1972).

The defects in the mutated plastids of the plastom mutants *en:alba-1* and

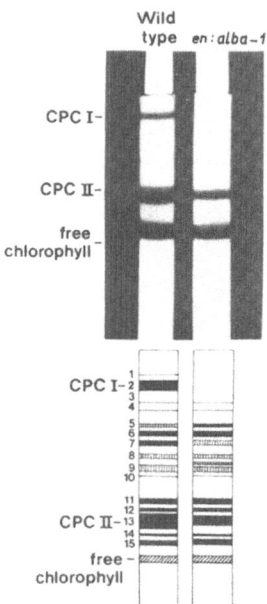

Fig. 1. Electrophoretic separation of pigment-protein complexes (CPC) and electropherograms of chloroplast lamellar proteins of wild-type and mutant *en:alba-1* of *Antirrhinum majus*.

en:viridis-1 are caused by mutations of the plastid DNA. (The genetic information of the nucleus is not changed in these mutants.) The absence or decrease of definite protein bands of the plastid lamellar proteins in particular mutants indicates that the plastid DNA controls the formation of these proteins.

The white plastids of the periclinal-chimerical cv. 'Mrs. Parker' of *Pelargonium zonale* represent a white plastome mutant. Comparisons between green and pure white branches of variegated F_1-plants (with equal nuclear genetic information, but differences in the genetic information of the plastids) revealed marked differences in the RNA content (BÖRNER et al. 1972). From normal green cells of *Pelargonium zonale* four bands of high molecular weight ribosomal RNA can be isolated: 25S and 18S RNA of the cytoplasmic ribosomes and 23S and 16S RNA of plastid ribosomes. The mutation of the plastid DNA in the plastids of 'Mrs. Parker' causes an altered RNA pattern: The 23S and the 16S RNA of the plastid ribosomes cannot be detected in polyacrylamide gels (whereas 25S and 18S RNA are present) (Fig. 2). The cells of white leaves contain numerous plastids, which are smaller than normal chloroplasts; the formation of normal internal membrane structures is blocked. In mutated plastids of 'Mrs. Parker' ribosomes cannot be detected in electron micrographs.

The absence of the high molecular weight ribosomal RNA of the plastids as a

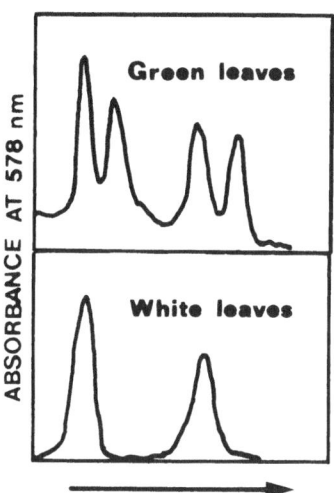

Fig. 2. Electrophoretic separation of high molecular weight ribosomal RNA of green leaves and white leaves with 'Mrs. Parker' plastids. Densitograms.

consequence of a plastome mutation demonstrates a direct or indirect control of the plastid DNA on the plastid ribosomal RNA (23S and 16S).

From these findings it follows that protein synthesis cannot be performed in mutated plastids of 'Mrs. Parker'. The multiplication of the plastids — and presumably also the replication of plastid DNA — is not impaired by the deficiency in plastid protein synthesis. Protein synthesis within the plastids is necessary for full development and differentiation of the chloroplasts, although an essential part of the plastid proteins are synthesized on cytoplasmic ribosomes (for a recent review see BÖRNER 1973). Further results of our investigations on plastome mutants of higher plants have been reported by BÖRNER et al. (1973), HAGEMANN et al. (1973), HAGEMANN & SCHOLZE (1974), HERRMANN et al. (1974) and KNOTH et al. (1974).

Summary

The function of genetic information in plastids was investigated by studying plastid (= plastome) mutants. Mutations in the plastid DNA cause defects in the chloroplast lamellar proteins in two mutants of *Antirrhinum* and plastid ribosome deficiency in a mutant of *Pelargonium*.

REFERENCES

BÖRNER, T., *Biol. Zentralbl.* 92: 545, 1973.
BÖRNER, T., HERRMANN, F. & HAGEMANN, R., *FEBS Lett.* 37: 117, 1973.
BÖRNER, T., KNOTH, R., HERRMANN, F. & HAGEMANN, R., *Theor. appl. Genet.* 42: 3, 1972.
HAGEMANN, R., *Biol. Zentralbl.* 90: 409, 1971.
HAGEMANN, R., BÖRNER, T. & KNOTH, R., *Genetics* 1973 (2, Part 2): 103, 1973.
HAGEMANN, R. & SCHOLZE, M., *Biol. Zentralbl.* 93: 625, 1974.
HERRMANN, F., *Photosynthetica* 5: 258, 1971a.
HERRMANN, F., *FEBS Lett.* 19: 267, 1971b.
HERRMANN, F., *Exp. Cell Res.* 70: 452, 1972.
HERRMANN, F. H., MATORIN, D., TIMOFEEV, K., BÖRNER, T., RUBIN, A. B. & HAGEMANN, R., *Biochem. Physiol. Pflan.* 165: 393, 1974.
KNOTH, R., HERRMANN, F. H., BÖTTGER, M. & BÖRNER, T., *Biochem. Physiol. Pflan.* 166: 129, 1974.

THE RELATIONSHIP OF CHLOROPHYLL BIOSYNTHESIS TO PROTEIN AND RNA SYNTHESIS IN GREENING AND GREEN LEAVES

A. A. SHLYK, INNA V. PRUDNIKOVA, GALINA E. SAVCHENKO, NATALIYA G. AVERINA, NATALIYA N. KOSTYUK, LARISSA K. KAMYSHENKO, LARISSA I. VLASENOK. V. I. GAPONENKO, ELIZABETA F. BALEVA, TATYANA K. PARAMONOVA, TATYANA V. LOSITSKAYA & A. Yu. VEZITSKIĬ

Institute of Photobiology, Academy of Sciences of the B.S.S.R., Minsk, U.S.S.R.

Biosynthesis of chlorophyll (Chl) is closely associated with other processes of plant metabolism, of which the most important is protein biosynthesis. The latter supplies the enzymatic apparatus necessary for the construction of new Chl molecules and the carrier on which pigment molecules are mounted. It is known that the control of Chl biosynthesis takes place to a considerable degree at the stage of formation of δ-aminolevulinic acid (ALA). In many cases the formation and regulation of the activity of ALA synthetase is a key point in the whole process (GRANICK 1959; LASCELLES 1959; MARSH et al. 1963; SISLER & KLEIN 1963; KLEIN & BOGORAD 1964; KALER & PODCHUFAROVA 1965; SCHIFF & EPSTEIN 1965). However, Chl biosynthesis may be regulated in a more complex way, and, in particular, it may be carried out by a multienzyme complex, chlorophyll-synthetase, and may be directed both through changes in the number of acting centres of Chl biosynthesis and through changes within the centres (SHLYK et al. 1969a).

One of the experimental approaches to the investigation of the regulation of Chl biosynthesis is the application of inhibitors and stimulators of protein synthesis, acting at different stages in the carrying out of genetic information. It is then taken into account that the chloroplast contains its own DNA and synthesizes some of its own proteins, but in addition the chloroplasts are influenced by gene mutations, which means that their essential activities are directed in a fairly complex manner. Probably this complexity can explain why the results have not led to an unified concept, although problems of the relationship of Chl biosynthesis to the biosynthesis of protein and other metabolic processes have long been a subject of discussion. Moreover, the discussion is usually limited to experimental data obtained for postetiolated seedlings in which accumulation of Chl takes place together with the development of lamellar and granular structure, and the separation of these two aspects of the single process of transformation of etioplast into chloroplast is extremely difficult.

It seemed advisable, therefore, to study the significance of protein synthesis in the formation of new Chl molecules in mature chloroplasts of the normal green leaf and to consider the observed quantitative regularities on the basis of the concept of centres of Chl biosynthesis (SHLYK et al. 1969a). In addition to the usual tests, based on the recording of changes in the accumulation of protochlorophyllide (P) or total Chl, particular attention has been paid to the ratio between different forms of P and the transformation of chlorophyll a (Chl_a) to chlorophyll b (Chl_b). It was believed that these indexes would facilitate the differentiation of observed effects into those characteristics of the system of centres of Chl biosynthesis that were actually changed by inhibitors and those that remained unchanged.

Changes in Volume of the System of Acting Centres of Chlorophyll Biosynthesis

Our earlier published data show that in experiments with green leaves containing mature chloroplasts, the introduction of chloramphenicol brings about the suppresion of Chl biosynthesis, as revealed by P accumulation in darkness (SHLYK et al. 1969b). On the other hand, it was found that stimulation of protein synthesis by kinetin not only maintains or reestablishes the Chl level in old leaves (KURSANOV et al. 1964) but increases synthesis of P both in normal green and in etiolated leaves (SHLYK & AVERINA 1969; SHLYK et al. 1970b; AVERINA & SHLYK 1972). This means that the level of protein synthesis is the limiting factor determining the general productivity of the system of centres of Chl biosynthesis. Fig. 1 gives an example of effects of chloramphenicol and kinetin on P accumulation in green leaves.

It is probable that the stimulating action of kinetin is determined by its effects on the formation of RNA participating in this process (KULAEVA 1967; SELIVANKINA et al. 1972). On the other hand, the inhibition of RNA synthesis reduces Chl synthesis: Fig. 2 shows the content of P before and after darkening

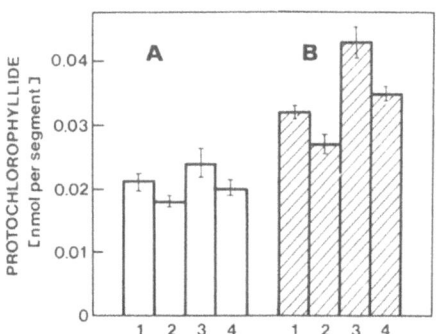

Fig. 1. Content of protochlorophyllide in middle segments of green leaves of barley incubated for 24 h in light (A) and then for 18 h in darkness (B) on water (1) or solutions of chloramphenicol (2) or kinetin (3), and also on a solution containing both kinetin and chloramphenicol (4). (From SHLYK & AVERINA 1973.)

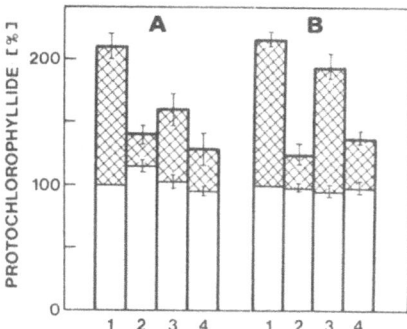

Fig. 2. Relative content of protochlorophyllide, before darkening of postetiolated and green barley leaves (light columns) and its increase in darkness (dashed columns) in the presence of inhibitors of RNA synthesis. 7 to 8-day-old etiolated and green leaves of barley were kept for 20 h in darkness on 0.2 M solutions of sucrose with or without inhibitor, then the etiolated leaves were illuminated for 1 min (2 klx, 25 °C) and green leaves for 30 min (2 klx, 0 °C), and again transferred to darkness and left on the same solutions: postetiolated for 3 h, green for 24 h. The protochlorophyllide content before darkening in the samples kept on water is taken as 100%. A – postetiolated; B – green; 1 – water; 2 – actinomycin; 3 – 5-fluorouracil; 4 – 8-azaguanine. (From SHLYK et al. 1973b.)

of postetiolated and green seedlings of barley which had previously been incubated for 20 h on solutions of inhibitors of RNA synthesis. Before the transfer of plants to darkness the level of P in the samples with and without inhibitors was similar, since under illumination the inactive form of the pigment is almost the only one present. In darkness, active P accumulates as well, and this accumulation is considerably decreased by all inhibitors. While similar observations have already been made for postetiolated leaves and have served as an indication that new RNA molecules must be formed in order that the greening process may proceed, for green leaves they have now apparently been made for the first time.

These results indicate that RNA(s) and protein(s) necessary for Chl biosynthesis are subject to turnover and are capable of ensuring the normal process of pigment formation in the green leaf only if their loss is constantly made up by new synthesis. The latter is just as necessary in the green as in the greening leaf.

Inadequate RNA and protein synthesis in green leaves treated with actinomycin D and chloramphenicol leads to a decrease both in the rate and in the maximum level of P accumulation (Figs. 3 and 4). When the inhibitor is added a sufficiently long time before darkening, both these values decrease in such a way that the ratio of initial rate to maximum level of P accumulation in the presence of inhibitors is of the same order as in the control without inhibitor. In other words, whatever the interpretation of the mechanism of the inhibitor effect, there is always a close connection between the volume which can be filled by P and the activity of the biosynthetic apparatus: a unit of activity corresponds to a unit of

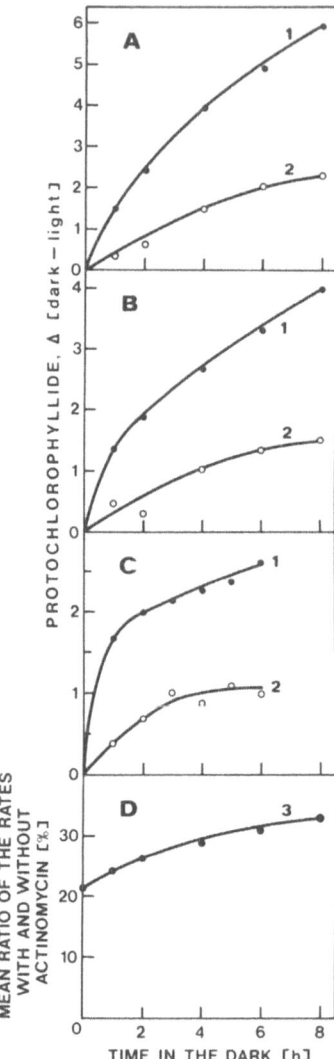

Fig. 3. Kinetics of accumulation of protochlorophyllide in green barley leaves that were previously kept in darkness for 20 h on 0.2 M solutions of sucrose with (2) or without (1) actinomycin D, then illuminated for 30 min (2 klx, 0 °C) and again transferred to darkness on the same solutions. The samples were taken during this last period of darkness. The quantity of protochlorophyllide accumulated is expressed as the difference between its content after and before the second darkening. The initial sections of the curves of dark accumulation of protochlorophyllide in three experiments (A, B, C) and the ratio of the rates of this process at different times (D) with and without actinomycin D are shown. The rates were evaluated as the tangent of the angle of slope of the curve at the corresponding time. 1 − control without actinomycin, 2 − with actinomycin, 3 − ratio between rates of resynthesis of protochlorophyllide with and without actinomycin [%]. (From SHLYK et al. 1973b.)

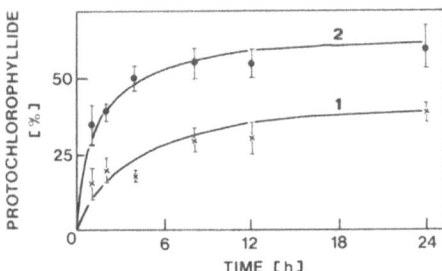

Fig. 4. Kinetics of accumulation of protochlorophyllide in the dark in the upper parts of green leaves of barley infiltrated with solutions of chloramphenicol (1) or dextromycetin (2) (1 mg l^{-1}), kept for 24 h in light and afterwards placed in darkness. The quantity of protochlorophyllide accumulated is expressed as the ratio (in percent) of the difference in its content after and before darkening to the content in light (%). (From SHLYK et al. 1969b.)

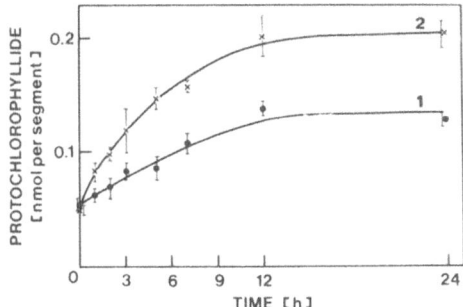

Fig. 5. The change in the content of protochlorophyllide in one segment of the leaf when the middle parts of green leaves of barley are darkened, having first been held for 48 h in light on the surface of water (1) or a solution of kinetin (20 mg l^{-1}) (2).

volume. This conclusion also agrees with the nature of the action of kinetin on the kinetics of dark accumulation of P in green barley leaves (Fig. 5).

Such a simple relationship between the two parameters could hardly be observed if they were not internally related and determined by a common mechanism. Obviously they should be considered as two coupled characteristics of some elementary structural-functional unit, within which both the formation and the accumulation of P take place. While under certain conditions the number of these units is greater, and under others smaller, the total volume and total activity of the system will differ accordingly. Thus, the concept of the localization of the apparatus responsible for Chl formation in special centres of biosynthesis, which has been developed by us over a number of years, receives additional substantiation. The number of acting centres is the variable most affected by the application of inhibitors.

We have already mentioned above that the maintenance of the activity of the system of centres of Chl biosynthesis in the green leaf requires constant replenishment of RNA(s) and protein(s) subject to turnover. The insufficient production of protein creates a deficiency of certain essential components necessary for the operation of centres of Chl biosynthesis, and the number of active centres decreases. Then the general form of the kinetic curve of P accumulation for the whole leaf, which is lower in conditions of inhibition, is still determined in the first place by internal characteristics of the active centres.

The most probable cause of the stoppage of P accumulation is the switching off of the biosynthetic chain at the step of ALA formation by a feedback mechanism triggered by P (GRANICK 1959; LASCELLES 1959; MARSH et al. 1963; SISLER & KLEIN 1963; KLEIN & BOGORAD 1964; KALER & PODCHUFAROVA 1965; SCHIFF & EPSTEIN 1965). From this model it may be assumed that inhibition of protein synthesis interferes, first of all, with the normal process of turnover of ALA synthetase. The effects of chloramphenicol or actinomycin D on the initial rate of P resynthesis are not exactly equal to but somewhat greater than their effects on the maximum level. Thus, after a pretreatment of green leaves of barley with chloramphenicol (1 mg l^{-1}) for 24 h the ratio of the initial rate to the final level (from which the amount of P present in the light before darkening was previously subtracted) was equal to 41% per hour while without the inhibitor it corresponded to 58% per hour. In experiments with actinomycin D the initial rate was only 21% of that observed in the control and, as pigment was accumulated, this value gradually increased (Fig. 3D). The difference between the final amount of P accumulated after 24 h of darkening and its starting level in the light in the variant with actinomycin D corresponds to 35% of the difference in the control without the inhibitor. Consequently, the decrease in the volume of the system of active centres of Chl biosynthesis caused by inhibition of RNA and protein synthesis is accompanied by a reduction in the enzyme activity of those centres that are still working. This may be associated with the particular sensitivity of ALA synthetase.

Such special sensitivity of ALA synthetase is also revealed by experiments with kinetin. Though this stimulator of RNA and protein synthesis causes an increase in both initial rate and maximum level of P accumulation (Fig. 5), the initial rate is stimulated with kinetin 2.5 times, while the final increase in the P content (i.e. the dark minus light amount) is augmented only 1.8 times.

Thus the main result of a change in the rate of protein synthesis in either a positive or negative direction is a corresponding change in the volume of the system of acting centres of Chl biosynthesis, while changes in some internal properties of those centres that remain working (such as the activity of ALA synthetase) represent an additional effect.

The constancy of several other characteristics of centres of Chl biosynthesis also attracts attention. Earlier we showed that the increase in P accumulation under the influence of kinetin does not lead to a change in the ratio between forms of pigment capable and incapable of photoreduction (SHLYK et al. 1970b). The work with postetiolated leaves showed that if we compared the content of the inactive form of P which remained after a short period of illumination with that after another illumination period (separated from the first by a dark interval), a partial activation of the pigment took place during the dark period. The extent of this activation was not changed by the presence of kinetin (AVERINA & SHLYK 1972). The proportion of P capable of phototransformation was also unchanged in experiments where either actinomycin D, 8-azaguanine or 5-fluorouracil were introduced into green leaves and 8-azaguanine and 5-fluorouracil into postetiolated leaves. And although we observed in chloramphenicol treated green seedlings or actinomycin D treated postetiolated ones some decrease in the percentage of the pigment active in phototransformation (in addition to a decrease in P accumulation) this decrease was only of a secondary importance. Hence an exact description of the phenomena requires taking into account some internal disordering of centres of biosynthesis, but nevertheless, the main conclusion from all these observations is that characteristics of this type are approximately constant for a given system.

The degree of activity of P is not determined by ALA synthetase, but depends on the placing and condition of the pigment in the centres of biosynthesis. However, this enzyme seems to be a part of a multienzyme complex carrying out the whole sequence of reactions of the biosynthetic chain, i.e. chlorophyll-synthetase (SHLYK et al. 1969a). Chlorophyll-synthetase itself is closely associated with other active and structural elements of the biosynthetic centre. Therefore the constancy of activity of P accumulated at different rates of protein synthesis signifies that different components of the centre of biosynthesis, and in particular of chlorophyll-synthetase, which behave in many respects as a whole, are jointly dependent on its level.

Does Inhibition of RNA Synthesis Actually Affect the Transformation of Chlorophyll a into Chlorophyll b?

The conclusion that Chl synthesis is regulated in such cases by preferential action on a number of centres capable of reaction, with a less essential effect on particular internal characteristics of a single centre, is confirmed if we compare the action of inhibitors of RNA synthesis on the biosynthesis of Chl_a and Chl_b. So far information about the action of this group of inhibitors on the transformation of Chl_a into Chl_b is lacking, since the determination of extremely small quantities of Chl_b at the very beginning of greening requires the use of special procedures that have only recently been developed (FRADKIN et al. 1966; RUDOÏ et al. 1968; SHLYK et al. 1970a; ZEN'KEVICH & LOSEV 1970; THORNE & BOARDMAN 1971). Nor has anyone attempted to study it with green leaves. Certainly in this case quantitative pigment determinations alone would be almost useless for such a

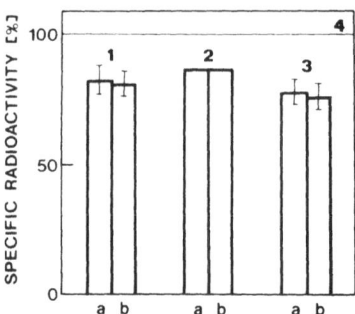

Fig. 6. Relative specific radioactivity of chlorophylls *a* (a) and *b* (b) in green barley leaves which assimilated $^{14}CO_2$ for a short time and were then incubated on solutions of actinomycin D (18 h in darkness) (1), 8-azaguanine (2) or 5-fluorouracil (3) (18 h in darkness, 24 h illumination with 1.2 klx). Specific radioactivity of the corresponding chlorophyll in the parallel control without an inhibitor is taken as 100% (4). (From SHLYK et al. 1972.)

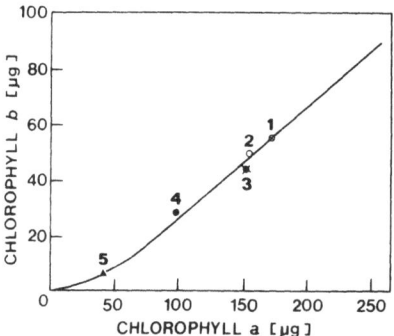

Fig. 7. The relationship between contents of chlorophyll *b* and chlorophyll *a* in the process of greening of etiolated leaves of barley in the course of the first 24 h of illumination (2 klx) on 0.2 M solution of sucrose (continuous line) and the amounts of the two pigments after 12 h of illumination on the same solution without (1) or with 5-bromuracil (2), 5-fluorouracil (3), 8-azaguanine (4), or actinomycin D (5). Before illumination the leaves were kept for 20 h in darkness on a solution with or without an inhibitor. (From SHLYK et al. 1973a.)

purpose. In green leaves the effect of actinomycin D, 5-fluorouracil and 8-azaguanine has been investigated by following ^{14}C incorporation. The decrease in specific radioactivities caused by the inhibitors was the same for both pigments (Fig. 6). During the illumination of etiolated seedlings actinomycin D, 5-haloiduracils* and 8-azaguanine delay the accumulation of both Chl's to the same extent, so that

---

* The action of 5-haloiduracil and 8-azaguanine should be especially emphasized, since GASSMAN & BOGORAD (1967) and NADLER & GRANICK (1970) did not observe any effects which, apparently, are especially dependent on experimental conditions.

their quantities correspond simply to a lower degree of greening (Fig. 7).

There is a danger that such data will be incorrectly interpreted if we do not take into account the fact that the rate of formation of Chl_b may be decreased even without a specific effect of an agent on the transformation of Chl_a into Chl_b (SHLYK 1971) but rather as a consequence of the abnormal synthesis of Chl_a. An extreme example of this is the absence of Chl_b in etiolated leaves that have been found recently to be capable of performing this reaction, but could not make use of this ability only because of the absence of Chl_a as the precursor: if exogenous Chl_a is added to a homogenate of etiolated leaves which have never been illuminated, then Chl_b appears even in the dark (SHLYK et al. 1971). Obviously, under illumination the quantity of Chl_b formed should depend on the quantity of accumulated Chl_a. The factor linking the accumulation of the two pigments may also be the development of the ultrastructure, where the appropriate biosynthetic apparatus is localized and where molecules are fixed and stabilized as they arise. It is not an accident that in Fig. 7 the points reflecting the dependence of accumulation of Chl_b on the accumulation of Chl_a after treatment with inhibitors lie on the curve characterizing the mutual relationship between the two Chl's in the case without an inhibitor. This clearly shows that inhibitors of RNA synthesis have no direct effect on the Chl_a to Chl_b transformation, but affect the accumulation of Chl_b indirectly as a result of the suppression of Chl_a synthesis.

Clear evidence for such a conclusion was obtained when the synthesis of Chl_b and the synthesis of Chl_a were purposely separated – in the study of dark accumulation of Chl_b in postetiolated leaves of rye. Etiolated seedlings were kept in darkness for 20 h on solutions of inhibitors or water, then illuminated for 1 min so that P was photoreduced giving quantities of Chl_a similar in all variants, and then the plants were again kept for 24 h in darkness, during which time part of the Chl_a, as is known (RUDOÏ et al 1968; SHLYK et al. 1970a), is transformed into Chl_b. During the dark period the amount of Chl_b formed was more than 20% of Chl_a, and this value was the same both without and with inhibitor (Fig. 8).

Thus the effect of inhibitors of RNA synthesis is manifested in a decrease in ^{14}C incorporation in Chl_a and Chl_b in green plants, in the suppression of accumulation of both Chl's in the greening process, but does not concern the transformation of Chl_a to Chl_b as shown by unchanged ratios of Chl_b to Chl_a in postetiolated leaves and of specific radioactivities of the pigments in green leaves. These facts indicate again that the inhibitors of RNA synthesis mainly reduce the total volume of the system of active centres of Chl biosynthesis, while the activity of any individual centre that remains active (in particular the proportion of Chl_a molecules transformed into Chl_b molecules) remains unchanged.

Of course, a change in the volume of the system of working centres of Chl biosynthesis does not always clearly dominate other manifestations of the action of one factor or another such as the inhibition of RNA and protein synthesis. In particular, changes in Chl_b content do not always reflect the behaviour of Chl_a so simply. In our experiments with p-chloromercuribenzoate and α-iodoacetamide, which act on SH-groups and disrupt protein structure, Chl_b synthesis was more strongly inhibited than the synthesis of Chl_a. The same effect was observed with ethionine, which interferes with the specific use of methionine for Chl synthesis. These data indicate the importance of SH-groups for the activity of the enzyme apparatus which transforms Chl_a into Chl_b. Such direct action on this transformation reflects not only changes in the volume of the biosynthetic apparatus, but also important internal changes in the centres of biosynthesis.

Importance of Internal Changes within Centres of Chlorophyll Biosynthesis

In separating effects into those due to changes in the volume of the system of centres of Chl biosynthesis and those associated with changes within individual centres, we are, of course, somewhat formalizing the discussion of the question. From the phenomenological point of view, the first category may also be characterized as changes in the number of open entrances into the centres. This is correct when the entrance is the limiting stage in the whole process. Moreover, such expression should imply that simultaneously with changes in the entrance other changes may be included, provided that the latter do not proceed to such a high degree that the limiting role of the entrance begins to lose its importance, or provided that the nature of the changes is such that they are not revealed by the test used. Of course, changes may well take place within the centre simultaneously

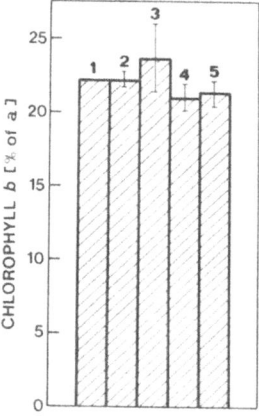

Fig. 8. Ratio of chlorophyll b to chlorophyll a after darkening of postetiolated leaves of rye treated with inhibitors of RNA synthesis (7 to 8-day-old etiolated leaves were kept for 20 h in darkness on 0.2 M solutions of sucrose with or without inhibitor, then illuminated for 1 min (2 klx, 25 °C) and again transferred for 24 h to darkness, on the same solutions). The values for the samples with inhibitors in each experiment were normalized to the mean value in the water variant before being averaged. 1 – water; 2 – actinomycin D; 3 – 5-fluorouracil; 4 – 5-bromuracil; 5 – 8-azaguanine. (From SHLYK et al. 1973a.)

with changes in the entrance, and they would be accentuated in those cases where they are distinctly reflected on the behaviour of the system.

We have already discussed the widespread view according to which Chl biosynthesis is limited at the stage of ALA formation, which may be considered as an entrance into the biosynthetic centre. Inasmuch as this limitation may be removed by the artificial addition of ALA from the outside, it was of interest to use this procedure in order to bring out limitations of the second order, which may in certain situations materially supplement the value of the entrance.

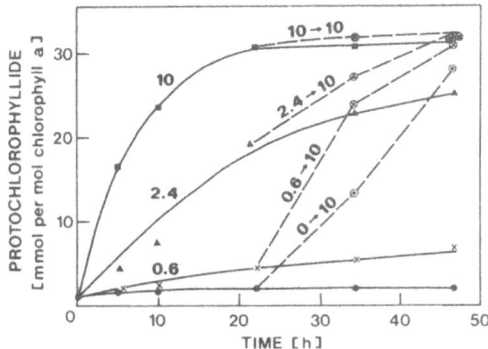

Fig. 9. The change in the content of protochlorophyllide during darkening of green leaves of barley in the presence of different concentrations of δ-aminolevulinic acid (mM – figures on the curves) and after transfer of part of the plants of each variant to a fresh solution of 10 mM δ-aminolevulinic acid. (From SHLYK & KOSTYUK 1972b.)

Fig. 9 shows the accumulation of P during the darkening of green leaves of barley when ALA is present in different concentrations. In the water variant the accumulation practically ceases after 5 h of darkening with a P content three times the initial level in light. With an increase in the concentration of ALA, the accumulation of P increases. With an ALA concentration of 10 mM it stops after 22 h of darkness, when the quantity of P increased 50 times. This is considerably more than the tenfold increase usually described for etiolated leaves with the same concentration of ALA (GRANICK 1962; KALER & PODCHUFAROVA 1965). The addition of fresh ALA solution (dashed curves) after the cessation of accumulation of P does not lead in this case to a new rise in its content. The existence of a limit in the accumulation of P is confirmed when some plants of all variants are transferred to a 10 mM ALA solution; this leads to the attainment of a common P level, always in 20 h, and the higher the P concentration at the moment of transfer, the lower the rate of its further accumulation.

Thus, the usual explanation of the regulation of chlorophyll biosynthesis by means of inhibition of further ALA formation by accumulated P may be now expanded by taking into account the existence of a blocking action of excess P to its further formation even from abundant ALA. This new point of switching off lies after the formation of ALA and apparently the role of a reserve regulation mechanism is performed here. Its sensitivity to the accumulation of P is only about 1/20th of the sensitivity of regulation through ALA synthetase. If the potentialities of chlorophyll-synthetase at all stages after ALA are normally so great that they permit it to respond to the addition of ALA by a proportional (SHLYK & KOSTYUK 1972b) increase in the initial rate of P formation, then the excess accumulation of the latter, even with an abundant supply of ALA, is permitted only up to a certain limit.

By removing the limiting action of ALA synthetase through exogenous addition of ALA we can understand the role of protein synthesis in ensuring that further stages in the chain of Chl formation may take place. We have already shown that chloramphenicol inhibits the accumulation of P even with an excess of exogenous ALA, although to a lesser extent than without this extra supply (SHLYK & KOSTYUK 1972a). This is also reflected in a lower P level after darkening of green seedlings, when ALA was given to them during chloramphenicol treatment (Fig. 10). Consequently we see that Chl biosynthesis requires normal turnover not only of ALA synthetase but also of other components of chlorophyll-synthetase.

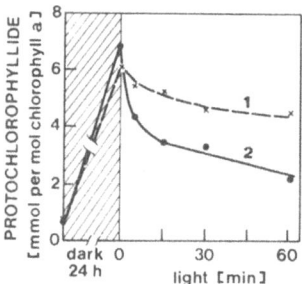

Fig. 10. Change in the content of protochlorophyllide in green leaves of barley which were first kept in a solution of chloramphenicol (1) or water (2), then transferred to darkness for 24 h in the presence of δ-aminolevulinic acid, and subsequently illuminated (70 lx). (From SHLYK & KOSTYUK 1972b.)

The right-hand side of Fig. 10 shows still another aspect of the action of chloramphenicol — the effect of the inhibitor on the phototransformation of P formed from exogenous ALA. Considering the ease with which such P might be bleached, the work was carried out under low illumination of 70 lux. Insufficient synthesis of protein leads not only to a smaller accumulation of P in darkness but also to a lower capacity of the formed P for phototransformation, and as a result the content of inactive P remaining in light is higher. An increase in the level of inactive P in light was observed earlier to be induced by chloramphenicol without the addition of exogenous ALA (SHLYK et al. 1969b). Loss of the photochemical activity of P indicates the absence of the normal binding of this molecules to the protein carrier and/or the hydrogen donor (which may also be a carrier). This means that even in the green leaf there exists a requirement for continuous protein synthesis in order to ensure both the normal activity of the enzymatic apparatus carrying out the biosynthesis of Chl and the normal state of the products formed. And in this the coupling of different features of the activity of centres of Chl biosynthesis is manifested.

Summary

The regulation of chlorophyll (Chl) formation is mediated mainly by changes in the number of acting centres of Chl biosynthesis and by changes within individual centres. Continuous protein synthesis is important for both types of regulation in greening and green leaves. Actinomycin or chloramphenicol decrease and kinetin increases similarly both the initial rate and the final level of protochlorophyllide (P) accumulation upon darkening green leaves. Thus, as a result of a change in RNA and protein turnover rate, the number of acting centres becomes smaller or larger. The additional action of these agents on the events within centres is apparent from the fact that the response of the initial rate of P accumulation is somewhat stronger than that of the final level. The effect of chloramphenicol is only partly overcome by feeding a green leaf δ-aminolevulinic acid which causes formation of additional P. The limitation at a later step of biosynthesis is also revealed in the saturation of accumulation of this P (even in the case of a continuous feeding) and in its lower reducibility. Actinomycin and other inhibitors of RNA synthesis decrease ^{14}C-incorporation in Chl_a and Chl_b in green leaves as well as Chl_a and Chl_b accumulation during greening of etiolated ones, but in both cases they do not affect the transformation of Chl_a into Chl_b. Ratio $Chl_b:Chl_a$ is independent of them when etiolated leaves have been illuminated for 1 min and kept 24 h in the dark.

REFERENCES

AVERINA, N. G. & SHLYK, A. A., *Fiziol. Rast.* 19: 487, 1972.
FRADKIN, L. I., SHLYK, A. A. & KOLYAGO, V. M., *Dokl. Akad. Nauk SSSR* 171: 222, 1966.
GASSMAN, M. & BOGORAD, L., *Plant Physiol.* 42: 774, 1967.
GRANICK, S., *Plant Physiol. 34* (Suppl.): XVIII, 1959.
GRANICK, S., in: Mechanism of Photosynthesis, International Biochemical Congress, Symposium VI, p. 184, Moskva 1962.
KALER, V. L. & PODCHUFAROVA, G. M., in: Fiziologo-biokhimicheskie Issledovaniya Rastenii, p. 15, Nauka i Tekhnika, Minsk 1965.
KLEIN, S. & BOGORAD, L., *J. Cell Biol.* 22: 443, 1964.

KULAEVA, O. N., *Uspekhi sovrem. Biol.* 63: 28, 1967.
KURSANOV, A. L., KULAEVA, O. N., SVESHNIKOVA, I. N., POPOVA, E. A., BOLYAKINA, Yu. P., KLYACHKO, N. L. & VOROB´EVA, I. P., *Fiziol. Rast.* 11: 838, 1964.
LASCELLES, J., *Biochem. J.* 72: 508, 1959.
MARSH, H. V. Jr., EVANS, H. J. & MATRONE, C., *Plant Physiol.* 38: 632, 1963.
NADLER, K. D. & GRANICK, S., *Plant Physiol.* 46: 240, 1970.
RUDOÏ, A. B., SHLYK, A. A. & VEZITSKIĬ, A. Yu., *Dokl. Akad. Nauk SSSR* 183: 215, 1968.
SCHIFF, J. A. & EPSTEIN, H. T., in: LOCKE, M. (ed.): Reproduction: Molecular, Subcellular, and Cellular, Symp. Soc. Developm. Biol. 24: 131, Academic Press, New York – London 1965.
SELIVANKINA, S. Yu., KUROEDOV, V. A. & KULAEVA, O. N., *Fiziol. Rast.* 19: 508, 1972.
SHLYK, A. A., *Annu.Rev. Plant Physiol.* 22: 169, 1971.
SHLYK, A. A. & AVERINA, N. G., *Dokl. Akad. Nauk SSSR* 186: 1209, 1969.
SHLYK, A. A. & AVERINA, N. G., *Fiziol. Rast.* 20: 725, 1973.
SHLYK, A. A., GAPONENKO, V. I., VLASENOK, L. I., BALEVA, E. F. & PARAMONOVA, T. K., *Dokl.Akad.Nauk SSSR* 207: 1002, 1972.
SHLYK, A. A. & KOSTYUK, N. N., *Dokl. Akad. Nauk SSSR* 202: 707, 1972a.
SHLYK, A. A. & KOSTYUK, N. N., *Dokl. Akad. Nauk SSSR* 206: 1002, 1972b.
SHLYK, A. A., PRUDNIKOVA, I. V., FRADKIN, L. I., NIKOLAYEVA, G. N. & SAVCHENKO, G. E., in: METZNER, H. (ed.):Progress in Photosynthesis Research, Vol. 2, p. 572, Tübingen 1969a.
SHLYK, A. A., PRUDNIKOVA, I. V., KAMYSHENKO, L. K., LOSITSKAYA, T. V., MITSUK, Z. I. & GROZOVSKAYA, M. S., *Dokl.Akad. Nauk SSSR* 208: 472, 1973a.
SHLYK, A. A., PRUDNIKOVA, I. V. & MALASHEVICH, A. V., *Dokl. Akad. Nauk SSSR* 201: 1481, 1971.
SHLYK, A. A., PRUDNIKOVA, I. V., SAVCHENKO, G. E., KAMYSHENKO, L. K., GROZOVSKAYA, M. S., MITSUK, Z. I. & LOSITSKAYA, T. V., *Dokl. Akad. Nauk SSSR* 211: 744, 1973b.
SHLYK, A. A., RUDOÏ, A. B. & VEZITSKIĬ, A. Yu., *Photosynthetica* 4: 68, 1970a.
SHLYK, A. A., SAVCHENKO, G. E., VEZITSKIĬ, A. Yu. & KARAKO, P. S., *Dokl.Akad. Nauk SSSR* 88: 718, 1969b.
SHLYK, A. A., VAL´TER, G., AVERINA, N. G. & SAVCHENKO, G. E., *Dokl.Akad.Nauk SSSR* 193: 1429, 1970b.
SISLER, E. C. & KLEIN, W. H., *Physiol. Plant.* 16: 315, 1963.
THORNE, S. W. & BOARDMAN, N. K., *Plant Physiol.* 47: 252, 1971.
ZEN´KEVICH, E. I. & LOSEV, A. P., *Zh. prikl. Spektroskop.* 13: 1032, 1970.

GENETIC CONTROL OF CHLOROPHYLL BIOSYNTHESIS AND FORMATION OF ITS FORMS IN VIVO

YU. S. NASYROV, YU. E. GILLER & P. D. USMANOV

Institute of Plant Physiology and Biophysics, Tajik Acad. Sci., Dushanbe, U.S.S.R.

Introduction

At present we seem to have a clear view of the pathways for the biosynthesis of chlorophylls and other plastid pigments (GODNEV 1963; GOODWIN 1965; SMIT 1965; SHLYK 1965, 1971; GRANICK 1967; SHLYK et al. 1972, 1975). It has been stated that light-induced synthesis of chlorophylls runs simultaneously with the formation of chloroplast lamellar structure.

Particular consideration is now devoted to metabolic and genetic regulatory mechanisms of the plastid pigments' biosyntheses. We do not exactly know the contribution of nuclear and plastid genetic factors to chlorophyll biosynthesis and the formation of their forms in vivo. It is supposed that synthesis of the key enzymes of chlorophyll biosynthesis is determined by the nucleus, whereas the nonlimiting enzymes are generated in chloroplasts and are controlled by the chloroplast transcription-translation system. In addition, the aggregation and packing of pigments in thylakoids are determined by lamellar proteins, their formation being under dual control (NASYROV 1972).

Biochemical methods of inhibition of different chains of reactions coupled with experimental mutagenesis and followed by genetic analysis seem to be the perspective for understanding the nuclear-plastid control of chlorophyll biosynthesis and formation of the forms in vivo.

It has been previously shown that formation of δ-aminolevulinic acid (ALA) synthetase is a limiting factor for chlorophyll biosynthesis (GRANICK 1967). Subsequently it was established that synthesis of this enzyme is inhibited by cycloheximide, a specific inhibitor of protein synthesis on 80S cytoplasmic ribosomes, and that exogenous δ-aminolevulinic acid can remove the action of this inhibitor (NADLER & GRANICK 1970). Hence the key enzyme of chlorophyll biosynthesis is formed in the cytoplasm and is – according to NADLER & GRANICK (1970) under the control of nuclear genes.

Induction of pigment mutations in higher plants and algae by different physical and chemical agents enables us to study the influence of nuclear genes on the biosynthesis of plastid pigments. A wide spectrum of chlorophyll mutants of *Arabidopsis thaliana* and *Gossypium hirsutum* from 'albino' up to 'dark-green' has been obtained in our Institute. Genetic analysis of these mutants showed they belonged, mainly, to monohybrid-segregating recessives, and there were also cases

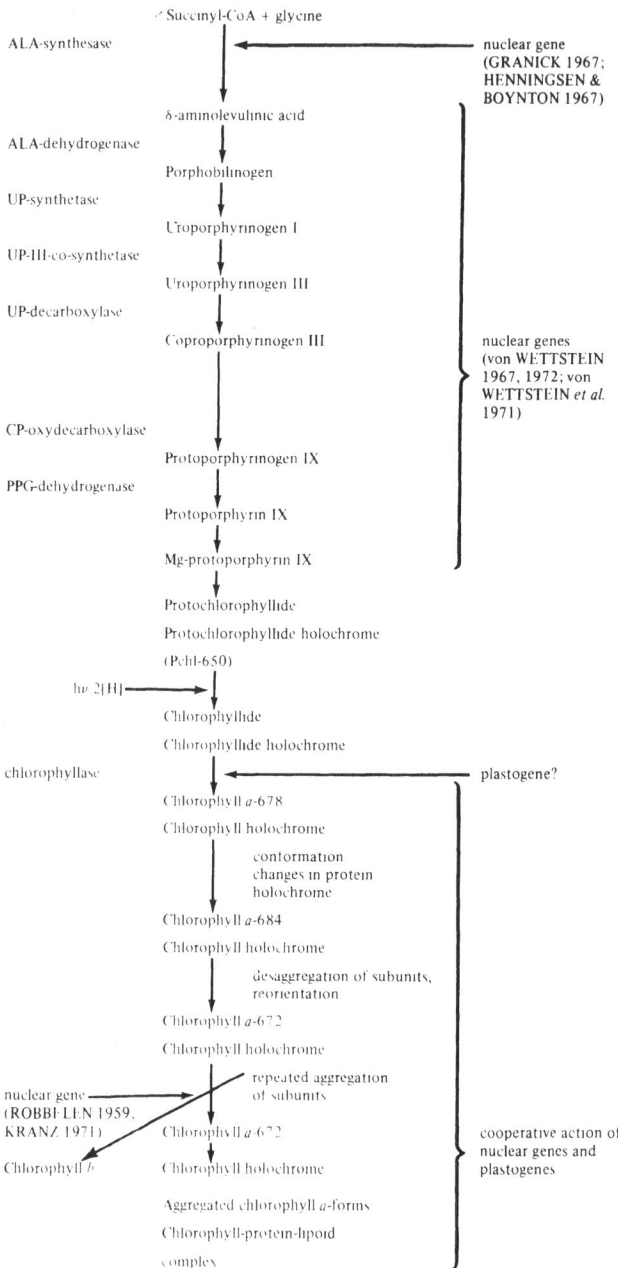

Fig. 1. Scheme of genetic regulation of chlorophyll biosynthesis and the initial stages of formation of its native forms.

of chromosome changes affecting the chloroplast pigment system (KAS'YA-NENKO & NASYROV 1968; BATALOV et al. 1972; BIKASIYAN et al. 1972). Today 60 nonallelic genes which influence the synthesis of chlorophylls and carotenoids are known for *Arabidopsis thaliana* (RÖBBELEN 1963; KAS'YA-NENKO 1967; NASYROV et al. 1971; USMANOV et al. 1971). 176 types of mutations with a genetic block in the biosynthesis of plastid pigments have been reported for *Pisum sativum* (GOTTSCHALK 1964), 100 for maize (WEIER 1952), and 200 for barley (NILAN 1967; VON WETTSTEIN 1967).

Analysis of the genetic regulatory systems of the biosynthesis of plastid pigments revealed that synthesis of key enzymes catalysing this process is controlled by nuclear genes. According to the possible scheme (Fig. 1) of genetic control of chlorophyll biosynthesis (NASYROV 1972) and the initial stages of formation of its forms in vivo (HENNINGSEN 1971), the nuclear genes exercise control over the reactions of ALA transformation up to protochlorophyllide (GRANICK 1967; HENNINGSEN & BOYNTON 1967; VON WETTSTEIN 1967, 1972; VON WETTSTEIN et al. 1971) as well as formation of chlorophyll *b* from chlorophyll *a* (RÖBBELEN 1959).

Recently KRANZ (1971) reported the possibility of biosynthesis of chlorophyll *b* directly from protoporphyrin through protochlorophyllide *b* which pathway is controlled by nuclear genes. Other researchers have claimed to have induced mutants of plants with a block in the synthesis of chlorophyll *b* (HIGHKIN 1950; RÖBBELEN 1959; HIRONO & RÉDEI 1963). Mutations of genes coding synthesis of the enzymes responsible for the reactions between ALA and protochlorophyllide leads to the accumulation of porphyrins. Among 86 nuclear genes which control chloroplast development in barley, five were found to be structural genes regulating different stages of uroporphyrinogen transformation to protochlorophyllide, and three genes were identified as the regulatory ones. In contrast to structural genes, two regulatory genes demonstrated similar dominating effects in the wild type and alleles of mutants in heterozygote (VON WETTSTEIN 1972).

On the other hand, inhibition of chlorophyll formation by D-threo-chloramphenicol (SMILLIE et al. 1971) indicates the participation of proteins synthesized on chloroplast ribosomes in chlorophyll synthesis. Some nonlimiting enzymes of intermediate stages of chlorophyll biosynthesis are probably determined by plastogenes, while chlorophyllase represents the plastid enzyme. Defects in the plastid DNA cause the plastid mutations (RYZHKOV 1965; KIRK & TILNEY-BASSETT 1967; WALLES 1971). Because of this, variegated plastom mutants of *Antirrhinum majus* status *albomaculata* were found to contain heteroplastid cells ('Mischzellen') with normal and mutant 'bleached' chloroplasts (DÖBEL & HAGEMANN 1963). We observed heteroplastid cells (Fig. 2) in leaves of the mutant *Arabidopsis thaliana* 40/3, where abnormalities of the pigment system were due to a block in leucine biosynthesis as a result of nuclear mutation (ABDULLAEV et al. 1972). These informations confirm the complex character of the genetic control of chlorophyll biosynthesis performed both by the genome of chloroplasts and the inheritance factors of the nucleus and cytoplasm.

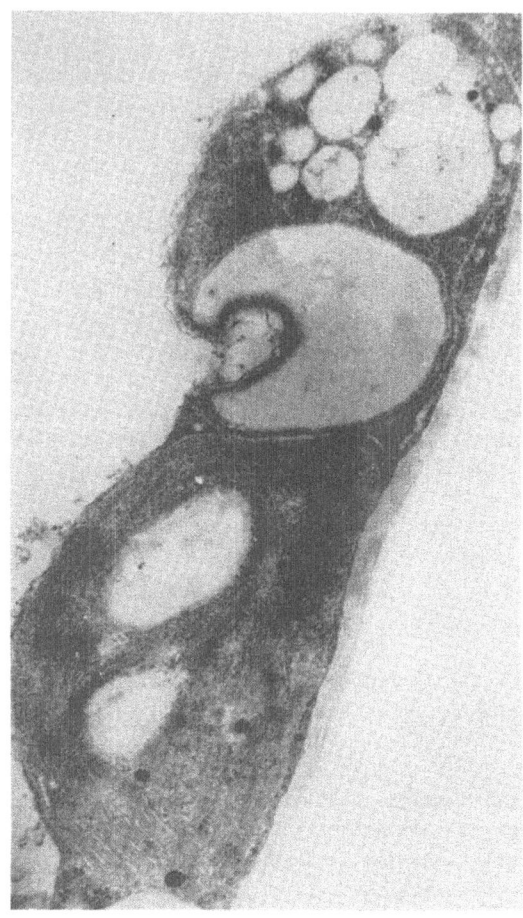

Fig. 2. Heteroplastids in leaf cells of the mutant *Arabidopsis thaliana* 40/3. Bottom: normal chloroplast; top: greatly vacuolized plastid with a very reduced membrane system.

The Amount of Chlorophyll and the Presence of its Forms in Vivo

Current concepts on the mechanism of photosynthesis suggest that normal activity of the photosynthetic apparatus requires not only the accumulation of a certain quantity of chlorophyll but also a specific structural organization of chlorophyll into the system of native forms (SMITH & FRENCH 1963; BROWN 1972) differing in spectral properties, which reflect their molecular organization, and in the functional activity during primary radiant energy conversion in photosynthesis. Naturally, the complex investigations of genetic regulation of the formation and activity of the photosynthetic apparatus have given rise to the problem of genetic control of formation of the chlorophyll forms in vivo.

The important aspect of this problem is the role of pigment amount in the formation of its discrete spectral forms. Acknowledging the role of this parameter, expressed in the modelling of the native state of pigments using their solutions, monolayers and films (KRASNOVSKIĬ 1962, 1965, 1970; LITVIN 1965; BYSTROVA 1972), it has been established that the presence, structure and state of specific proteins and lipoproteids – the carriers of the chlorophyll forms in vivo – are the decisive factors of molecular organization of pigment-containing structures in higher plants, algae and photosynthetic bacteria. This conclusion is supported by experimental data on the role of proteins and lipids in the configuration of different types of packing of chlorophyll molecules in native structures of chloroplasts (OSTROVSKAYA 1972) and chromatophores (EROKHIN 1969, 1972; EROKHIN & SINEGUB 1970a,b; SINEGUB & EROKHIN 1971) and in model systems of synthetic complexes of chlorophyll with proteins and lipoproteids (GILLER 1968, 1969, 1972; GILLER et al. 1968, 1970, 1975; SEMICHAEVSKIĬ 1972; DINANT & AGHION 1973).

A secondary role of the quantity of pigment in the formation of its aggregated forms is also confirmed by the presence of these forms in the first hours of greening in the etiolated leaves when the total concentration of the pigment is far from normal (LITVIN & BELYAEVA 1971a, b). In addition, in the pathway of chlorophyll biosynthesis (Fig. 1) variations in the spectral properties of the pigment resulting from the intensification or reduction of pigment-pigment interaction (aggregation or disaggregation) are related to the conformational changes of protein holochrome and disaggregation and reorientation of subunits. There are mutants either lacking the pigment-protein complex of Photosystem 1 with normal pigment content but lacking the most long-wave forms of chlorophyll (the corresponding protein fraction was also absent) (GREGORY et al. 1971) or, on the contrary, there are some mutants containing the abnormal aggregated form of chlorophyll a-745 in plastids (VON WETTSTEIN et al. 1971).

Algae Mutants

Analysis of the chlorophyll state in the plastids of mutants of higher plants and unicellular algae provided additional evidence for the independence of molecular organization of the chloroplast pigment structures on the concentration of chlorophyll. The comparison of fluorescence spectra at 77 K of cells from the wild strain B and two mutant forms of *Chlorella* (KVITKO & KHROPOVA 1963; GILLER et al. 1971) containing the same amount (3.3 μg per 10^9 cells, 8.4% of the wild type) of differently bound chlorophyll in chloroplast (Fig. 3, left) shows that the relatively low amount of chlorophyll may be distributed between the spectral forms as in the wild type (mutant 37-5), or may be concentrated in the short-wave and the abnormally long-wave forms and almost contain none of the

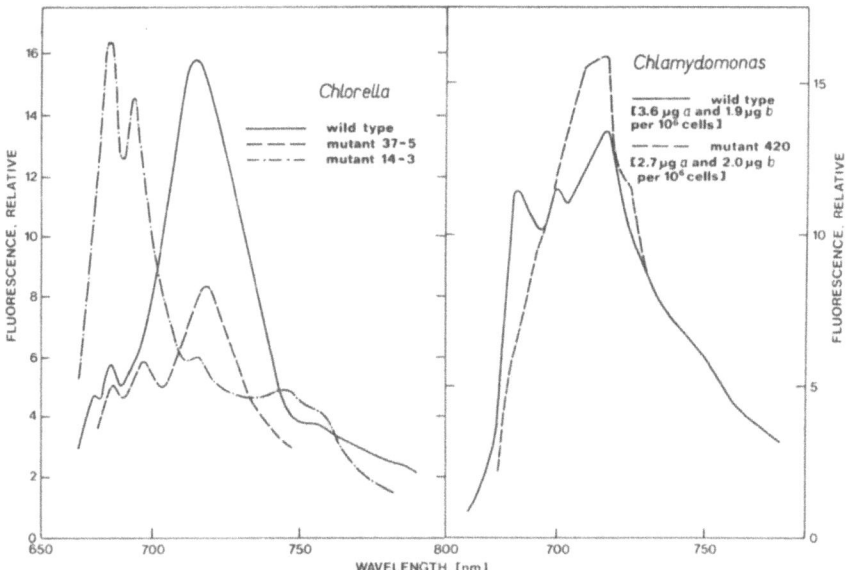

Fig. 3. Fluorescence spectra of *Chlorella* and *Chlamydomonas* cells (wild type and mutants) at 77 K.

form with the maximum of fluorescence at 715-720 nm, which is the main band of the wild type (mutant 14-3). Both these mutants are capable of photosynthesis, but the mutant 37-5 demonstrates autotrophic growth and division soon after the start of illumination, pigment composition of the plastids being constant; mutant 14-3 starts greening in the light accumulating an amount of chlorophyll close to normal (KVITKO & KHROPOVA 1963) and only then begins to develop.

In another case almost all chlorophyll may be concentrated in only one spectral form (Fig. 3, right). These mutant cells contain more chlorophyll in long-wave (aggregated) forms, though the total chlorophyll content is 25% lower than in the wild type cells.

Cotton Mutants

The most convincing results were obtained on variegated cotton mutants produced in our Institute. These mutants were registered as 826:8 (initial cv. 108-F) and 814:16 (initial cv. C-4727)[1]. Their progeny consisted always of four phenotypic classes: green, variegated, pale-green and yellow. Seedlings of the last two classes died early in the cotyledon stage. Variegated seedlings developed into variegated plants, produced normal bolls, and their seeds gave the same four

1 First figures stand for number of the family; second figures for doses of irradation of M_1 seeds in kR (G.P. BIKASYAN & P.D. USMANOV).

classes of seedlings. Green seedlings gave normal plants. Thus these variegated mutants represent the class of constant segregation. The ratio between green, variegated, pale-green and yellow seedlings from individual mutant families was similar, clearly indicating the genotypic similarity of these mutants. The ratio between normal and mutant plants in the progeny greatly deviated from the expected 3:1 (Table 1). The genetic nature of these mutants is not yet established, but according to the above data the mutations may be in both the nucleus and the extrachromosome systems (see also RYZHKOV 1933; HAGEMANN 1964, 1967).

Table 1. Frequency of different phenotypes in the progeny from variegated cotton plants.

Strain		Number of families studied	Colour of cotyledon leaves							
			green		varie-gated		light-green		yellow	
			[amount]	[%]	[amount]	[%]	[amount]	[%]	[amount]	[%]
826:8	(108-F)	2	456	84	17	3	21	4	48	9
814:16	(C-4727)	7	1 881	82	83	4	96	4	238	10
		Found	2 337				503			
		Expected (3:1)	2 130				710			

$$\chi^2 = 80.5; p < 0.01$$

Fig. 4. Variegated mutant of *Gossypium hirsutum*.

In the pale sections of leaves from variegated mutants (Fig. 4) an almost equal amount of chlorophyll may occur in different states — practically as a monomer (Fig. 5 B) or close to normal (Fig. 5 A). These results support the postulated independence of the chlorophyll state in vivo to its content in the plastids, and — coupled with the preliminary data on the genetic nature of mutants — indicate the complex nature of the genetic regulation of the pigment system's molecular organization in the photosynthetic apparatus.

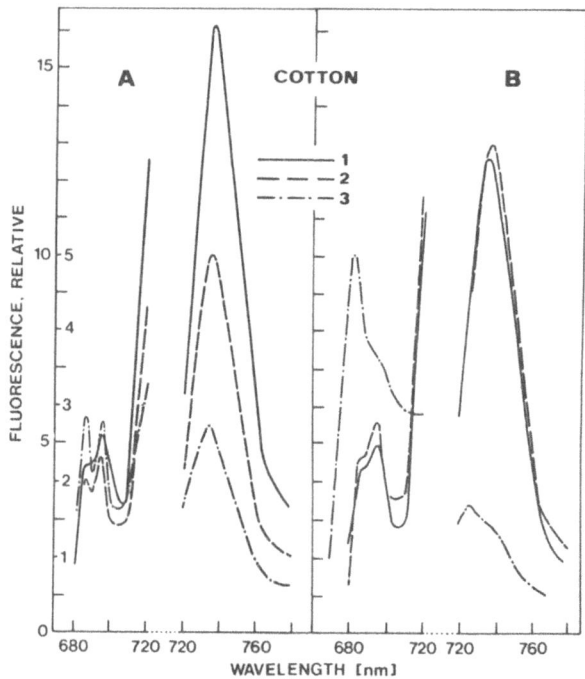

Fig. 5. Fluorescence spectra of cotton leaves at 77 K. A: 1. cv. 108-F (chl.a – 2 000 µg g^{-1} fresh weight, chl.b – 660 µg g^{-1}), 2. mutant 826:8, green section of a leaf (chl.a – 1 100 µg g^{-1}, chl.b – 480 µg g^{-1}), 3. mutant 826:8, light section of a leaf (chl.a – 43 µg g^{-1}; chl.b – 30 µg g^{-1}). B: 1. cv. C-4727 (chl.a – 1 230 µg g^{-1}; chl.b – 420 µg g^{-1}). 2. mutant 814:16, green section of a leaf (chl.a – 1 020 µg g^{-1}; chl.b – 430 µg g^{-1}). 3. mutant 814:16, light section of a leaf (chl.a – 48 µg g^{-1}; chl.b – 23 µg g^{-1}).

The Effect of Inhibitors of Protein Synthesis on Chlorophyll in vivo

As has been previously stated, the decisive role of protein or protein-lipoid matrix in the molecular organization of chloroplast pigment structures, well demonstrated on model systems (GILLER 1968, 1969, 1972; GILLER et al. 1968, 1970, 1971, 1975), is the basis of studies of the genetic control of the formation of chlorophyll forms in vivo. The relative independence of the chlorophyll state upon its amount in plastids agrees well with this assumption, whereas the effects of inhibitors of protein synthesis on the pigment state in lamellar membranes may be regarded as a direct indication of the decisive role of protein matrix in vivo. Cotyledons from cotton seedlings were treated with actinomycin D and cycloheximide at concentrations and times chosen so that the amounts of chlorophyll synthesised were equal. Changes in the state of the pigment were observed (Fig. 6). The pigment state in the seedlings treated with cycloheximide is close to normal, whereas in the variant with actinomycin D the fluorescence yield of all forms (and particularly of the short-wave ones) is increased, and the unusual form with fluorescence maximum at 722 nm appears. Taking into account the specificity of the antibiotics used, the formation of in vivo forms of chlorophylls in chloroplast membranes must be closely related to the protein synthesis controlled by genetic factors in the nucleus.

When cotton seedlings were grown on media with actinomycin D an insufficient amount of chlorophylls was formed. The presence of chlorophyll was de-

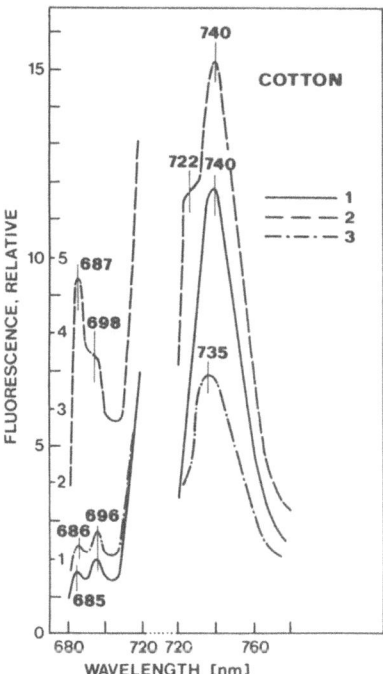

Fig. 6. The effects of inhibitors of protein synthesis on the fluorescence spectra of cotton seedlings at 77 K; 1. control (chlorophyll 100%); 2. actinomycin D (25 mg l^{-1}) – 76 h of darkness + 76 h of light, antibiotic (chlorophyll 12%): 3. cycloheximide (10 mg l^{-1}) – 24 h of darkness, antibiotic + 96 h of light, antibiotic (chlorophyll 20%).

tected only by the absorption spectrum of the acetone extract. The seedlings after 2 days illumination contained one of the primary forms of chlorophyll with the fluorescence maximum at 683 nm (LITVIN & BELYAEVA 1971a, b). Repacking of pigment molecules took place through the abnormal forms 705, 757, and even 770 nm, and finally a new appearance of chlorophyll a-683 and the long-wave forms 725 and 733 nm (close to normal chlorophyll a-740) was observed (Fig. 7).

There are well-known facts of the incomplete inhibition of protein biosynthesis by actinomycin D in germinating seeds of cotton due to the presence of long-living 'masked' mRNA's (KARIMOV & KUDINOVA 1969; KARIMOV & DONTSOVA 1970) or due to sites in the nuclear DNA resistant to the antibiotic (MARRÉ et al. 1965). So the changes in the low-temperature fluorescence spectra (Fig. 7) may be explained by changes in chlorophyll aggregation apparently related to the processes of protein biosynthesis.

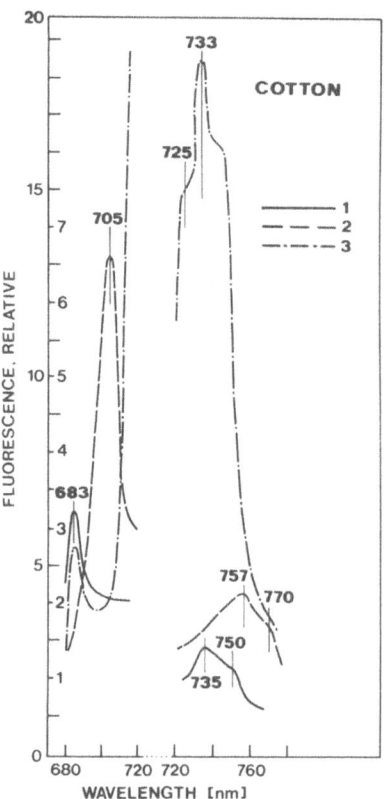

Fig. 7. Changes in fluorescence spectra at 77 K of cotton seedlings grown on medium with actinomycin D (25 mg l^{-1}); 1. 24 h of darkness, 48 h of light; 2. 24 h of darkness, 96 h of light; 3. 24 h of darkness, 168 h of light.

The Interaction of Genetic Factors of Nucleus and Chloroplasts in Chlorophyll Formation

The above data constitute good evidence that the biosynthesis and formation of in vivo forms of chlorophyll in photosynthetic membranes are controlled by the cooperative interaction of genetic factors of the nucleus and chloroplasts. Synthesis of the key enzymes of chlorophyll formation is coded by nuclear genes, but its realization and regulation occur on the translational level. Synthesis of ALA-synthetase proceeds on 80S cytoplasmic ribosomes. Other enzymes of pigment biosynthesis are probably formed on 70S plastid ribosomes with the participation of mRNA of nuclear origin. So the block of nuclear gene transcription by actinomycin D inhibits the formation of chlorophyll forms in vivo. Formation of lamellar proteins of photosynthetic membranes is also controlled by nuclear-plastid genes. Mutations of these genes may lead to a mixed population of plastids and the loss of photosynthetic capacity in the membranes.

Summary

The biosynthesis and formation of native forms of chlorophyll is under the dual genetic control of the genome and plastome. Synthesis of the key-enzymes of chlorophyll formation is coded in nuclear genes, but its realization and regulation occur on the translation level. The state of chlorophyll is relatively independent of its content in the plastids; the processes of protein biosynthesis influence the packing patterns of the pigment in lamellar membranes. The protein or the protein-lipoid matrix of native complexes play a directive role in the molecular organization of the chloroplast pigment structures, which is genetically controlled by the nuclear-plastid regulation of the biosynthesis of the lamellar proteins.

REFERENCES

ABDULLAEV, Kh.A., USMANOV, P. D. & NASYROV, Yu. S., *Dokl. Akad. Nauk Tadzh. SSR* 15 (8): 48, 1972.
BATALOV, R. B., KVITKO, K. V. & USMANOV, P. D., *Arabidopsis Inform. Service* 1972 (9) : 25, 1972.
BIKASIYAN, G. R., MUSTAFAEV, A. & NEGMATOV, M., in: Geneticheskie Aspekty Fotosinteza, Tezisy Dokladov, p. 55, Donish, Dushanbe 1972.
BROWN, J. S., *Annu. Rev. Plant Physiol.* 23: 73, 1972.
BYSTROVA, M. I., in: KIRICHENKO, E. B. (ed.): Metody Issledovaniya Struktury Fotosinteticheskogo Apparata, p. 81, Pushchino-na-Oke 1972.
DINANT, M. & AGHION, J., *Photochem. Photobiol.* 17: 25, 1973.
DÖBEL, P. & HAGEMANN, R., *Biol. Zentralbl.* 82: 149, 1963.
EROHKIN, Yu. E., in: Vtoroï Vsesoyuznyï Biokhimicheskiï S"ezd, Tezisy Dokladov, p. 278, FAN, Tashkent 1969.
EROKHIN, Yu. E., in: KIRICHENKO, E. B. (ed.): Metody Issledovaniya Struktury Fotosinteticheskogo Apparata, p. 155, Pushchino-na-Oke 1972.

EROKHIN, Yu. E. & SINEGUB, O. A., *Mol. Biol. (Moskva)* 4: 401, 1970a.
EROKHIN, Yu. E. & SINEGUB, O. A., *Mol. Biol. (Moskva)* 4: 541, 1970b.
GILLER, Yu. E., *Biofizika* 13: 1006, 1968.
GILLER, Yu. E., in: Mekhanizm Fotosinteza, Tezisy Dokladov II Vsesoyuznogo Biokhimicheskogo S"ezda, Sektsiya Problemy Fotosinteza, p. 40, FAN, Tashkent 1969.
GILLER, Yu. E., in: IV Mezhdunarodnyï Biofizicheskiï Kongress, Tezisy Sektsionnykh Dokladov, Vol. I, p. 368, Moskva 1972.
GILLER, Yu. E., KRASICHKOVA, G. V. & SAPOZHNIKOV, D. I., *Dokl. Akad. Nauk SSSR* 182: 1230, 1968.
GILLER., Yu. E., KRASICHKOVA, G. V. & SAPOZHNIKOV, D. I., *Biofizika* 15: 38, 1970.
GILLER, Yu. E., STOLBOVA, A. V., VAKHIDOVA, L. R. & KVITKO, K. V., *Biofizika* 16: 67, 1971.
GILLER, Yu. E., VAKHIDOVA, L. R., YUKHANANOVA, L. N. ABDULLAEVA, S. K., LIPKIND, B. I., KRASICHKOVA, G. V. & YUSUPOVA, G. A., in: NASYROV, Yu. S. & ŠESTÁK, Z. (ed.): Genetic Aspects of Photosynthesis, p. 271, Junk, The Hague 1975.
GILLER, Yu. E., YUKHANANOVA, L. N. & ABDULLAEVA, S. K., *Dokl. Akad. Nauk SSSR* 207: 1475, 1972.
GODNEV, T. N.: Khlorofill, ego Stroenie i Obrazovanie v Rastenii, Izd. Akad. Nauk BSSR, Minsk 1963.
GOODWIN, T. W. (ed.): Chemistry and Biochemistry of Plant Pigments, Academic Press, London-New York 1965.
GOTTSCHALK, W.: Die Wirkung mutierter Gene auf die Morphologie und Funktion pflanzlicher Organe, VEB G. Fischer Verlag, Jena 1964.
GRANICK, S., in: GOODWIN, T. W. (ed.): Biochemistry of Chloroplasts, Vol. II, p. 373, Academic Press, London-New York 1967.
GREGORY, R. P. F., RAPS, S. & BERTSCH, W., *Biochim. biophys. Acta* 234: 330, 1971.
HAGEMANN, R., Plasmatische Vererbung, VEB G. Fischer Verlag, Jena 1964.
HAGEMANN, R., *Biol. Zentralbl.* 86 (Suppl.): 163, 1967.
HENNINGSEN, K. W., *J. Cell Sci.* 7: 587, 1971.
HENNINGSEN, K. W. & BOYNTON, J. E., *Stud. biophys.* 5: 89, 1967.
HERRMANN, F. & HAGEMANN, R., *Biol. Zentralbl.* 86 (Suppl.): 181, 1967.
HIGHKIN, H. R., *Plant Physiol.* 25: 294, 1950.
HIRONO, G. & RÉDEI, G. P., *Nature* 197: 1324, 1963.
KARIMOV, Kh. Kh. & DONTSOVA, S. V., *Dokl. Akad. Nauk Tadzh. SSR* 13 (8): 1970.
KARIMOV, Kh. Kh. & KUDINOVA, S. V., *Fiziol. Rast.* 16: 730, 1969.
KAS'YANENKO, A. G., in: Issledovaniya po Fotosintezu, p. 77, Akad. Nauk Tadzh. SSR, Dushanbe 1967.
KAS'YANENKO, A. G. & NASYROV, Yu. S., *Fiziol. Rast.* 15: 422, 1968.
KIRK, J. T. O. & TILNEY-BASSETT, R. A. E.: The Plastids. Their Chemistry, Structure, Growth and Inheritance, W. H. Freeman and Co., London-San Francisco 1967.
KRANZ, A. R., *Theor. appl. Gen.* 41: 45, 1971.
KRASNOVSKIĬ, A. A., in: Mekhanizm Fotosinteza, Trudy V Mezhdunarodnogo Biokhimicheskogo Kongressa, Simp. VII, p. 196, Izd. Akad. Nauk SSSR, Moskva 1962.
KRASNOVSKIĬ, A. A., in: KRASNOVSKIĬ, A. A. et al. (ed.): Biokhimiya i Biofizika Fotosinteza, p. 26, Nauka, Moskva 1965.
KRASNOVSKIĬ, A. A., in: Funktsional'naya Biokhimiya Kletochnykh Struktur, p. 15, Nauka, Moskva 1970.
KVITKO, K. V. & KHROPOVA, V. I., *Vestn. leningrad. gos. Univ., Ser. biol.* 1963 (2): 150, 1963.
LITVIN, F. F., in: KRASNOVSKIĬ, A. A. et al. (ed.): Biokhimiya i Biofizika Fotosinteza, p. 96, Nauka, Moskva 1965.
LITVIN, F. F. & BELYAEVA, O. B., *Biokhimiya* 36: 615, 1971a.
LITVIN, F. F. & BELYAEVA, O. B., *Photosynthetica* 5: 200, 1971b.

MARRE, E., COCUCCI, S. & STURANI, E., *Plant Physiol.* 40: 1162, 1965.
NADLER, K. D. & GRANICK, S., *Plant Physiol.* 46: 240, 1970.
NASYROV, Yu. S., *Zh. obshch. Biol.* 33: 683, 1972.
NASYROV, Yu. S., KAS'YANENKO, A. G. & ABDURAKHMANOVA, Z. N., in: Biokhimiya i Biofizika Fotosinteza, p. 121, Irkutsk 1971.
NILAN, R. A., *Abhandl. deut. Akad. Wiss. Berlin, Kl. Med.* 1967 (2): 5, 1967.
OSTROVSKAYA, K. L., in: KIRICHENKO, E. B. (ed.): Metody Issledovaniya Struktury Fotosinteticheskogo Apparata, p. 8, Pushchino-na-Oke 1972.
RÖBBELEN, G., *Z. Indukt. Abstamm. Vererbungslehre* 90: 503, 1959.
RÖBBELEN, G., in: VELEMÍNSKÝ, J. & GICHNER, T. (ed.): Induction of Mutations and the Mutation Processes, p. 42, Publ. House Czech. Acad. Sci., Praha 1965.
RYZHKOV, V. L.: Mutatsii i Bolezni Khlorofillovogo Zerna, Sel'khozgiz, Moskva-Leningrad 1933.
RYZHKOV, V. L., *Dokl. Akad. Nauk SSSR* 162: 1177, 1965.
SEMICHAEVSKII, V. D., *Biofizika* 17: 530, 1972.
SHLYK, A. A.: Metabolizm Khlorofilla v Zelenom Rastenii, Nauka i Tekhnika, Minsk 1965.
SHLYK, A. A., *Annu. Rev. Plant Physiol.* 22: 169, 1971.
SHLYK, A. A., PRUDNIKOVA, I. V., SAVCHENKO, G. E., AVERINA, N. G., KOSTYUK, N. N., KAMYSHENKO, L. K., VLASENOK, L. I., GAPONENKO, V. I., BALEVA, E. F., PARAMONOVA, T. K., LOSITSKAYA, T. V. & VEZITSKII, A. Yu., in: NASYROV, Yu. S. & ŠESTÁK, Z. (ed.): Genetic Aspects of Photosynthesis, p. 119, Junk, The Hague 1975.
SHLYK, A. A., RUDOI, A. B. & VESITSKY, A. Y., in: FORTI, G., AVRON, M. & MELANDRI, A. (ed.): Photosynthesis, Two Centuries after Its Discovery by Joseph Priestley, Vol. 2, p. 2291, Junk, The Hague 1972.
SINEGUB, O. A. & EROKHIN, Yu. E., *Mol. Biol.* (Moskva) 5: 472, 1971.
SMILLIE, R. M., BISHOP, D. G., GIBBONS, G. G., GRAHAM, D., GRIEVE, A. M., RAISON, J. K. & REGER, B. J., in: BOARDMAN, N. K., LINNANE, A. W. & SMILLIE, R. M. (ed.): Autonomy and Biogenesis of Mitochondria and Chloroplasts, p. 422, North Holland Publ. Comp., Amsterdam-London 1971.
SMIT, J., in: Mekhanizm Fotosinteza, Trudy V Mezhdunarodnogo Biokhimicheskogo Kongressa, Simp. VII, p. 157, Izd. Akad. Nauk SSSR, Moskva 1965.
SMITH, J. H. C. & FRENCH, C. S., *Annu. Rev. Plant Physiol.* 14: 181, 1963.
USMANOV, P. D., KAS'YANENKO, A. G. & BATALOV, R. B., in: NASYROV, Yu. S. (ed.): Geneticheskie Aspekty Fotosinteza, p. 24, Donish, Dushanbe 1971.
WALLES, B., in: GIBBS, M. (ed.): Structure and Function of Chloroplasts, p. 51, Springer-Verlag, Berlin-Heidelberg-New York 1971.
WEIER, J., *Bibliogr. genet.* 14: 189, 1952.
VON WETTSTEIN, D., in: GOODWIN, T. W. (ed.): Biochemistry of Chloroplasts, Vol. I, p. 19, Academic Press, London-New York 1966.
VON WETTSTEIN, D., in: SAN PIETRO, A., GREER, F. A. & ARMY, T. J. (ed.): Harvesting the Sun, p. 153, Academic Press, New York-London 1967.
VON WETTSTEIN, D., in: Geneticheskie Aspekty Fotosinteza, Tezisy Dokladov, p. 21, Donish, Dushanbe 1972.
VON WETTSTEIN, D., HENNINGSEN, K. W., BOYNTON, J. E., KANNANGARA, G. C. & NIELSEN, O. F., in: BOARDMAN, N. K., LINNANE, A. W. & SMILLIE, R. M. (ed.): Autonomy and Biogenesis of Mitochondria and Chloroplasts, p. 205, North Holland Publ. Comp., Amsterdam-London 1971.

GENETIC CONTROL OF PHOTOSYNTHETIC CO$_2$ ASSIMILATION

PHOTOSYNTHETIC CO_2 ASSIMILATION IN MAIZE AND SPINACH LEAVES AND CHLOROPLASTS[1]

M. GIBBS, E. LATZKO[2], L. J. LABER[3] & G. HINES

Department of Biology, Brandeis University, Waltham, Massachusetts 02154, U.S.A.

Introduction

A considerable amount of data has been collected indicating that higher plants can be classified into two groups on the basis of their primary photosynthetic carboxylation reaction (HATCH & SLACK 1970). Species using the Calvin cycle and designated as C_3-plants utilize ribulose 1,5-diphosphate carboxylase while phosphoenolpyruvate carboxylase catalyzes the initial reaction involving atmospheric CO_2 in plants designated as C_4-types. An example of the former group is spinach and of the latter group is maize. In order to determine the validity of this point of view, we have compared some properties of photosynthetic CO_2 fixation in chloroplasts from spinach and maize. Furthermore, we have compared enhanced dark fixation following illumination in intact leaves of the two plants.

Studies with Intact Leaves

In these experiments, we have utilized a method described by several investigators for elucidating the path of carbon during photosynthesis in algae (FAGER et al. 1950, TOGASAKI & GIBBS 1967, HOGETSU & MIYACHI 1970). The method termed 'preillumination' involves identification of labeled compounds formed upon introduction of $^{14}CO_2$ to photosynthetic tissues which have been illuminated for a period of time in the absence of CO_2. Also, it is possible to observe changes in the levels of photosynthetic intermediates during preillumination and following introduction of CO_2 (WILSON & CALVIN 1955).

Both spinach and maize leaves exhibit the phenomenon of light enhanced dark CO_2 fixation (Fig. 1). The enhanced rates are comparable in the two leaves

---

1. This research was supported by grants from the National Science Foundation and from the United States Atomic Energy Commission AT (11-1) 3231.
2. Permanent address: Chemisches Institut, Technische Hochschule, München Weihenstephan, Germany.
3. A post-doctoral trainee (2 TO1 GM 1586) of the National Institutes of Health. Permanent address: Department of Botany and Plant Pathology, University of Maine, Orono, Maine, U.S.A.

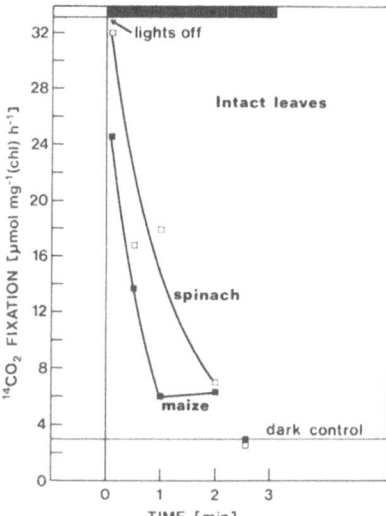

Fig. 1. The effect of preillumination upon the rate of $^{14}CO_2$ fixation. Intact maize (■) and spinach (□) leaves were illuminated for 30 min at 38 klx under N_2 + 1.5% O_2. Gaseous $^{14}CO_2$ was added for 1 min at the indicated points. The leaves were removed and killed in boiling 80% (v/v) ethyl alcohol.

though the rate decays more rapidly in spinach. After 30 minutes of preillumination, the enhanced dark rate was about 30 μmol CO_2 per mg chlorophyll per hour for spinach leaves and about 22 μmol for maize leaves. These rates are about 30% of photosynthetic CO_2 fixation determined with comparable leaves.

The bulk of the enhanced CO_2 fixation ceases in less than one minute of exposure to CO_2. These results are most easily interpreted in terms of the accumulation of a precursor in the light, in the absence of CO_2 which rapidly disappears in darkness. Thus, it seems likely that we are observing the result of a carboxylation of an accumulated CO_2 acceptor.

Transient changes in the levels of ribulose 1,5-diphosphate (RuDP), glycerate-3-phosphate (G3P) and phosphoenolpyruvate (PEP) in spinach and maize leaves were measured enzymatically during preillumination without CO_2, after a pulse of CO_2 given in the light and a pulse of CO_2 after the lights were switched off. Malate was also analyzed in the maize leaves under these conditions. If PEP were the primary acceptor in maize under these conditions, we would expect that the level of this compound would parallel that of RuDP in plants possessing the Calvin cycle only.

In spinach (Fig. 2), a linear increase in the level of RuDP was observed when the intact leaves were illuminated in the absence of CO_2. After a pulse of CO_2, either in the light or in the subsequent dark period, its concentration fell sharply

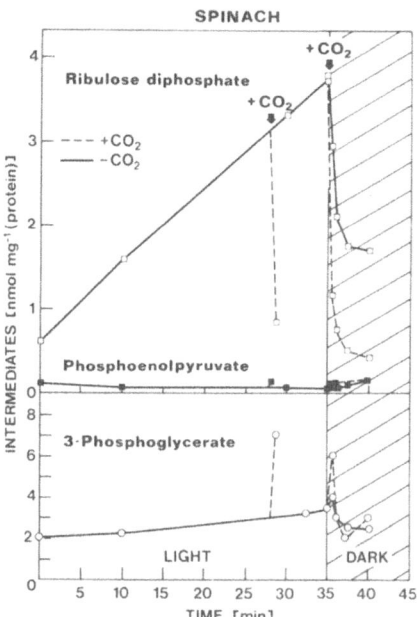

Fig. 2. Changes in the levels of the intermediates after $^{12}CO_2$ pulses and during light to dark transitions in spinach leaves. Two spinach leaves were placed in each of ten 100 ml flasks and treated as described for Fig. 3.

and the level of G3P showed a concomitant increase. When CO_2 was not present during the dark period, RuDP decreased more gradually to a higher steady level. Presumably, these transients are due to CO_2 generated within the spinach leaf. The level of PEP remained consistently low and was unaffected by either CO_2 pulses or light-dark transitions.

The levels of RuDP and G3P in maize leaves responded to light-dark transitions in the same manner as did spinach leaves (Fig. 3). After a pulse of CO_2, the level of RuDP fell sharply and the level of G3P increased. In contrast to spinach, PEP increased in a manner similar to G3P. Malate decreased rapidly during the preillumination period and rose sharply during the CO_2 pulse in the light and in the dark (Fig. 4). Clearly, the transient change in malate is far higher (20-30 fold) than the other intermediates studied. Presumably malate is converted to oxaloacetate and this dicarboxylic acid is oxidized via the Krebs cycle or decarboxylated to CO_2 and pyruvate. If the latter reaction does occur, the released CO_2 is not sufficient to stop the linear formation of RuDP.

The divergent changes of the concentrations of RuDP and of G3P during light-dark transitions and after pulses of CO_2 served as a major part of evidence in establishing the reductive pentose phosphate cycle in photosynthetic tissue (WIL-

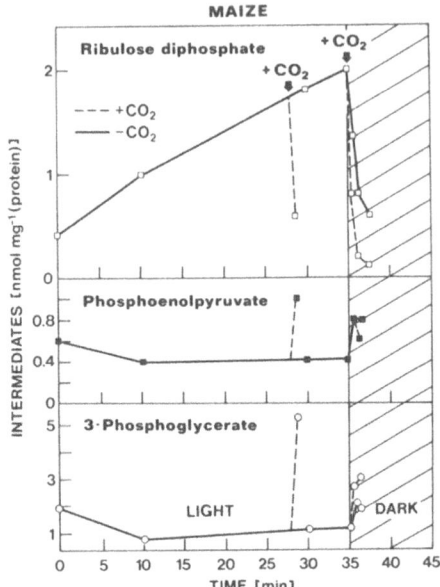

Fig. 3. The effect of preillumination under N_2 +1.5% O_2, addition of $^{12}CO_2$, and light to dark transitions upon the levels of intermediates in maize leaves. Each point on the graph represents the analysis of one flask containing 5 primary leaves from 8-day-old maize seedlings. After 28 min of preillumination gaseous $^{12}CO_2$ was added to one flask for one min. Gaseous $^{12}CO_2$ was also administered to the indicated samples 5 s after turning the lights out (35 min). Metabolism was stopped by immersion in liquid nitrogen. Dotted lines indicate addition of CO_2. Solid line in dark represents no addition of CO_2.

SON & CALVIN 1955). Reasoning in an analogous fashion we conclude that in maize under the conditions of our preillumination experiments, PEP behaved as a product of CO_2 fixation rather than as a primary acceptor of CO_2. It would appear that carbon in maize leaves can flow from RuDP to G3P and then onto PEP. Finally, we propose that RuDP can serve as a primary acceptor of CO_2 in both C_3 and C_4 plants.

Demonstration of this route of carbon flow in maize leaves does not contradict the kinetic labeling experiments in plants possessing a C_4 pathway or Crassulacean Acid Metabolism (CAM) (HATCH & SLACK 1966). Neither does our view conflict with the concept of compartmentation of the two carboxylases in different regions of the plant (EDWARDS et al. 1970). We propose that photosynthetic carbon flow can be initiated in C_4 or CAM plants by either RuDP or PEP carboxylase. Under steady state conditions, these plants maintain in their cytoplasm a high level of PEP and its carboxylase. Therefore, under most conditions, tracer is seen first in the dicarboxylic acids. In both C_4 and CAM plants, the dicarboxylic

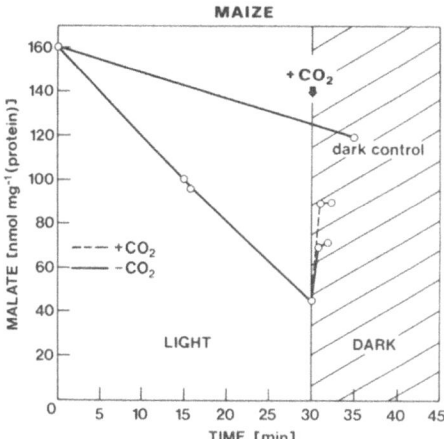

Fig. 4. Changes of the malate level in maize leaves during light-dark transition. Conditions as for Fig. 3, except that the gaseous CO_2 pulse in the light was applied after 15 min of illumination and the light was removed after 30 min. Dotted lines indicate addition of CO_2. Solid line in dark represents no addition of CO_2.

acids are storage depots for CO_2. These organic acids are decarboxylated to CO_2 which enters the chloroplast and is assimulated to carbohydrates.

Studies with Chloroplasts

Recently we have succeeded in isolating intact chloroplasts from 6-day-old maize leaves capable of assimilating 45 μmol CO_2 per mg chlorophyll per h (O'NEAL et al. 1972). Photosynthetic carbon metabolism of these plastids appears to be similar to that observed in isolated spinach and pea chloroplasts (GIBBS 1971). Although the rate of CO_2 fixation in maize was lower, most of the differences are quantitative rather than qualitative, such as percentage of stimulation by sugar phosphates, percentage of inhibition by oxygen, and product distribution.

The major products of assimilation by the maize chloroplasts were G3P, sugar phosphates, dihydroxyacetone-P, and starch. Only traces of isotope were located in the organic acid fraction which has traces of aspartate, malate and glycolate (Table 1). One major difference between the preparation of maize and spinach chloroplasts is the need of a strong reducing agent such as dithiothreitol in the maize reaction mixture.

The products formed by the chloroplasts are consistent with the enzyme profile. The maize chloroplast preparation contains RuDP carboxylase but no PEP carboxylase (GIBBS et al. 1970). This enzyme profile study also showed that PEP carboxylase was present in the supernatant fraction from which the chloroplasts

Table 1. Fixation of $^{14}CO_2$ during 10 (20) min in light into photosynthates [% of total counts per min] in mixed chloroplast (C) and supernatant (S) fractions from 6-day-old maize leaves. The amount of chloroplast (left) or supernatant (right) fraction remained stable.

Compound	Chloroplast fraction				Supernatant fraction			
	0.2 ml C	0.1 ml S 0.2 ml C	0.2 ml S 0.2 ml C	0.4 ml S 0.2 ml C	0.1 ml C	0.2 ml S 0.1 ml C	0.2 ml S 0.2 ml C	0.2 ml S 0.4 ml C
Glycerate-3-P	77(66)	32(36)	39(24)	16(10)	71(67)	38(22)	43(31)	38(20)
Sugar phosphate	10(13)	15(15)	13(17)	7(11)	19(15)	12(12)	14(13)	15(17)
Aspartate	0(0)	11(12)	18(22)	41(39)	0(0)	18(20)	13(22)	14(23)
Malate	0(0)	6(7)	6(9)	12(15)	0(0)	7(10)	7(12)	9(18)
Origin	2(5)	4(3)	2(2)	2(2)	3(3)	0.9(1.8)	0.6(0.8)	1(1)
Other	11(16)	31(28)	22(27)	21(24)	7(15)	24(34)	23(22)	23(20)
Rate [μmol mg^{-1} (chl) h^{-1}]	9(7)	9(7)	14(12)	16(14)	4(3)	11(8)	9(7)	10(8)

were spun down. When supernatant was combined with chloroplasts (Table 1 left), the percentage of isotope in G3P decreased and there was a concomitant increase in aspartate and malate. When the addition of supernatant was increased, over 50% of the ^{14}C was found in aspartate and malate. In addition, the ratio of aspartate to malate widened as the ratio of supernatant to chloroplast was raised. Finally, the percentage of isotope in the sugar phosphate fraction was little affected.

Table 1 (right) tabulates data when the chloroplastic content of the reaction was increased and the supernatant fraction remained fixed. While a great deal of the glycerate-3-P was converted to aspartate and malate, nevertheless, the amount was smaller than that illustrated in the left part of the Table. Also, with increasing chloroplast amount the ratio between aspartate and malate tended toward unity.

Table 2. Fixation of $^{14}CO_2$ during 20 min in the dark into metabolites [% of total counts per min] in mixed chloroplast (C) and supernatant (S) fractions from 6-day-old maize leaves.

Compound	0.1 ml S	0.1 ml S 0.1 ml C	0.4 ml S 0.2 ml C
Aspartate	72	69	75
Malate	13	13	20
Citrate	15	18	5

This distribution of isotope is dependent upon light. In the dark (Table 2), aspartate, malate and possibly citrate accounted for the bulk of the fixed $^{14}CO_2$. Note that in the absence of light, the ratio of aspartate to malate is about 6 or at least 3 times that found in the light.

Another indication that we are dealing with a photosynthetic process is the result of the addition of DCMU to the reaction mixture. With increasing concen-

Table 3. Fixation of $^{14}CO_2$ during 20 min in light into photosynthates [% of total counts per min] in mixed chloroplast (C) and supernatant (S) fractions from 6-day-old maize leaves without and under the addition of DCMU.

Compound	0.2 ml C	0.2 ml S 0.2 ml C	+DCMU [2.5 x 10^{-7} M] 0.2 ml S 0.2 ml C	+DCMU [2.5 x 10^{-5} M] 0.2 ml S 0.2 ml C
Glycerate-3-P	55	14	9	0
Sugar phosphate	21	16	6	0
Aspartate	0	31	55	67
Malate	0	13	21	21
Origin	3	1	0	0
Other	21	25	9	12
Rate [μmol mg^{-1}(chl) h^{-1}]	7	8	4	3

tration of DCMU, the level of isotope in G3P and sugar phosphates fall off until eventually only the products of dark fixation are seen on the chromatograms (Table 3).

The present interest in the relationship between carbonic anhydrase and photosynthesis prompted us to study the effect of sodium azide on photosynthetic CO_2 fixation. Sodium azide up to 100 μM is known not to affect photosynthetic electron transport but to stop carbonic anhydrase. Furthermore, AVRON & GIBBS (unpublished data) have observed that 50 μM azide eliminates completely CO_2 fixation in the isolated spinach chloroplast. Apparently, the inhibition is due to carbonic anhydrase. In contrast, photosynthesis in the maize chloroplast is unaffected up to 500 μM (Table 4). On addition of 50 mM azide, the products resemble that of dark fixation or fixation in the presence of DCMU. Apparently, the photochemistry is affected at the elevated level of azide. The significance of this difference with respect to azide between the two plastids is unknown but the K_m of each chloroplast for bicarbonate was found to be roughly identical.

The plastids used in this experiment were isolated from young plants. Electron

Table 4. Fixation of $^{14}CO_2$ during 10 (20) min in light into photosynthates [% of total counts per min] in mixed chloroplast (C) and supernatant (S) fractions from 6-day-old maize leaves under the action of sodium azide.

Compound	0.2 ml C	0.2 ml S 0.2 ml C	+Azide [5 x 10^{-4} M] 0.2 ml S 0.2 ml C	+Azide [5 x 10^{-2} M] 0.2 ml S 0.2 ml C
Glycerate-3-P	62(53)	23(11)	30(17)	3(1)
Sugar phosphate	20(16)	18(21)	16(21)	0(0)
Aspartate	0(0)	17(22)	22(20)	66(62)
Malate	0(0)	8(12)	10(14)	15(18)
Origin	4(3)	2(2)	2(1)	0(0)
Other	14(28)	29(30)	16(23)	15(21)
Rate [μmol mg^{-1}(chl) h^{-1}]	8(6)	11(8)	8(6)	8(5)

Table 5. Fixation of $^{14}CO_2$ into photosynthates [% of total counts per min] in maize seedlings of various age (5 to 10 d).

Age [d]	Origin	G3P Sugar-P	Aspartate	Malate	Alanine	Other
5	0	29	15	35	4	13
6	5	48	17	25	1	4
7	5	39	23	31	1	2
8	6	42	13	36	1	2
10	5	49	21	18	0	3

microscopy analysis indicated that the plants had not fully differentiated into a bundle sheath and mesophyll celltype. All of the plastids were of the agranal type whereas it is known that in older maize, the mesophyll cells contain more fully developed grana. We have investigated the distribution of isotope following photosynthesis in the developing seedling (Table 5). Product distribution was apparently independent of fully developed sheath and mesophyll components.

Conclusions

Perhaps the most important results of the present investigations lie in their relationship to a better understanding of the nature of the CO_2 fixation pathway in C_4-plants. The preillumination findings and the data obtained with the isolated plastids establish that a C_4-plant can assimilate CO_2 via the reductive pentose phosphate cycle without the necessity of a functional C_4-cycle or a decarboxylation of malate or aspartate. Unlabeled malate or aspartate did not inhibit the photosynthetic rate in the presence of $^{14}CO_2$. Indeed, aspartate and malate were not synthesized until supernatant was combined with the chloroplast indicating that the enzymes responsible for the conversion of glycerate-3-P to aspartate and malate are more likely localized in the cytoplasm rather than in the chloroplast. This is not to suggest that malate and aspartate play no important roles in the C_4-plant. These compounds serve as means for storing CO_2 that is subsequently released within the cell and reduced in the Calvin cycle. If this speculation is correct, then the metabolism of CO_2 by the maize would be comparable to that described for the crassulacean type plants.

REFERENCES

EDWARDS, G. E., LEE, S. S., CHEN, T. M. & BLACK, C. C., *Biochem. biophys. Res. Commun.* 39: 389, 1970.
FAGER, E. W., ROSENBERG, J. L., GAFFRON, H., *Fed. Proc.* 9: 535, 1950.
GIBBS, M., in: GIBBS, M. (ed.): Structure and Function of Chloroplasts, p. 169, Springer-Verlag, Berlin-Heidelberg-New York 1971.

GIBBS, M., LATZKO, E., O'NEAL, D. & HEW, C.-S., *Biochem. biophys. Res. Commun.* 40: 1356, 1970.
HATCH, M. D. & SLACK, C. R., *Biochem. J.* 101: 103, 1966.
HATCH, M. D. & SLACK, C. R., *Annu. Rev. Plant Physiol.* 21: 141, 1970.
HOGETSU, D. & MIYACHI. S., *Plant Physiol.* 45: 178, 1970.
O'NEAL, D., HEW, C. S., LATZKO, E. & GIBBS, M., *Plant Physiol.* 49: 607, 1972.
TOGASAKI, R. K. & GIBBS, M., *Plant Physiol.* 42: 991, 1967.
WILSON, A. T. & CALVIN, M., *J. amer. chem. Soc.* 77: 5948, 1955.

CARBON DIOXIDE TRANSFER AND PHOTOCHEMICAL ACTIVITIES AS FACTORS OF PHOTOSYNTHESIS DURING ONTOGENESIS OF PRIMARY BEAN LEAVES

Z. ŠESTÁK, J. ČATSKÝ, JARMILA SOLÁROVÁ, HELENA STRNADOVÁ & INGRID TICHÁ

Institute of Experimental Botany, Czechoslovak Academy of Sciences, 160 00 Praha 6, Flemingovo n.2, Czechoslovakia

Introduction

The maximum photosynthetic rate which can be attained by plants under optimum conditions is limited in principle by two genetically controlled factors, i.e. by the maximum activity of the photochemical systems and by the maximum conductance of leaf tissues for carbon dioxide transfer from ambient air to carboxylation sites. Both the physical, CO_2 transfer part and the biochemical, energy conversion part of photosynthesis (see schemes on Figs. 1 and 2) are composed of

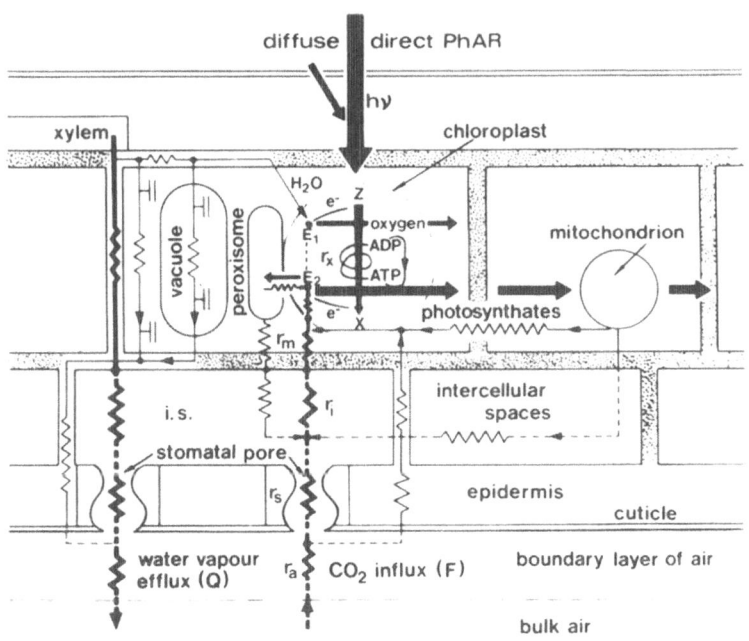

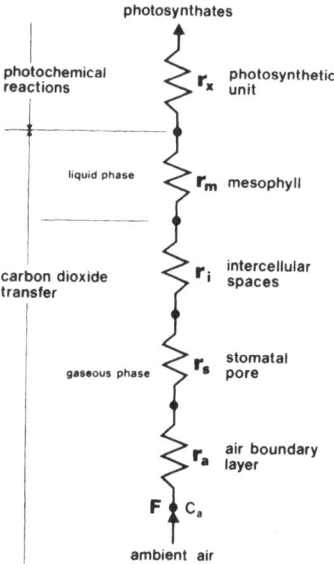

Figs. 1 and 2. Schemes of mass and energy transfers in photosynthesis and related processes. Carbon dioxide transfer in the gaseous phase is controlled by the resistances in the boundary layer of air (r_a), in the stomatal pores (r_s) and cuticle (r_c), and the intercellular spaces (r_i). CO_2 transfer in the liquid phase, i.e. the transfer from the outer surface of the photosynthesizing cell to the carboxylation site, is limited by the 'mesophyll' resistance, r_m. A value reflecting the activity of all photochemical and biochemical reactions involved in photosynthesis, expressed in dimensions of a diffusive resistance, is called 'carboxylation resistance', r_x (for details see ŠESTÁK et al. 1971 and JARVIS 1971). This hypothetical r_x may enable a comparison of the importance of different segments of the CO_2 pathway as rate-limiting factors of photosynthesis.

several catenary segments. The activity of each individual segment changes in varying degree during the development of a leaf and a plant, resulting finally in ontogenetic changes of net CO_2 influx or dry matter increment. The changes in these generally used measures of photosynthetic rate are often reported in the literature (for review see ŠESTÁK & ČATSKÝ 1967).

To show the importance of the above mentioned segments of the CO_2 pathway and their limiting effect on the photosynthetic rate, diffusion and quasi-diffusion resistances in the stomata and leaf mesophyll were measured during the ontogenesis of primary bean leaves, simultaneously with the Hill reaction and photophosphorylation rates.

Material and Methods

Seedlings of the French bean (*Phaseolus vulgaris* L. cv. Jantar) were grown in an air-conditioned chamber (CEL 37-14; Sherer, Marshall, Mich., U.S.A.) in pots with coarse sand and nutrient solution. Controlled cultivation conditions were: irradiance (400-700 nm), 300 µeinstein $m^{-2} s^{-1}$, i.e. *ca.* 65 W m^{-2}, a 16 h-day including two 30 min-twilight periods, day/night air temperature and relative humidity 24/16 °C and 60/80%, respectively. The primary leaves unfolded on the 5th day of cultivation.

Net photosynthetic rate (P_N) was measured in an open system (ČATSKÝ & TICHÁ 1974) as carbon dioxide influx by means of an infra-red gas analyser (Infralyt III, Junkalor, Dessau, G.D.R., provided with Grubb Parsons, Newcastle/Tyne, U.K., filters for determining CO_2 in the presence of water vapour); the difference in CO_2 concentration at the inlet and at the outlet of the assimilation chamber was measured. The assimilation chamber was ventilated with two fans (r_a for $CO_2 < 0.4$ s cm^{-1}); the effective CO_2 concentration was controlled to a constant value by mixing CO_2-free air and pure CO_2 before the assimilation chamber by means of a mixing pump (H. Wösthoff, Bochum, F.R.G.) and a precision needle valve (Edwards High Vacuum, Crawley, Sussex, U.K.). Irradiance (400-700 nm) at the leaf level: up to 3 150 µeinstein $m^{-2} s^{-1}$, i.e. *ca.* 680 W m^{-2}; leaf temperature 28 ± 1 °C at saturating irradiances.

Leaf temperature was measured by a chromel-constantan (0.08 mm) thermocouple inserted into the leaf mesophyll. Irradiance was measured with an LI-170 Quantum/Radiometer/Photometer (Lambda Instruments Co., Lincoln, Nebraska, U.S.A.) and a compensated thermopile CA 1 (Kipp & Zonen, Delft, Holland) with filters WG 1 or RG 8 (Schott, Jena, G.D.R.) and microvolt ammeter 150 B (Keithley, Cleveland, Ohio, U.S.A.).

The total resistance to CO_2 transfer was calculated from P_N and the gradient of CO_2 concentration in ambient air and at carboxylation sites (see JARVIS 1971 for details). The resistances in the boundary layer of air and in the stomata, r_a and r_s, were calculated for water vapour, i.e. from the measurements of evaporation rate from a leaf model (moist green blotting paper) and of transpiration rate, respectively, and then converted to resistances for CO_2 according to THOM (1968) and to COWAN & MILTHORPE (1968), respectively:

$$r_{a,CO_2} = 1.37 r_{a,H_2O}; \quad r_{s,CO_2} = 1.54 r_{s,H_2O}$$

The intracellular resistance (i.e. the sum of 'mesophyll' and 'carboxylation' resistances, $r_m + r_x$) was calculated as residual resistance. In preliminary experiments, CHARTIER's model (CHARTIER 1966; CHARTIER et al. 1970) was used for the separation of r_m and r_x. Stomatal diffusive resistance was also measured separately in adaxial and abaxial epidermis with a 'diffusion porometer' of the van Bavel type with correction for the temperature of the LiCl sensor (DJAVANCHIR 1970).

Leaf area was determined on leaf copies by a polar planimeter. Dry matter was weighed after drying the material for 24 h at 85 °C. Leaf thickness was measured on fresh free-hand transverse sections using a calibrated eyepiece.

For the isolation of chloroplasts, 0.01 M KH_2PO_4, 0.01 M ascorbic acid and bovine serum albumin (10 mg per 10 g fresh leaves) were added to the basic 0.05 M Tris − 0.35 M saccharose buffer (pH 7.5). Chloroplast fraction, separated from the homogenate by centrifugation at 200 × g for 5 min, was washed (600 × g for 20 min) in the basic buffer. For the spectrophotometric measurement of rates of the Hill reaction and non-cylic photophosphorylation, the following reaction mixture was used [μmol ml^{-1}]: Tris 50, saccharose 210, KH_2PO_4 0.55, ADP 0.53, $MgCl_2$ 1.2, either $K_3Fe(CN)_6$ 0.73 or DCPIP 0.03, and chloroplasts with $ca.$ 12.5 μg chlorophyll ml^{-1}. The temperature was 20 °C, irradiance (400-700 nm) 1 500 μeinstein m^{-2} s^{-1}, i.e. $ca.$ 325 W m^{-2}. The Hill reaction rate was measured at 410 nm ($K_3Fe(CN)_6$) or 590 nm (DCPIP). The rate of non-cyclic photophosphorylation was calculated from the decrease in inorganic phosphate measured at 750 nm by the modified molybdate method of MURPHY & RILEY (COLE & ROSS 1966). The spectrophotometric determination of the Hill reaction rate was checked by polarographic measurements of oxygen production with a special concentration Pt/Ag-electrode at the same reaction conditions, but with an increased $K_3Fe(CN)_6$ concentration (1.5 μmol ml^{-1}). All three methods of measurement of the Hill activity gave comparable results (see also STRNADOVÁ & ŠESTÁK 1974).

Amounts of chlorophyll a and b were determined spectrophotometrically in 85% acetone extracts using the equations of ARNON (see ŠESTÁK 1971).

The experiments were repeated five times yielding similar results. Hence the data from one characteristic experiment shown in this paper represent a typical response of leaves.

Results

Net photosynthetic rate, P_N, measured as CO_2 influx under saturating irradiance (Fig. 4, top) increased rapidly after the unfolding of the primary leaves (the 5th day) and reached a maximum between the 12th and the 15th day from sowing, i.e. in leaves having about two thirds of their final size (Fig. 3). P_N then decreased rapidly to $ca.$ 60% of the maximum value from the 16th to the 21st day, with a small shoulder around the 20th day. After a following slow decline of P_N the zero CO_2 influx was observed on the 31st day.

The equivalent (total) stomatal diffusive resistance, r_s (Fig. 4, bottom) measured under near-saturating irradiance, was constant till the 12th day; starting from the 15th day a considerably rapid increase was observed. This increase, corresponding to the more or less complete closure of the stomata, began sooner on the adaxial epidermis (on the 18th day) than on the abaxial side (on the 26th day).

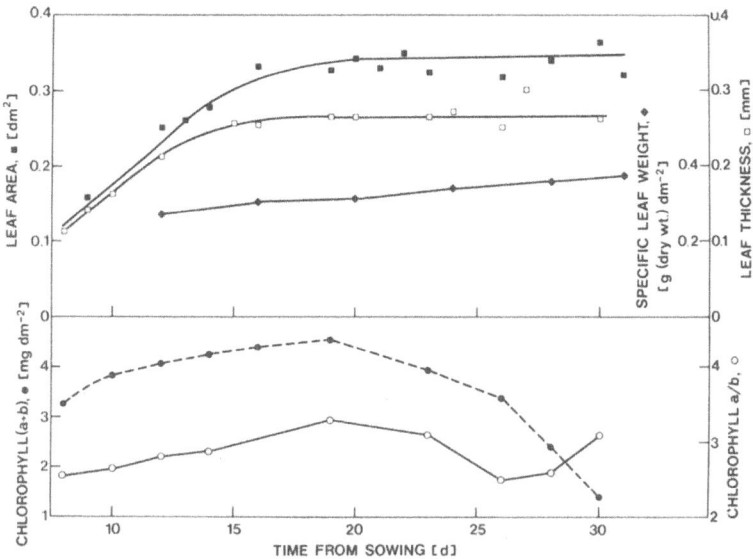

Fig. 3. Leaf characteristics (area, thickness and specific leaf weight), and chlorophyll $(a + b)$ amount and a/b ratio during the development of primary leaves of *Phaseolus vulgaris* L.

While chlorophyll $(a + b)$ amounts per unit leaf area gradually increased to a maximum value on the 19th to 20th day (Fig. 3), the initial higher rates of both the Hill reaction and non-cyclic photophosphorylation per unit amount of chlorophyll $(a + b)$ decreased to a minimum at about the 13th day. The second very sharp peak was observed on the 20th to 22nd day of plant development (Fig. 4, middle). The following rapid decline in the activities of photochemical processes was typical for the second phase of life of the leaf characterized by chlorophyll destruction (21st to 30th day). The third peak of photochemical activities per chlorophyll unit was observed just before the death of the leaf.

Discussion

The above experimental data and information published earlier were used to assess the relative importance of individual segments of the entire CO_2 pathway from the ambient air to photosynthates efflux as rate-limiting factors of photosynthesis during leaf development (Fig. 5). During the first half of leaf expansion the increase in photosynthetic rate seems to be controlled namely by the increase in conductance for CO_2 transfer in the liquid phase ($1/r_m$ on Fig. 5); the stomatal conductance ($1/r_s$) is high and constant and the rate of biochemical reactions

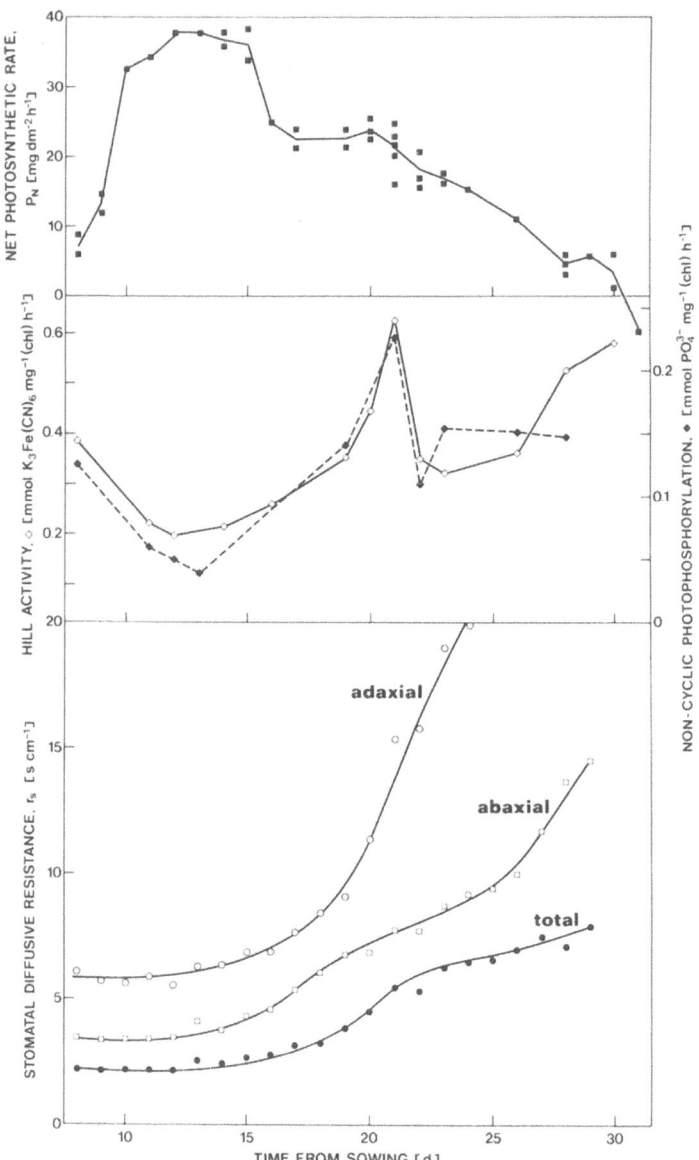

Fig. 4. Net photosynthetic rate [CO_2 concentration: 600×10^{-6} kg m^{-3}, i.e. *ca.* 330 v.p.m., irradiance (400-700 nm) 1 250 µeinstein m^{-2} s^{-1}, i.e. *ca.* 270 W m^{-2}, leaf temperature 28 ± 1 °C], Hill reaction rate (measured with ferricyanide), non-cyclic photophosphorylation, and stomatal diffusive resistance of adaxial and abaxial epidermis during the development of primary leaves of *Phaseolus vulgaris* L.

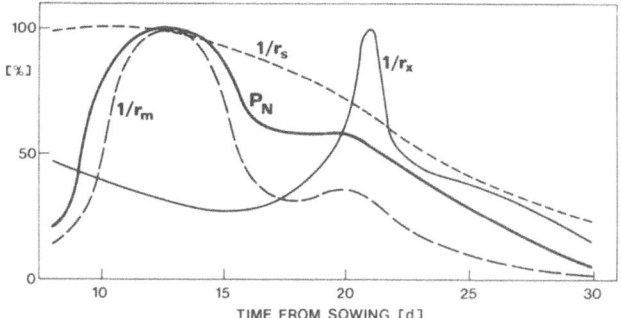

Fig. 5. Generalized scheme of the development of photosynthetic rate (P_N) and its main rate-limiting factors: stomatal ($1/r_s$) and intracellular, i.e. 'mesophyll' ($1/r_m$) and 'carboxylation' ($1/r_x$) conductances. $1/r_x$ is a hypothetical value calculated from the measured Hill reaction and non-cyclic photophosphorylation rates.

(here shown as the 'carboxylation conductance', $1/r_x$) steadily decreases. The second half of leaf expansion is characterized by a decline in photosynthetic rate due to a decrease in both stomatal and intracellular transfer conductances. At the end of the expansion period a rapid increase in rates of photochemical processes is responsible for the slowing down of the decline in photosynthetic rate and perhaps also for the shoulder in net CO_2 uptake. The small peak in 'mesophyll' conductance in this period need not be realistic. It may result from the presuppositions involved in the calculation models used, and from the rather small value of r_x as compared with r_m found in primary bean leaves when using the model of CHARTIER et al. (1970). Intracellular transfer conductances calculated from the average CO_2 transfer path and inner leaf surface according to RACKHAM (1966) and LAÏSK (1971) did not show any corresponding peak. After reaching the final leaf size (20th to 21st day) all parameters observed show a declining trend simultaneously with the chlorophyll ($a + b$) amount. This results in a decrease of P_N towards the compensation point.

The third peaks of photochemical activities (Fig. 4, middle) almost disappear in the calculation of $1/r_x$ (Fig. 5). This value is based on the activity of a unit leaf area containing, in the given phase of leaf ontogenesis, progressively decreasing amounts of the chlorophyll apparatus of a changing composition (see the chlorophyll ($a + b$) amounts and a/b ratios in Fig. 3). The ontogenetic changes in the contribution of individual factors to the limitation of photosynthesis reported here may explain the rather contradictory information available in the literature concerning the decisive rate-limiting factor of photosynthesis.

Summary

The relative importance of several genetically controlled (internal) factors limiting net photosynthetic rate (P_N) was studied during ontogenesis of primary leaves of *Phaseolus vulgaris* L. grown in a controlled environment. CO_2 exchange rate and main conductances for carbon dioxide transfer were correlated with the Hill reaction and non-cyclic photophosphorylation rates. The data obtained in these and earlier experiments indicate that the well-known ontogenetic changes of leaf P_N may be limited in different phases of leaf development by various factors: in young leaves by a rather low photochemical activity, in mature leaves by low conductances for CO_2 diffusion in stomata and intracellular spaces ($1/r_s$). The upper limit of P_N may be limited by the conductance for CO_2 transfer in the liquid phase, i.e. by the intracellular transfer conductance, $1/r_m$.

REFERENCES

ČATSKÝ, J. & TICHÁ, I., *Biol. Plant.* 16: 144, 1974.
CHARTIER, P., *Ann. Physiol. vég.* 8: 167, 1966.
CHARTIER, P., CHARTIER, M. & ČATSKÝ, J., *Photosynthetica* 4: 48, 1970.
COLE, C. V. & ROSS, C., *Anal. Biochem.* 17: 526, 1966.
COWAN, I. R. & MILTHORPE, F. L., in: ECKARDT, F. E. (ed.): Functioning of Terrestrial Ecosystems at the Primary Production Level, p. 107, Unesco, Paris 1968.
DJAVANCHIR, A., *Oecol. Plant.* 5: 301, 1970.
JARVIS, P. G., in: ŠESTÁK, Z., ČATSKÝ, J. & JARVIS, P. G. (ed.): Plant Photosynthetic Production. Manual of Methods, p. 566, Junk, The Hague 1971.
LAÏSK, A., in: Fotosintez i Ispol'zovanie Solnechnoï Energii, p. 97, Nauka, Leningrad 1971.
RACKHAM, O., in: BAINBRIDGE, R., EVANS, G. C. & RACKHAM, O. (ed.): Light as an Ecological Factor, p. 167, Blackwell sci. Publ., Oxford & Edinburgh 1966.
ŠESTÁK, Z., in: ŠESTÁK, Z., ČATSKÝ, J. & JARVIS, P. G. (ed.): Plant Photosynthetic Production. Manual of Methods, p. 672, Junk, The Hague 1971.
ŠESTÁK, Z. & ČATSKÝ, J., in: SIRONVAL, C. (ed.): Le Chloroplaste, Croissance et Vieillissement, p. 213, Masson & Cie., Paris 1967.
ŠESTÁK, Z., JARVIS, P. G. & ČATSKÝ, J., in: ŠESTÁK, Z., ČATSKÝ, J. & JARVIS, P. G. (ed.): Plant Photosynthetic Production. Manual of Methods, p. 1, Junk, The Hague 1971.
STRNADOVÁ, H. & ŠESTÁK, Z., *Photosynthetica* 8: 130, 1974.
THOM, A. S., *Quart. J. roy. meteorol. Soc.* 94: 44, 1968.

REGULATORY ACTION OF BLUE LIGHT ON THE ACTIVITY OF CARBOXYLATING ENZYMES AND ENZYMES OF THE GLYCOLLATE PATHWAY IN BROAD BEAN AND MAIZE PLANTS

NATALIYA P. VOSKRESENSKAYA, NATALIYA M. POYARKOVA, A. KHODZHIEV & INNA S. DROZDOVA

Institute of Plant Physiology, Academy of Sciences of the U.S.S.R., Moscow, and *Institute of Plant Physiology and Biophysics, Tajik Academy of Sciences, Dushanbe, U.S.S.R.*

Regulatory Role of Light

The concept of a regulatory role of light in the vital activity of green plants arose for the first time in connection with the effect of light on morphogenesis and growth motions of plants. At present such ideas may be applied also to photosynthesis. In other words, light in photosynthesis may carry out two functions: (1) the role of a source of energy; and (2) it can act as a regulatory factor of photosynthesis. This was first shown in studies of photosynthetic carbon metabolism (VOSKRESENSKAYA 1953; KOWALLIK 1962; KROTKOV 1964). The regulatory function of light is apparently effected through absorption by various photoreceptors. Having absorbed light, the chromophore-carrying molecules undergo conversions and affect the subsequent course of metabolic events in the plant. Metabolism can be controlled by light in two different ways (VOSKRESENSKAYA 1972; ZUCKER 1972): (a) A fast, reversible action of light on the activity of an enzyme resulting mainly in conformational changes of the enzyme. The role of photoreceptor during this process may be performed by a chromophore group which is a constituent of the enzyme itself (SCHMID 1970). Conformational changes based on cooperative interactions of light-excited molecules with enzyme molecules of the nearest environment seem also to be widespread (KONEV et al. 1970). Finally, allosteric or substrate regulation of an enzymatic activity by light may be connected with light-induced alterations in the substrate accessibility and environmental conditions (pH, ion composition of media, etc.). (b) The prolonged action of light on metabolism is connected with the induction of additional biosynthesis of an enzyme (as compared with darkness). In this case light seems to influence the protein-synthesizing system of the cell.

General Effects of 'Blue' and 'Red' Light

The regulatory function of two regions of wavelengths, the so called 'red and blue light' on the vital activities of plants is rather well-known at present. The effect of 'red light' (RL) is connected with the transition of the chromoprotein phytochrome into the physiologically active state (MOHR 1969), and the action of 'blue light' (BL) with the excitation of flavins or photoconversion of carotenoids (see, e.g. VOSKRESENSKAYA 1972). Physiological and morphogenetic responses of plants to RL or BL are different. For instance, only BL induces two-dimensional growth of fern gametophytes (MOHR 1969). Under RL the enhancement in activities of photosynthetic enzymes in etiolated seedlings is connected with their additional biosynthesis mediated by the photoconversion of phytochrome (SMILLIE & SCOTT 1969; GRAHAM et al. 1971).

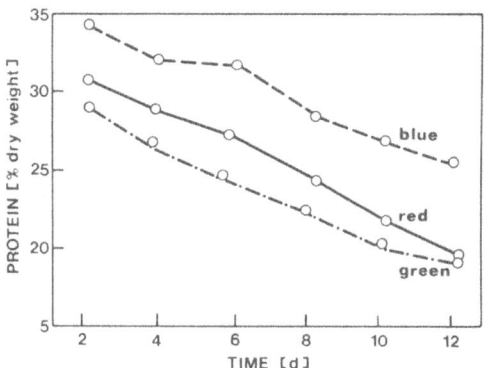

Fig. 1. Protein content in the first leaf of barley seedlings grown under equienergetic 'blue', 'red' or 'green' light.

However, the regulatory action of radiation on the biosynthesis of photosynthetic enzymes is probably not restricted only to RL. Under equienergetic conditions for photosynthesis BL essentially increases the protein content in plants as compared with RL and RL in comparison with 'green light' (Fig. 1). This fact is interpreted as evidence for the functioning of the active form of phytochrome under RL in contrast to 'green light' and as proof for the existence under BL of an additional regulatory photoreaction specifically inducing protein biosynthesis (VOSKRESENSKAYA 1953; KOWALLIK 1962; MOHR 1969). The response of plants previously grown under RL to BL is completely developed during two days of illumination under BL. The reaction is completely reversible when plants are returned to RL. The increase in the protein content in leaves under BL is accompanied by an increase in the RNA content (VOSKRESENSKAYA 1972; VOSKRESENSKAYA & NECHAEVA 1967), size of chloroplasts and their contents of protein and chlorophyll (BERGFELD 1964; MOHR 1969). BL accelerates (as compared with RL) the formation of chloroplast ultrastructure (VLASOVA & VOSKRESENSKAYA 1973) and retards their aging (VOSKRESENSKAYA et al. 1968): The ultrastructure of chloroplasts decays as the result of fast aging of leaves under RL, but may be restored if the detached leaves are illuminated for 48 h in BL. The retardation of leaf aging under BL may be coupled with the activation of protein biosynthesis or with the formation of some specific compounds, which are limiting under RL. There is also an analogy in the effects of BL and kinetin on the restoration of structures in aged barley leaves grown under RL. Kinetin had no effect on the leaves of plants grown under BL (VOSKRESENSKAYA et al. 1968).

'Blue' and 'Red' Light, Photosynthetic Rate and Carboxylating Enzymes

Growing plants under BL activates photosynthetic electron transfer (Hill reaction) and photophosphorylation (VOSKRESENSKAYA & OSHMAROVA 1969; VOSKRESENSKAYA 1972). Prolonged illumination of plants with BL also increases the CO_2-uptake. In both broad bean and maize plants (which differ in their type of carbon metabolism, sensitivity to oxygen and rate of photosynthesis with saturating light intensity) the rate of photosynthesis was much lower in plants grown under RL compared to those grown under BL for all light intensities tested (Fig. 2). At saturating illuminances the photosynthetic rate in maize was higher than in broad bean. Since the maximal differences in CO_2-exchange between the plants grown under RL or BL are observed with saturating illumination, the primary carboxylation reactions in both plant species may be activated by BL. Broad bean, as a C_3-plant, fixes CO_2 using ribulose diphosphate carboxylase

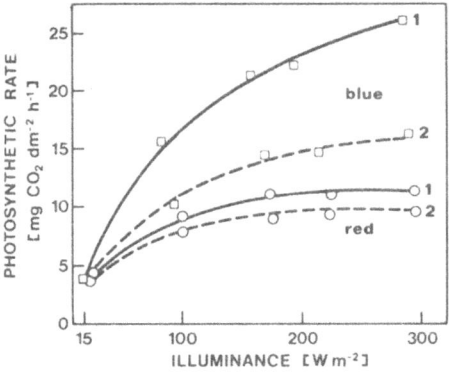

Fig. 2. Effect of 'white light' illuminance on the photosynthetic rate of maize (1) and broad bean (2) grown under 'blue' or 'red' light.

(RDC). In maize (a C_4-plant) the primary fixation of CO_2 proceeds with the participation of phosphoenolpyruvate carboxylase (PEPC) and the dicarboxylic acids thus formed serve as the source of CO_2 for the Calvin cycle (HATCH & SLACK 1970). Thus RDC plays a subordinate role in maize plants. The C_4-dicarboxylic acid pathway may, in turn, be considered as a mechanism additional to the reductive pentose phosphate cycle of CO_2 assimilation (FULLER 1971) in tropical plants. It is possible that, due to a continuous supply of CO_2 from this additional mechanism, these plants may photosynthesize effectively in conditions of very high illumination and very low CO_2 concentration. Broad bean and maize (differing in primary carboxylases and other organizational features of photosynthesis) may serve as good models for the elucidation of general and particular effects of BL on the enzymes of photosynthesis. The techniques for determination of the activities of the carboxylases and glyceraldehyde-3-phosphate dehydrogenase (GPD) were described in detail by POYARKOVA et al. (1973).

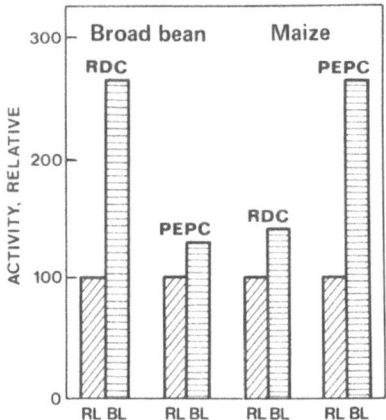

Fig. 3. Relative activity of RDC and PEPC in leaves of broad bean and maize grown under 'red' (RL) or 'blue' (BL) light.

The potential activities (determined in darkness with an excess of substrates) of RDC in bean and PEPC in maize (per unit fresh weight) were considerably higher in plants grown under BL (Fig. 3). Similar differences were observed when these enzyme activities were calculated per mg of supernatant protein in spite of the 25-35% increase in the total content of protein in leaves under BL compared with RL. Hence it seems that proteins of the enzymes responsible for primary carboxylation represent the main soluble proteins synthesized during BL action both in broad bean and maize. The activity of PEPC in broad bean, where this enzyme plays a subordinate role, represents less than 10% of the RDC activity under RL and it is only slightly increased by BL (Table 1); the ratio of the

Table 1. Activity of ribulose diphosphate carboxylase (RDC), phosphoenolpyruvate carboxylase (PEPC) and NADP-dependent glyceraldehyde-3-phosphate dehydrogenase (GPD) in plants grown under 'blue' (BL) light or 'red' (RL) light.

Number of experiment	Light	RDC	PEPC	PEPC/RDC	RDC	PEPC	GPD
		[μmol CO_2 min^{-1} mg^{-1} (protein)]			[% to red light]		
			Broad bean				
1	RL	0.296	0.0324	0.11			
	BL	0.612	0.0382	0.07	205	123	148
2	RL	0.405	0.0496	0.12			
	BL	0.817	0.0496	0.06	202	100	133
			Maize				
1	RL	0.378	0.382	1.05			
	BL	0.579	0.813	1.43	151	213	161
2	RL	0.340	0.382	1.13			
	BL	0.439	0.766	1.74	129	200	131

activities PEPC/RDC is thus low both under RL and BL. In maize where PEPC plays the predominant role in carboxylation the ratio PEPC/RDC remains high at any illuminance and is enhanced by BL. Under RL this ratio was always close to 1 which is unusually low for C_4-plants. Under BL it remained high due to the predominant increase in PEPC activity. Therefore it is possible that the presence of blue rays in the spectrum of 'white light' creates the necessary conditions for the typical PEPC/RDC ratios which, in maize, are always higher than 1.

In broad bean plants the enhancement of GPD activity induced by BL did not correspond to the increase of RDC activity and was always significantly lower (Table 1). Such disproportionate increases in RDC and GPD activities under BL indicate (in comparison with RL) a rising possibility for the removal of phosphoglyceric acid (PGA) from the Calvin cycle. OGASAWARA & MIYACHI (1971) suggest the use of PGA for PEP formation in *Chlorella* in connection with the specific activation of PEPC biosynthesis by BL; this metabolic shunt is not probable for broad bean since the BL activation of PEPC was insignificant. This plant apparently has another way for PGA utilization outside the Calvin cycle, i.e. for biosynthesis of alanine which may accumulate in great amounts in plants grown under BL (PAYER 1969). In maize plants the relative increases of RDC and GPD activities under BL were usually identical (Table 1).

Time Course of Enzyme Activation by 'Blue' Light

The time dependence of the variation in the enzyme activities induced by BL was ascertained as follows: Some plants grown under RL were placed under BL and, after definite time intervals, investigated together with plants which had continuously remained under RL (Fig. 4). During the first hours after transfer of broad bean plants to BL the activities of all investigated enzymes were much lower than in plants under continuous RL. The inhibition decreased with increasing exposure time and then changed to a stimulated activity (RDC was the most stimulated activity and GPD the least stimulated). In maize BL did not inhibit enzyme activities after transfer from RL. 3-6 h later carboxylases (especially PEPC) were activated. The different reactions of broad bean and maize to the transfer to BL may be explained by the different nature of the enzymes participating in primary carboxylation. Since oxygen uptake is activated by BL (in comparison with RL) and oxygen is a competitive inhibitor of RDC (OGREN & BOWES 1971) the transfer of broad bean plants from RL to BL provokes the inhibition of RDC by oxygen. However, after some time this inhibition is masked by progressively increasing biosynthesis of the enzyme and after 16-18 h the biosynthetic processes dominate. Primary carboxylation in maize plants is realized by PEPC and is not inhibited by oxygen. The possibility of a competitive inhibition of carboxylation of RuDP by oxygen in maize should be limited by the high concentrations of CO_2 resulting in chloroplasts from the carboxylation of PEP and subsequent decarboxylation of C_4-acids (HATCH & SLACK 1970). Certainly, all the above explanations for the differences in the time courses of enzyme activation by BL in maize and broad bean need additional examination.

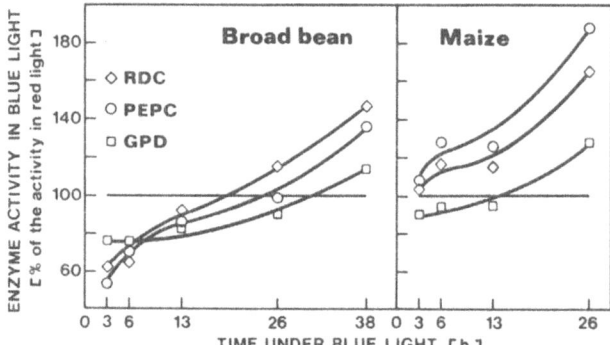

Fig. 4. The effect of exposure period in 'blue light' (22 W m^{-2}) on the enzyme activities in broad bean and maize leaves grown for 18 d under 'red light' (15 W m^{-2}). (From POYARKOVA et al. 1973.)

Effect of Blue Light on the Glycollate Pathway

Since C_3- and C_4-plants differ not only in the mode of primary carboxylation, but also in rates of reactions of the glycollate pathway (which is only slightly manifested in C_4-plants) we decided to study the effect of BL on the enzymes of this pathway (for methods see VOSKRESENSKAYA & KHODZHIEV 1973). The only literature data on light control of the glycollate pathway concerns the activation of biosynthesis of glycollate oxidase in etiolated seedlings of mustard which is mediated by phytochrome (POUCKE et al. 1970), and the increased glycollate formation in plants adapted to BL (HESS & TOLBERT 1967). In our experiments the prolonged action of BL as compared with RL exerts a positive after-effect on the activity of two peroxisome enzymes — glycollate oxidase and glyoxylate amino transferase. In spite of the significant differences in absolute activities of these enzymes in broad bean and maize (higher activities in broad bean), their activation by BL in both plants was very significant (by a factor of 1.5-3.0; more pronounced effect in maize). The stimulating effect of BL on the activity of the glycollate pathway is due to an additional biosynthesis of the enzymes since the differences in favour of BL were also observed when enzyme activities were calculated per unit amount of protein in crude extract. At present it is not clear whether the activation of the glycollate pathway might be explained simply by general enhancement of CO_2 assimilation induced by BL. It is not excluded that an independent activation of glycollate formation takes place which is not proportional to the enhancement of photosynthesis (e.g. increased oxidation of a C_2-constituent of keto-saccharides or increased competition of oxygen with CO_2 for RDC as a result of active oxygen uptake — see ANDREWS et al. 1973).

Conclusions

The photoreaction provoked by RL (and mediated by phytochrome) is not sufficient for revealing all possibilities of the protein-synthesizing system of green plant cells and chloroplasts. BL, as compared with RL, favours a better display of genetic peculiarities of plants with respect to biosynthesis of enzymes of primary carboxylation, the activities of which determine the level of photosynthesis. Certain features of the activation may be explained by different responses of carboxylating enzymes (RDC and PEPC) to oxygen concentration. Since the activating effect of BL is observed both for RDC and PEPC, it is apparent that common reasons for their activation do not lie in their specific properties. Under BL some factor seems to appear which provokes the extra biosynthesis of these enzymes. The extra biosynthesis especially of RDC may be provoked by an enhancement of accessibility of the substrate (CO_2) under BL, since BL diminishes protoplasm viscosity (VIRGIN 1952) and can apparently diminish the resistance to diffusion of CO_2. BL can also alter the state of chloroplast membranes and thus increase the accessibility of the substrate for the enzyme (LAUDENBACH & PIRSON 1969). In addition, BL can increase the entry of CO_2 by enhancing stomatal opening (for review see VOSKRESENSKAYA 1972). The main reason for the enhancement of carboxylation under BL may also be the increase in the rate of regeneration of the CO_2 acceptors (RuDP and PEP), since under the prolonged action of BL photosynthetic electron transfer and photophosphorylation are stimulated.

The use of radiation of different wavelengths permits elucidation of the principles of photosynthesis control by light. Knowing metabolic and energetic alterations provoked by long- and short-term irradiation of plants with light of different spectral compositions, one can obviously select the optimum irradiation for the control of photosynthesis, activities of different biosynthetic pathways, and growth of plants.

Summary

The growing of broad bean (C_3) and maize (C_4) plants under fluorescent red lamps results in a very low rate of photosynthesis, even when saturated with radiant energy, in comparison with 'blue light'. Increases in the potential capacity for photosynthesis after growing under 'blue light' may be explained by an activation of biosynthesis of carbon metabolism enzymes. In bean leaves 'blue light' principally stimulates the activity of RDC, in maize the activity of PEPC.

In bean and maize plants growing under 'red light' and subsequently transferred to 'blue light' the activities of carboxylating enzymes greatly increase after 20 h of 'blue light', probably due to the biosynthesis of additional enzyme proteins.

The reactions of the glycollate pathway are more pronounced in bean than in maize. However, the use of 'blue light' for growing plants activates the glycollate pathway in both plants.

It is suggested that 'blue light' is an obligatory constituent of the net flux of radiation for green plants. The presence and regulatory effects of 'blue light' on the activity of primary carboxylating enzymes provide an increased productivity of photosynthesis in higher plants with different types of carbon metabolism.

REFERENCES

ANDREWS, T. J., LORIMER, G. H. & TOLBERT, N. E., *Biochemistry* 12: 11, 1973.
BERGFELD, R., *Z. Naturforsch.* 19 B: 1076, 1964.
FULLER, R. C., in: SCHOFFENIELS, E. (ed.): Biochemical Evolution and the Origin of Life. Molecular Evolution, Vol. 2, p. 259, North-Holland Publ. Comp., Amsterdam 1971.
GRAHAM, D., GRIEVE, A. M. & SMILLIE, R. M., *Phytochemistry* 10: 2905, 1971.
HATCH, M. D. & SLACK, C. R., *Annu. Rev. Plant Physiol.* 21: 141, 1970.
HESS, J. L. & TOLBERT, N. E., *Plant Physiol.* 42: 1123, 1967.
KONEV, S. V., AKSENTSEV, S. L. & CHERNITSKIĬ, E.A., in: Kooperativnye Perekhody Belkov v Kletke, p. 202, Nauka i Tekhnika, Minsk 1970.

KOWALLIK, W., *Planta* 58: 337, 1962.
KROTKOV, G., *Trans. roy. Soc. Can., Sect.* 1-3, 2: 205, 1964.
LAUDENBACH, B. & PIRSON, A., *Arch. Mikrobiol.* 67: 226, 1969.
MOHR, H.: Lehrbuch der Pflanzenphysiologie, Springer-Verlag, Berlin-Heidelberg-New York 1969.
OGASAWARA, N. & MIYACHI, S., *Plant Cell Physiol.* 12: 675, 1971.
OGREN, W. L. & BOWES, G., *Nature – new Biol.* 230: 159, 1971.
PAYER, H. D., *Planta* 86: 103, 1969.
POUCKE, M., CERFF, R., BARTHE, F. & MOHR, H., *Naturwissenschaften* 57: 132, 1970.
POYARKOVA, N. M., DROZDOVA, I. S. & VOSKRESENSKAYA, N. P., *Photosynthetica* 7: 58, 1973.
SCHMID, G. H., *Ber. deut. bot. Ges.* 83: 399, 1970.
SMILLIE, R. M. & SCOTT, N. S., *Progr. mol. subcell. Biol.* 1: 136, 1969.
VIRGIN, H. I., *Physiol. Plant.* 5: 575, 1952.
VLASOVA, M. P. & VOSKRESENSKAYA, N. P., *Fiziol. Rast.* 20: 96, 1973.
VOSKRESENSKAYA, N. P., *Dokl. Akad. Nauk SSSR* 93: 911, 1953.
VOSKRESENSKAYA, N. P., *Annu. Rev. Plant Physiol.* 23: 219, 1972.
VOSKRESENSKAYA, N. P. & KHODZHIEV, A. Kh., *Fiziol. Rast.* 20: 309, 1973.
VOSKRESENSKAYA, N. P. & NECHAEVA, E. P., *Fiziol. Rast.* 14: 299, 1967.
VOSKRESENSKAYA, N. P., NECHAEVA, E. P., VLASOVA, M. P. & NICHIPOROVICH, A. A., *Fiziol. Rast.* 15: 890, 1968.
VOSKRESENSKAYA, N. P. & OSHMAROVA, I. S., in: METZNER, H. (ed.): Progress in Photosynthesis Research, Vol. III, p. 1669, Tübingen 1969.
ZUCKER, M., *Annu. Rev. Plant Physiol.* 23: 133, 1972.

LOCALISATION OF CARBON METABOLISM IN TWO ASSIMILATION TISSUES OF MAIZE LEAF

Yu. S. KARPILOV, TATYANA A. AVDEEVA & V. M. PERSANOV

Institute of Photosynthesis, Academy of Sciences of the U.S.S.R., Pushchino-na-Oke, Moscow region, U.S.S.R.

Introduction

The leaves of a great number of plants have chlorenchyma which consists of two specialized tissues: well developed bundle sheath cells, and mesophyll. In these plant species the photosynthetically fixed carbon is concentrated into malic and aspartic acids and after subsequent illumination it is translocated and bound into the products of the reductive pentose phosphate cycle and sugars. The scheme of 'co-operative' photosynthesis in species with the 'pathway of the C_4-dicarboxylic acids' (SLACK et al. 1969; KARPILOV 1969; KARPILOV & MALYSHEV 1970) was based on studies of the kinetics of photosynthates and localization of enzymes of carbon metabolism. In this scheme malic and aspartic acids, formed in the mesophyll cells as a result of carbon dioxide fixation by phosphoenolpyruvate carboxylase (PEPC), are assumed to be translocated to the parenchyma cells of the bundle sheath. Carbon dioxide formed during decarboxylation of these C_4-acids is refixed by ribulose-1,5-diphosphate carboxylase (RuDPC) located in bundle sheath chloroplasts. Decarboxylation of malic acid by NADP-dependant 'malic enzyme' in the plants which synthesize mainly this acid in mesophyll (e.g. maize) can partially provide reducing power for the lamellar plastids of parenchyma bundle sheath cells which are characterized by a low activity of photosystem 2 (KARPILOV et al. 1970). At the same time a portion of the phosphoglyceric acid (PGA) formed during RuDPC carbon dioxide fixation is translocated into mesophyll tissue together with alanine and pyruvate, the transport of which is required for closing the cycle and regeneration of phosphoenolpyruvate (PEP).

According to an alternative hypothesis the reductive pentose phosphate cycle operates in mesophyll chloroplasts only, while the reactions of the PEP-carboxylating system are located in the cytoplasm of these cells and serve as a 'pump' for carbon dioxide (COOMBS & BALDRY 1972). Bundle sheath plastids are considered to be amyloplasts synthesizing starch. This hypothesis is supported by data showing high activity of the reductive pentose phosphate cycle in isolated maize chloroplasts (GIBBS et al. 1970).

Investigations testing both these schemes were carried out with isolated tissues and cells. The methods applied, however, do not always ensure their complete separation. The analyses mentioned in this paper were carried out simultaneously on isolated tissues of mesophyll and parenchymal bundle sheath and on leaf strips of a maize chlorophyll mutant which contained neither chlorophyll nor developed chloroplasts in the bundle sheath cells, but whose mesophyll cells contained normal plastids.

Methods

The leaves of 30-d old plants were used for separation of assimilation tissues. The leaves were ground and the pulp was filtered through nylon sieves. The degree of tissue separation was monitored with the light microscope and by the chlorophyll *a/b* ratio. The first fraction of pure mesophyll and the last fraction of the bundle sheath cells were used for the analyses. The albino leaf strips and strips containing developed chloroplasts only in the mesophyll cells were cut from leaves of the maize mutant. Mesophyll chloroplasts isolated from the mutant strips and from the normal green leaves were similar in their ultrastructure, pigment composition and photochemical activity (BIL' et al. 1973).

Enzyme activity was determined in homogenates. A portion of leaf or tissue was ground with glass in 0.05 M Tris-HCl buffer, pH 8.3, containing 10^{-2} M $MgCl_2$, 2.7×10^{-4} M EDTA and mercaptoethanol, and was then centrifuged for 20 min at 22 000 x g. All operations were carried out at 4 °C. The activity of carboxylating enzymes was estimated from the amount of ^{14}C incorporated into acid-stable products (from sodium bicarbonate) during incubation of homogenates and substrates at 25 °C in reaction mixtures and with the incubation conditions recommended by ROMANOVA et al. (1968) and BJÖRKMAN & GAUHL (1969). The activities of 'malic' enzyme and GPD (glyceraldehyde phosphate dehydrogenase) were estimated by NADP reduction and NADPH oxidation (WALKER 1960; HEBER et al. 1963). Natural illuminance of about 80 klx and air with 0.3% CO_2 were used for radioactive CO_2 fixation. The photosynthates were separated, after elution with water-alcohol, by paper chromatography and their radioactivity was quantitatively evaluated (KARPILOV & MALYSHEV 1970).

Result and Discussion

The activity of RuDPC was measured using ribulose 1, 5-diphosphate (RuDP) or ribose-5-phosphate + ATP as substrate. In the latter case the combined activity of the three enzymes RuDPC, ribose-5-phosphate isomerase (RPI) and phosphoribulokinase (PRK) was estimated. With both methods, the activity of carboxylating enzymes of the reductive pentose phosphate cycle was 15-19 times higher in the bundle sheath fraction than in the mesophyll fraction on a chlorophyll basis and 5-7 times higher on a soluble protein basis (Table 1). The low values obtained when RuDP was used as substrate were apparently due to insufficient activity of the preparation.

Table 1. Activity of enzymes of carbon dioxide metabolism (μmol substrate per : g fresh maize leaves /I/, mg chlorophyll /II/, and mg soluble protein /III/).

Enzymes		green leaf	mutant meso-phyll	separated tissues	
				meso-phyll	bundle sheath
RuDPC	I	1.52	0.11	–	–
	II	0.69	0.39	0.44	8.51
	III	0.058	0.007	0.034	0.24
RuDPC + PRK + RPI	I	13.10	0.19	–	–
	II	6.01	0.62	1.93	28.4
	III	0.50	0.01	0.15	0.80
NADP-GPD	I	38.70	15.00	–	–
	II	17.40	50.40	16.65	12.90
	III	1.47	0.90	1.24	0.36
PEPC	I	13.70	15.30	–	–
	II	6.24	49.40	12.0	0.18
	III	0.52	0.90	0.90	0.005
'malic' enzyme	I	14.48	3.61	–	–
	II	6.48	12.20	0.74	14.16
	III	0.56	0.22	0.05	0.39

In mutant mesophyll the activity of RuDPC on a chlorophyll basis was similar to that of mesophyll tissue isolated from the green leaf, but considerably lower when expressed on a protein basis. This could be explained by an addition of proteins from bundle sheath cells which did not contain developed chloroplasts. RuDPC activity in mutant mesophyll, in comparison with green leaf and isolated bundle sheath, decreased when using ribose-5-P + ATP instead of RuDP as substrate. A similar difference was found in mesophyll tissue isolated from the green leaf. This could be due to a decreased activity of Calvin cycle enzymes regenerating RuDP (RPI and PRK) in mesophyll chloroplasts.

Fraction I protein (identical to RuDPC) isolated by electrophoresis on polyacrylamide gel was present in all fractions and in mutant mesophyll (Fig. 1), but it formed a much smaller portion of the total soluble proteins in mesophyll tissue than in the leaf and parenchyma bundle sheath.

The activity of one of the main enzymes of the Calvin cycle – NADP-dependant glyceraldehyde phosphate dehydrogenase (NADP-GPD) was similar in mutant mesophyll and isolated normal mesophyll tissue on a protein basis, and was three times greater than that of the bundle sheath. On a chlorophyll basis only the mutant mesophyll had a considerably increased activity.

On a protein basis the activity of PEPC was similar in both mesophyll tissues; this value was 180 times higher than that of parenchyma bundle sheath. On a chlorophyll basis the activity of PEPC of the mutant leaf strip (which did not

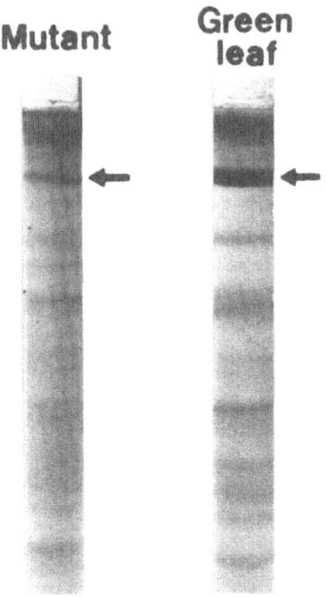

Fig. 1. Electropherograms of soluble proteins in a mutant strip containing neither chlorophyll nor developed plastids in bundle sheath cells, and in a normal green maize leaf.

contain developed chloroplasts in the bundle sheath cells) was four times higher than that of mesophyll isolated from the green leaf. This difference may be explained by the 4-6 times lower chlorophyll content in the mutant strip as compared to the green leaf; the activities of PEPC per g fresh weight of leaf were similar. These data suggest a complete or partial localization of this enzyme in the cytoplasm.

'Malic' enzyme is located mainly in the parenchyma bundle sheath where its activity is eight times higher on a protein basis and about 20 times higher on a chlorophyll basis compared to the mesophyll tissue. Mutant mesophyll contains a considerably higher activity of 'malic' enzyme when compared to normal mesophyll. Additional investigations showed (Table 2) that this difference was not induced by an increased enzyme activity in mutant mesophyll free of bundle sheath chloroplasts but by a lower retained enzyme activity of chlorophyll-free bundle sheath tissue.

On a fresh weight or leaf area basis the light-green strips of mutant leaf fixed CO_2 during steady-state photosynthesis 4-6 times slower than the normal green leaf. The difference was much smaller when the photosynthetic rate was calculated on a chlorophyll basis (Fig. 2).

Similar data were obtained from measurements of oxygen evolution, which

Table 2. 'Malic' enzyme activity in mutant albino tissue and etiolated maize leaves.

Tissue	Experiment no.	Chlorophyll [mg per g fresh weight]	Soluble protein	'Malic' enzyme activity [μmol]	
				per g fresh weight	per mg soluble protein
Albino	I	0.002	6.92	0.74	0.107
	II	0.003	8.27	1.03	0.075
Etiolated	I	–	25.5	1.66	0.065
	II	–	23.7	1.74	0.073

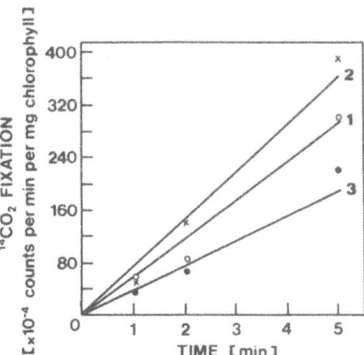

Fig. 2. Carbon dioxide fixation by normal maize leaf (1) and two different mutant mesophyll tissues (2, 3). The rates of mutant tissues calculated on a chlorophyll basis were lower or higher than those of normal tissues.

was 0.756 ml h^{-1} g^{-1} (fresh weight) in normal green leaf and 0.240 ml h^{-1} g^{-1} (f.w.) in mutant strips; the values per mg chlorophyll were 0.281 and 0.633 ml h^{-1} respectively. In spite of the similarity of PEPC activities (per g fresh weight) in the mutant strip (which does not contain developed plastids in bundle sheath tissue and some mesophyll cells) and in the normal green leaf, the rate of CO$_2$ fixation is slower in the mutant strip. It is apparently limited not only by the supply of free carbon compounds which are required for PEP regeneration, but also by photochemical reactions. This was shown by introduction of PEP and its precursors into mutant mesophyll: the rate of CO$_2$ fixation increased after incubation of leaf fragments in PEP solution but not in solutions of pyruvate, alanine and exogeneous RuDP (Fig. 3). The composition of ^{14}C photosynthates in the mutant mesophyll was similar to that formed in the albino tissue which had a fixation rate (per

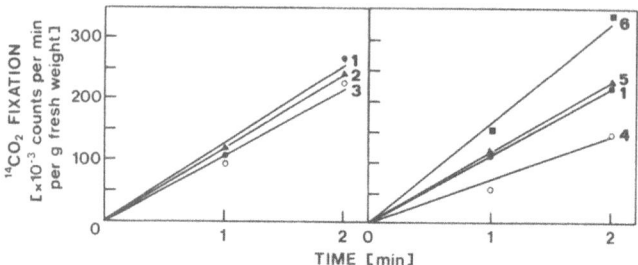

Fig. 3. Influence of carboxylation substrates and their precursors (1 – water control; 2 – pyruvate; 3 – alanine; 4 – RuDP; 5 – Ru-5-P + ATP; 6 – PEP) on the rate of carbon dioxide fixation in the mutant mesophyll.

g fresh weight) 12 times slower. ^{14}C was not present in phosphorylated compounds and sugars (Table 3). In only a few cases these compounds contained not more than 2% of fixed ^{14}C which was probably assimilated in the mutant strip in rare bundle sheath cells with developed chloroplasts.

Thus, RuDPC present in a mutant strip homogenate does not take part in CO_2 fixation. This may be explained by the low activity of enzymes regenerating RuDP in the mesophyll tissue and the resulting lack of this substrate. However,

Table 3. ^{14}C distribution among the photosynthates of albino leaves and of mutant strips of the leaf which contains neither chlorophyll nor developed plastids in the bundle sheath cells.

Duration of photosynthesis in $^{14}CO_2$ and $^{12}CO_2$ [min]	Radioactivity [% of total]					
	PGA + SPE* + sugars	Malate	Aspartate	Alanine	Glutamate	Citrate + succinate
		Albino				
5	–	67.2	28.4	1.4	3.0	–
5 + 10 in $^{12}CO_2$	–	54.4	5.9	3.8	16.4	9.5
		Mutant mesophyll				
0.25	–	71.1	28.9	–	–	–
1	0.4	72.0	27.0	–	–	–
2	0.2	71.0	28.8	–	–	–
5		65.5	32.3	0.7	1.5	–
5 + 10 in $^{12}CO_2$	2.1	60.3	13.9	12.0	8.3	3.4
	Mutant mesophyll unseparated from leaf					
0.25 + 3 in $^{12}CO_2$	18.6	52.1	11.6	13.9	3.8	–
0.25 + 10 in $^{12}CO_2$	50.0	30.1	5.1	12.2	2.5	–

* SPE – sugar phosphate esters.

the products of the reductive pentose phosphate cycle and sugars were not labelled after feeding exogeneous RuDP into the leaves (Table 4). Either the activity of RuDP carboxylase in mesophyll tissue chloroplasts is blocked, or the enzyme is located in chlorophyll-free proplastids of bundle sheath cells. In the latter case this enzyme may be inactive in the intact tissue as well as in etiolated leaves which do not fix CO_2 by RuDPC.

Table 4. Influence of carboxylation substrates on ^{14}C distribution among photosynthates. Fragments were kept for 3 h on the surface of solutions containing 15 μmol substrate in 6 ml water; illuminance 60 klx.

Compounds introduced	Exposure time [min]	Compound radioactivity [%]				Alanine	Sugars
		Acids					
		aspartic	malic	citric	glutamic		
Water	2	20.5	58.9	9.9	5.7	5.0	–
	5	19.1	54.9	10.0	10.2	5.8	–
PEP	2	18.5	69.8	7.1	2.8	1.8	–
	5	15.8	59.2	8.5	10.9	5.0	0.6
RuDP	2	65.4	34.6	–	–	–	–
	5	44.3	49.7	2.2	3.8	–	–

Mutant strips of the leaf lacking chloroplasts in the bundle sheath cells are characterized by a regular time-course of CO_2 fixation. Hence the question is: how is the CO_2 acceptor (PEP) regenerated? Kinetics of photosynthates (Table 3) indicate an increase in the relative radioactivity of alanine, glutamate, citrate and succinate with prolongated illuminance.

Appearance of ^{14}C in the acids of the Krebs cycle was more rapid after a 3 h incubation of mutant strips on a water surface (Table 4). Sugars and perhaps also alanine (from the nearest leaf parts with normally developed tissues) actively enter the mutant mesophyll tissue (Table 3). The feed-back transport of C_4-acids can possibly also take place. Hence pyruvate formed during oxidation of sugars may enter the mutant strip and be partially used for the synthesis of PEP.

Though chloroplasts of mutant mesophyll did not fix carbon dioxide by means of RuDPC they actively reduced PGA into phosphotrioses and transformed these into sucrose (Tabel 5). The reaction rate of the reductive pentose phosphate cycle in mutant mesophyll (as well as in the green leaf) was limited only by the entry of exogenous substrates since after exposure to light there were no labelled phosphorylated compounds in the tissues in either case. The intact tissue is responsible for the high activity of NADP-GPD in mesophyll chloroplasts. The activities of this enzyme and of those transforming phosphoglyceric aldehyde into sucrose are not blocked in plastids co-operating with the cytoplasm.

The obtained data support the recently proposed scheme of co-operative inter-

Table 5. Metabolism of phosphorylated compounds in leaf and mutant strip not containing developed bundle sheath chloroplasts.

Compound	Introduced extract [% of total]	Radioactivity					
		Mutant mesophyll			Mutant leaf		
		[% of total]	[counts per min × 10^{-3}]		[% of total]	[counts per min × 10^{-3}]	
			per mg chloro-phyll	per g fresh weight		per mg chloro-phyll	per g fresh weight
Insoluble fraction [% of total leaf activity]	–	13.5	266	89	25.1	107	214
PGA	41.2	–	–	–	–	–	–
SPE	51.3	–	–	–	–	–	–
Sucrose	–	84.6	1446	482	89.4	287	574
Serine	1.6	6.7	114	38	6.9	22	44
Glyceric acid	4.0	–	–	–	–	–	–
Alanine	1.9	8.7	148	49	3.7	12	24

action of mesophyll and parenchyma bundle sheath tissues in carbon dioxide metabolism (KARPILOV & MALYSHEV 1970).

Summary

On the basis of studies of the products of photosynthesis and the localization of the enzymes of carbon metabolism in maize a scheme of 'co-operative' photosynthesis has been suggested.

A hypothesis was advanced that the reductive pentose phosphate cycle is functioning only in mesophyll chloroplasts, and the reactions of the PEP carboxylating system are localized in the cytoplasm. The plastids of bundle sheath cells play the role of amyloplasts synthesizing starch. The investigations on isolated tissues of mesophyll, parenchyma, bundle sheath layers, and leaves of a maize mutant (in which bundle sheath cells contained neither chlorophyll nor developed chloroplasts while mesophyll cells contained normal plastids) confirm the scheme of co-operative interaction of the mesophyll tissues and parenchyma layer in carbon metabolism.

REFERENCES

BIL´, K. Ya., ARKHIPOV, V. N., KARPILOVA, I. F. & MOSHKOV, D. A., in: Fotosintez Kukuruzy, p. 7, Pushchino 1973.

BJÖRKMAN, O. & GAUHL, E., *Planta* 88: 197, 1969.
COOMBS, J. & BALDRY, C. W., *Nature-new Biol.* 238: 268, 1972.
GIBBS, M., LATZKO, E., O'NEAL, D. & HEW, C.-S., *Biochem. biophys. Res. Commun.* 40: 1356, 1970.
HEBER, U., PON, N. G. & HEBER, M., *Plant Physiol.* 38: 355, 1963.
KARPILOV, Yu. S., Tr. mold.nauch.-issled. Inst. orosh. Zemled. Ovoshchevod. 11: 3, 1969.
KARPILOV, Yu. S., BIL', K. Ya., MALYSHEV, O. G. & KARNAUKHOV, V. N., in: Kooperativnyï Fotosintez Kserofitov. *Tr. mold. nauch.-issled. Inst. orosh.Zemled. Ovoshchevod.* 11 (3): 25, 1970.
KARPILOV, Yu. S. & MALYSHEV, O. G., in: Kooperativnyï Fotosintez Kserofitov, *Tr. mold. nauch.-issled. Inst. orosh. Zemled. Ovoshchevod.* 11 (3): 33, 1970.
ROMANOVA, A. K., VEDENINA, I. Ya. & DOMAN, N. G., *Izv. Akad. Nauk SSSR, Ser. biol.* 1968 (3): 363, 1968.
SLACK, C. R., HATCH, M. D. & GOODCHILD, D. J., *Biochem. J.* 114: 489, 1969.
WALKER, D. A., *Biochem. J.* 74: 216, 1960.

GENETIC NATURE OF PHOTOSYNTHETIC MUTATIONS AND THEIR PHENOGENESIS

MUTATION VARIABILITY OF CHLOROPLASTS IN ARABIDOPSIS THALIANA (L.) HEYNH.

P. D. USMANOV, H. A. ABDULLAEV, OLGA V. USMANOVA & Sh. SOKHIBNAZAROV

Institute of Plant Physiology and Biophysics, Tajik Acad. Sci., Dushanbe, U.S.S.R.

Introduction

It is now universally recognized that the formation and activity of the photosynthetic apparatus is controlled by chromosome and extrachromosome systems (KIRK 1970, SAGER & RAMANIS 1970, HAGEMANN 1971, D. VON WETTSTEIN et al. 1971, NASYROV 1972). This extends our knowledge of the mechanisms of action of genes which regulate the photosynthetic process. However, to study the genetics of photosynthesis we must produce mutation systems with different degrees of complexity: first 'mutation-character' and then 'gene-character' systems (KVITKO 1972).

The present paper reviews the work of the Laboratory of Physiological Genetics on the selection of mutation systems of the 'mutation-character' type in *Arabidopsis thaliana* (L.) HEYNH. The first section analyses the frequency and phenotypic spectra of photosynthetic mutants after treating seeds of *A. thaliana* with chemomutagens and ionizing radiation. The second part describes the results of the phenotypic study of mutant plants with light and electron microscopes.

I. Quantitative Laws of Induction of Chloroplast Mutants

A. Frequency of Chlorophyll Mutations Induced by N-Methyl-N-Nitrosourea

The experimental plant *A. thaliana* is a suitable object for genetic research, on which the method of 'embryonic test' may be applied. The 'embryonic test' proposed by MÜLLER (1963) registers different classes of mutations, including chlorophyll mutations of the immature seeds in siliquae of M_1 plants (Fig. 1).

Using this method we have estimated the relative occurence of different classes of mutations induced by the chemomutagen N-methyl-N-nitrosourea (MNU) both in the embryonic stage of seed development and during germination (Fig. 2). Chlorophyll mutants constitute only 3% of the total mutation variability of *A. thaliana* in the embryonic stage (Fig. 2a), but they occur five times as often among the seedlings (Fig. 2b). This observation might be explained by the fact that some chlorophyll mutant types – bleaching, variegated, dark-green and even

Fig. 1. Open siliqua of *Arabidopsis thaliana*. Segregation of the sublethal mutation '*xantha*' is observed.

viridis – cannot be readily identified in the embryonic stage, their colour being practically normal.

The curve of effect of MNU concentration on the frequency of mutant seedlings in M_2 (Fig. 3) differs from the classical linear correlation expected for mutations according to the target theory (TIMOFEEFF-RESSOVSKY et al. 1972). Interpretation of such curves, often appearing in radiation-genetic investigations, is vague, although there is a number of tempting hypotheses and explanations in the literature (EMMONS & HOLLAENDER 1939, RÖBBELEN 1957, ZIMMER 1961, HUG & KELLERER 1966, VATTI et al. 1967, IVANOV et al. 1969, DUBININ 1970, USMANOV et al. 1971). Nevertheless, we have made an attempt to find the proper interpretation.

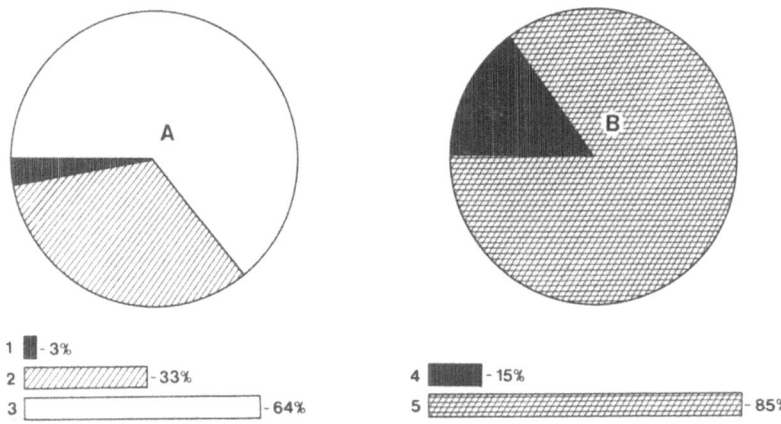

Fig. 2. Relative yield of different classes of mutations in the embryonic stage of seed development (A) and the ratio of chlorophyll mutants in the siliquae, embryos and seedlings (B). The seeds were soaked for 18 h in 125 μM MNU at 20 °C. A: 1 – chlorophyll mutants in the embryonic stage; 2 – recessive, embryonic (estimated as frequency of under-developed seeds in the siliquae; 3 – haplophase mutations (estimated as frequency of non-fertilized seed-buds in the siliquae); B: 4 – relative yield of chlorophyll mutants embryos and seedlings; 5 – in the stage of cotyledon leaves.

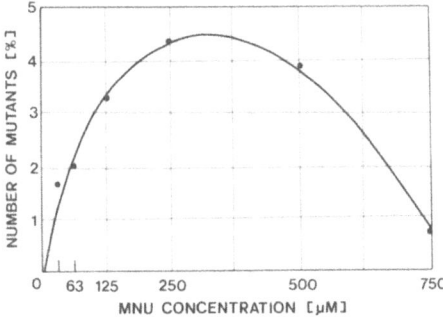

Fig. 3. Frequency of chlorophyll mutants in seedlings after soaking the seeds in different concentrations of MNU.

Only two ($P_{10}/43$ and 99) of the seven heterozygote strains analyzed for segregation in the lethal mutations '*xantha*' (Table 1) displayed Mendel's type of segregation in both the siliquae and the seedlings; in two strains (36 and 72) deficit of the recessive mutation (designated as +) was detected in both the siliquae and the seedlings; in all other cases segregation corresponding to that theoretically expected was found only in the embryonic stage.

Table 1. Deficit of the recessive mutations revealed while estimating segregation of heterozygotes in the lethal mutations 'xantha'. + = statistically significant deficit of recessive ($p \leqslant 0.05$).

Heterozygote strain	Deficit of recessive mutations	
	in siliquae	in seedlings
$P_{10/43}$		
72	+	+
36	+	+
127		+
79		+
27		+
99		

The 'embryonic test' in the siliquae can reveal practically all occurring chlo phyll mutants, but only a negligible number of these are capable of forming via seedlings with chlorophyll defects. Thus the deficit of recessive mutations in strains 27, 29 and 127 might be explained by a selective death of some le mutants already in the stage of embryonic development. This selective death r also explain the absence of the classical linear correlation between the mutai yield and the dose of chemomutagen. At MNU concentrations of 250 μM and higher the increasing toxicity of the mutagen probably accounts for the low chance of the mutant embryo deficient in chlorophyll to survive.

B. Phenotypic Spectra of Chlorophyll Mutations Induced by Chemomutagens MNU and EMS and Ionizing Radiation

The phenotypic spectrum for different concentrations of MNU shows that not all doses of the mutagen produce a wide spectrum of mutations (Table 2). Maximal spectra have been observed at concentrations of 63 and 125 μM; at other doses we failed to find some individual classes of mutations within the phenotypic spectrum.

The dose of 250 μM, noted for the absence of one phenotypic class corresponds to the bend of the 'dose-effect' curve in Fig. 3. This concentration is apparently critical, as some additional effects of the mutagen have been observed. The increasing toxicity of high concentrations of the mutagen affects the physiology of development of the mutants and hence some mutant classes may be missing in the phenotypic spectrum for these concentrations. The data in Table 2 thus lead to the conclusion that, to obtain the phenotypic spectrum, we should use doses of mutagens which do not influence the viability of the studied species.

At the 'optimal', 'subvital' concentrations, 63-125 μM, we detect a similar distribution of mutants in the spectrum and a similar sum of mutants per this mutagen. From such data we might compose phenotypic spectra and use them

Table 2. Spectrum of chlorophyll mutants in *Arabidopsis thaliana* after soaking seeds in different concentrations of MNU. (Analysis of families.)

Concentration [μM]	Distribution of mutants in phenospectrum [%]			
	albino	xantha	viridis	different
32	10	4	86	0
63	7	20	71	2
125	19	14	66	1
250	11	5	84	0
500	0	14	86	0
750	0	50	50	0
* Average for each concentration ration	11	11	77	1

* Ratio between the sum of mutant families per each phenotypic spectrum and the total sum of mutant families for all concentrations. (The seeds were treated by MNU solutions for 18 h at 20 °C.)

with certain precautions for characterizing the specificity of a mutagen action (D. VON WETTSTEIN et al. 1959, NILAN 1967, USMANOV et al. 1971). We have composed such spectra for six mutagens (Fig. 4). For X- and γ-irradiation a single spectrum has been represented due to their great similarity in physical nature and interaction with living matter. On the diagram the phenospectra are arranged according to the increasing number of *viridis* mutations, i.e. in the fol-

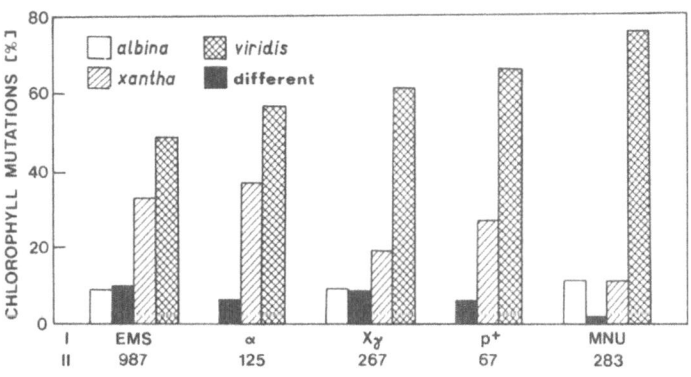

Fig. 4. Spectra of *A. thaliana* chlorophyll mutants induced by different mutagens. I: EMS – ethyl methanesulphonate; α-particles, X- and γ-irradiation, p$^+$-protons, MNU - N-methyl-N-nitrosourca. II: Number of mutant families studied.

lowing order: MNU > protons > X and γ-irradiations > α-particles > ethyl methanesulphonate (EMS). For mutations of the *xantha* type this sequence looks somewhat different: α-particles > EMS > protons > X- and γ-irradiation > MNU.

Heavy particles did not induce mutations of the *albino* type. This is assumed to result from a specified action of protons and α-particles on the genetic apparatus of cells.

Thus MNU is more suitable for inducing a maximal number of mutations of the *viridis* class, whereas heavy particles and EMS can be recommended for obtaining *xantha* mutations. *Albino* mutations are induced mainly by MNU, X- and γ-radiation.

II. Phenoanalysis and Phenogenesis of Chloroplast Mutants

A. *Genome Control of Chloroplast Size and Number*

As mentioned in the Introduction, the main aim of investigations on the genetics of photosynthesis is to assess the role of different components of idiotype in the process of chloroplast formation and activity.

In this connection we have studied the influence of genotype (number of chromosomes) on the size and number of chloroplasts in mesophyll cells of *Arabidopsis* species. We have examined chloroplasts from the first pair of leaves (YUNUSOV et al. 1969a, b, USMANOV et al. 1970) in five species of the genus *Arabidopsis* growing in Tajikistan (*A. thaliana*, 2n = 10; *A. wallichii*, 2n = 16; *A. mollissima*, 2n = 26; *A. pumila*, 2n = 32; *A. korschinsky*, 2n = 48) (Fig. 5) and 14 mutant strains of the race Enkheim.

The number of chloroplasts in mesophyll cells in the species studied increases with their chromosome number (Fig. 6); the respective regression coefficient $b = 0.38 \pm 0.098$ is different from zero (at $t_3 = 4.00; p = 0.025$). The regression coefficient for chloroplast diameters $f = 0.014 \pm 0.013$ ($t_3 = 1.08; p = 0.38$) indicates no changes in the size of chloroplasts with the chromosome number. Hence the size of the chloroplast is under the control of extrachromosome systems; cf. the results of F. VON WETTSTEIN (1924) and RHOADES (1959) for polyploid strains of mosses and maize.

No correlation was found between the number of chloroplasts in the cells and chloroplast diameter. The coefficient of range correlation $r_3 = 0.6$ is not statistically significant. Therefore the variability of chloroplast number and size is regulated by different mechanisms.

The study of 14 *Arabidopsis* mutant strains confirmed that nuclear mutations significantly influence the chloroplast number and size (see the data for chloroplast size in Fig. 7 – $b = 0.070 \pm 0.005; t_{13} = 14.7; p = 0.0002$).

Thus on the basis of our results (and literature data giving values for wheat, barley, peas, maize, sugar beet, etc.) we may conclude that the average number of chloroplasts in the cell is controlled by genetic factors located in the nuclear

Fig. 5. Species of *Arabidopsis* growing in Tajikistan. Top: (from left to right): *A. thaliana* – Tajik race, race Enkheim; *A. korshinsky*; *A. pumilla*; bottom: *A. wallichii*; *A. mollissima*.

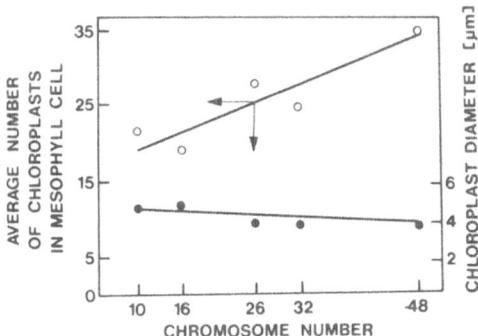

Fig. 6 Correlation between chromosome number, diameter of chloroplast (dark circles) and average number of chloroplasts in mesophyll cells (light circles) in five species of *Arabidopsis*.

apparatus (F. VON WETTSTEIN 1924, RHOADES 1959, VECHER 1961, LADYGIN 1965, BUTTERFASS 1967, USMANOV et al. 1970).

Significant disagreement in literature on the localization of genetic systems controlling the size of chloroplasts prevents us from reaching any final conclusion.

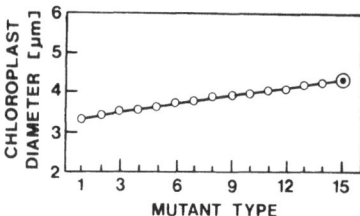

Fig. 7. Diameter of chloroplasts in 14 chlorophyll mutants of *A. thaliana* in comparison to the wild type (15). Mutants 1-14 are distributed according to increasing average chloroplast diameter.

B. Genome Control of Ultrastructural Organization of the Photosynthetic Apparatus

Analysis of a great number of strains of *Arabidopsis thaliana* showed that, in the mutation variability of chloroplasts, the only forms of membrane structure revealed are those which correspond to the main stages of chloroplast morphogenesis described by MÜHLETHALER & FREY-WYSSLING (1959), RÖBBELEN (1959) and D. VON WETTSTEIN (1959).

However, unusual membrane system morphotypes which are absent in the available schemes of chloroplast morphogenesis have also been observed (Fig. 8). Detection and analysis of such abnormalities help to clarify the mechanism of the photosynthetic membrane system formation.

Systematization and analysis of the morphotypes reveals the strict sequence of perfection and complexity of space organization of the photosynthetic membranes (Fig. 9).

It is interesting that characteristics of plastid membranes of the induced mutant morphotypes of *Arabidopsis thaliana* are almost the same as those of some taxons already existing in nature (Fig. 9). Occurence of such similar morphotypes of the chloroplast membrane systems of different plants representing different stages of the evolution process probably indicates the monophyletic origin of the chloroplast.

Elaboration of the optimal morphofunctional models of the photosynthetic apparatus is an important basis for formulating the theory and principles of optimization of plant photosynthetic activity. It is therefore necessary to define the limits of possible variability of different components of chloroplast ultrastructure.

Correlations between different chloroplast components occuring in the mutation process may be estimated by electron microscopic morphometry (WEIBEL 1969).

In ten chlorophyll mutants of *A. thaliana* we studied the limits of the possible variations in matrix, lamellae, plastoglobules, and starch grains (ABBULLAEV et al. 1972) and calculated their optimal ratios (Fig. 10). Such morphological models

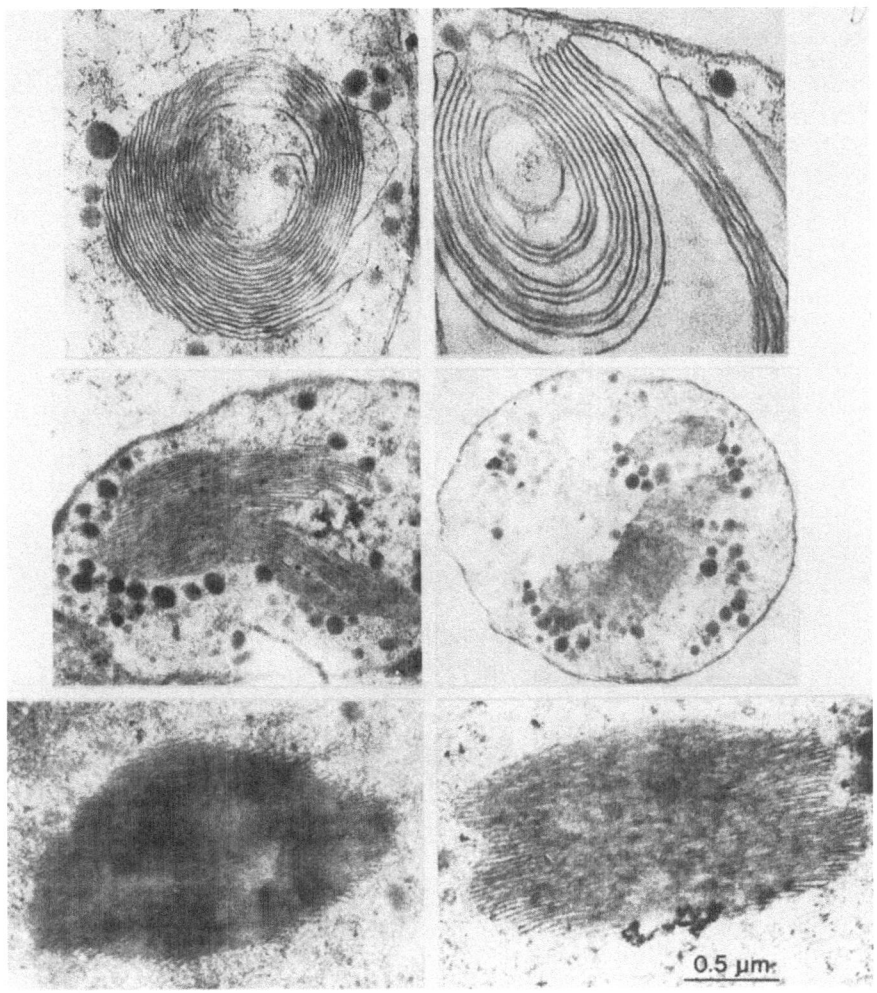

Fig. 8. Different configuration of thylakoids in the plastids of mutants of *A. thaliana*.

are regarded as an important step in creating the more complex morphofunctional models of chloroplasts of different plants. Their analysis would help to find the most optimal variants of the photosynthetic structure which might be applied in the selection of highly productive forms of plants.

By means of morphometry we can also estimate the rate of interaction and mutual conditionality of different components of the chloroplast. Having the

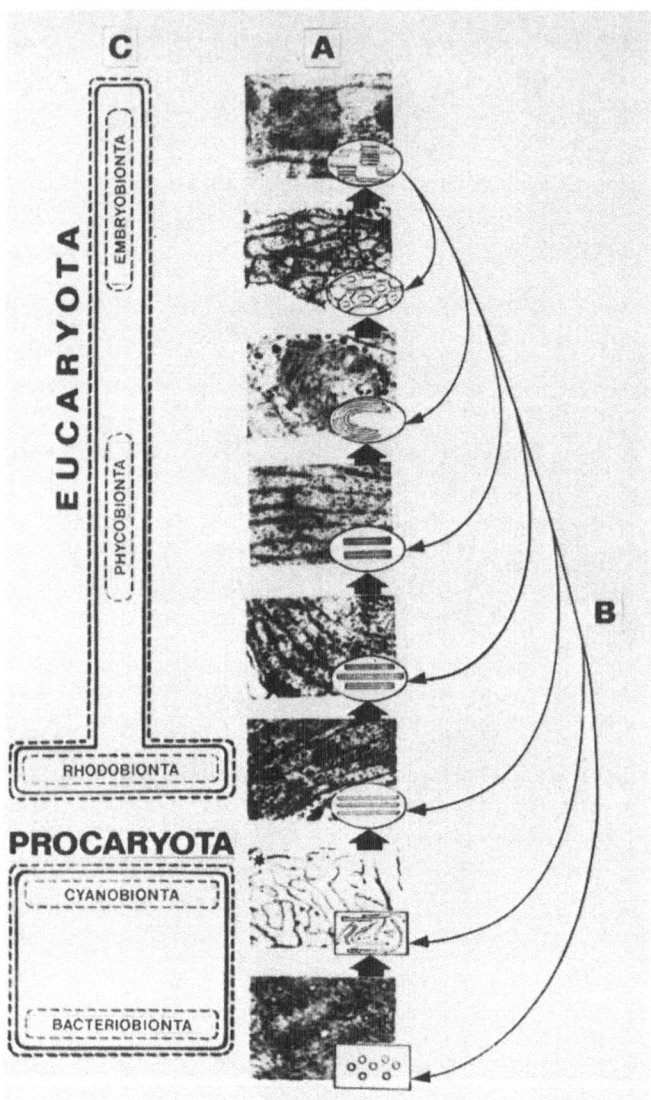

Fig. 9. The main morphotypes of the membrane systems of photosynthesizing organisms. A: The supposed direction of the successive perfection of space organization of photosynthetic membranes. B: Morphotypes of membrane systems in chloroplast mutants of *A. thaliana*. Membrane systems of plastids of the *Oenothera* type (SCHÖTZ 1970). C: Representatives of some taxons systematized according to TAKHTADZIIYAN (1973). Macrosystem of living organisms (in dark frame – kingdom, thick dotted – overkingdom, thin dotted – underkingdom).

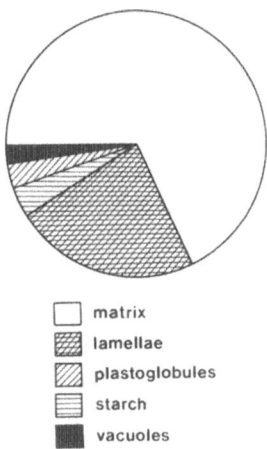

Fig. 10. Supposed optimal ratios of partial space occupied by components of chloroplasts from *A. thaliana*.

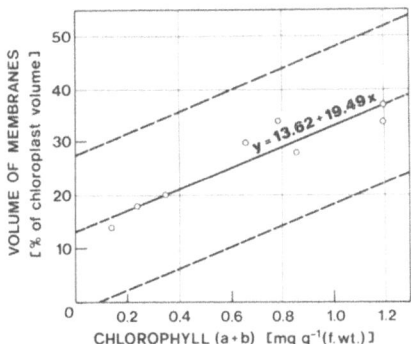

Fig. 11. Relationship between the volume of photosynthetic membranes and amount of chlorophylls ($a + b$).

quantitative data for the membrane systems of chloroplasts and knowing the amounts of chlorophylls we were able to show that the degree of correlation between these two values is strict (Fig. 11).

The coefficient of the range correlation was $r = 0.83 \pm 0.11$. The straight line of this correlation crosses the ordinate noticably above the zero point. This suggests that even in the absence of chlorophylls the chloroplasts have a certain number of membranes. However, calculations of the regression coefficient and its error ($b = 19.49 \pm 4.58$) and the significance of the regression line (Fig. 11, broken lines) show that such conclusions cannot be drawn from our results. Our experimental data are probably consistent with the opinion that photosynthetic membranes cannot be formed in the absence of chlorophyll.

Our data support the idea of WADDINGTON (1962) that there exists a reversible correlation between pigment content and lamellar structure; i.e. the formation of lamellar structures requires the presence of chlorophylls, whereas synthesis of chlorophylls in turn requires a certain quantity of lamellae.

Summary

An attempt was made to study the complex nature of photosynthetic mutations in *Arabidopsis thaliana* using mutation analysis and the light and electron microscopes. The yield of photosynthetic mutations after treatment of seeds of *A. thaliana* with N-methyl-N-nitrosourea, ethyl methanesulphonate, α-particles, protons, X- and γ-irradiation was calculated. The deviations of the 'dose-effect' curve from the classical linear dependence were shown to be induced by the selective death of chloroplast mutants in different phases of development. This selective death of chlorophyll mutants results from defects in the morpho-functional characteristics of chloroplasts resulting from the action of nuclear genes. The average number of chloroplasts per cell is controlled by genetic factors located in the nucleus. By electron microscopic morphometry the limits of variability of the chloroplast structural organization in *A. thaliana* were found. The variability of the chloroplast membrane system in the plant kingdom is homologous.

REFERENCES

ABDULLAEV, H. A., USMANOV, P. D., TAGEEVA, S. V. & NASYROV, Yu. S., *Arabidopsis Inform. Service* 9: 26, 1972.
BUTTERFASS, T., *Planta* 76: 75, 1967.
DUBININ, N. P.: Obshchaya Genetika, p. 273, Nauka, Moskva 1970.
EMMONS, C. W. & HOLLAENDER, A., *Amer. J. Bot.* 26: 467, 1939.
HAGEMANN, R., *Biol. Zentralbl.* 90: 409, 1971.
HUG, O. & KELLERER, A. M.: Stochastik der Strahlenwirkung, Springer-Verlag, Berlin – Heidelberg – New York 1966.
IVANOV, V. I., SANINA, A. V., TIMOFEEVA-RESSOVSKAYA, H. A., USMANOV, P. D. & HABERER, G., *Stud. biophys.* 13: 111, 1969.
KIRK, D. T. O., in: Funktsional'naya Biokhimiya Kletochnykh Struktur, p. 39, Nauka, Moskva 1970.
KVITKO, K. V., in: KIRICHENKO, E. B. (ed.): Metody Issledovaniya Struktury Fotosinteticheskogo Apparata, p. 119, Pushchino-na-Oke 1972.
LADYGIN, V. G., *Genetika* 6: 127, 1965.
MUHLETHALER, K. & FREY-WYSSLING, A., *J. biophys. biochem. Cytol.* 6: 507, 1959.
MÜLLER, A. J., *Biol. Zentralbl.* 82: 133, 1963.
NASYROV, Yu. S., *Zh. obshch. Biol.* 33: 683, 1972.
NILAN, R. A., *Abhandl. deut. Akad. Wiss. Berlin, Kl. Med.* 1967 (2): 5, 1967.

RHOADES, M. M., in: Maize Genetics Cooperation News Letter 33, Indiana Univ., Bloomington, Indiana 1959.
RÖBBELEN, G., Z. indukt. Abstamm. Vererbungslehre 88: 189, 1957.
RÖBBELEN, G., Z. Vererbungslehre 90: 503, 1959.
SAGER, R. & RAMANIS, Z., Proc. nat. Acad. Sci. U.S.A. 65: 593, 1970.
SCHÖTZ, F., in: MILLER, P. L. (ed.): Control of Organelle Development, Symp. Soc. exp. Biol. 24: 39, 1970.
TAKHTADZHYAN, A. L., Priroda 1973 (2): 22, 1973.
TIMOFEEFF-RESSOVSKY, N. V., IVANOV, V. I. & KOROGODIN, V. J.: Die Anwendung des Trefferprinzips in der Strahlenbiologie, VEB G. Fischer Verlag, Jena 1972.
USMANOV, P. D., ABDULLAEV, Kh. A. & BOBODZHANOV, V. A., Izv. Akad. Nauk Tadzh. SSR, Otd. biol. Nauk 1970 [3 (40)]: 49, 1970.
USMANOV, P. D., KAS'YANENKO, A. G. & BATALOV, R. B., in: NASYROV, Yu. S. (ed.): Geneticheskie Aspekty Fotosinteza, p. 24, Donish, Dushanbe 1971.
USMANOV, P. D., LOGINOV, M. A., ISRAFILOVA, U., AKHMEDOV, A. Ya. & YUNUSOV, S. Yu., Arabidopsis Inform. Service 7: 34, 1970.
VATTI, K. V., VIKTOROVA, G. V. & BELYATSKAYA, O. Ya., Issled. Genet. (Leningrad) 3: 40, 1967.
VECHER, A. S.: Plastidy Rasteniï, Ikh Svoïstva, Sostav i Stroenie, Izd. Akad. Nauk BSSR, Minsk 1961.
WADDINGTON, G. H.: New Patterns in Genetics and Development, Columbia Univ. Press, New York – London 1962.
WEIBEL, E. R., Int. Rev. Cytol. 26: 235, 1969.
VON WETTSTEIN, D., J. Ultrastruct. Res. 3: 235, 1959.
VON WETTSTEIN, D., GUSTAFFSON, A. & EHRENBERG, L., Mutationsforschung und Züchtung, Arbeitsgemeinschaft Forsch. Landes Nordrhein – Westfalen 73: 7, 1959.
VON WETTSTEIN, D., HENNINGSEN, K. W., BOYNTON, J. E., KANNANGARA, G. C. & NIELSEN, O. F., in: BOARDMAN, N. K., LINNANE, A. W. & SMILLIE, R. M. (ed.): Autonomy and Biogenesis of Mitochondria and Chloroplasts, p. 205, North-Holland Publ. Comp., Amsterdam – London 1971.
VON WETTSTEIN, F., Z. Indukt. Abstamm. Vererbungslehre 33: 3, 1924.
YUNUSOV, S. Yu., KAS'YANENKO, A. G. & USMANOV, P. D., Izv. Akad. Nauk Tadzh. SSR, Otd. biol. Nauk 1969 [2 (35)]: 3, 1969a.
YUNUSOV, S. Yu., KAS'YANENKO, A. G. & USMANOV, P. D., Arabidopsis Inform. Service 6: 3, 1969b.
ZIMMER, K. G.: Studies on Quantitative Radiation Biology, Oliver and Boyd, Edinburgh – London 1961.

THE RELATION OF PIGMENT COMPOSITION IN ALGAE MUTANT CELLS TO THEIR RESISTANCE TO INHIBITORS OF PHOTOPHOSPHORYLATION

B. T. MUKHAMADIEV* & K. V. KVITKO**

Institute of Plant Physiology and Biophysics, Academy of Sciences of the Tajik S.S.R., Dushanbe, U.S.S.R.*, and Leningrad State University, Leningrad, U.S.S.R.**

Introduction

As reported recently by various authors, mutants of photosynthesizing organisms can be successfully used for studying the mechanism of photosynthesis (LEVINE 1968, BISHOP 1971). Selection of appropriate mutants is the most complicated part of such investigations. Altered photosynthetic capacity and defects in pigmentation are the usual criteria for selecting mutants. Pigment mutants have often been used in investigations on the biosynthesis of chlorophylls and carotenoids and their role in the formation of chloroplast structure. But little is known (TAGAWA et al. 1963, SAGROMSKY 1975) on the contribution of certain pigments, or their combinations, to partial reactions of photosynthesis. This question has been the topic of our studies.

Growth of Algae Mutants on Mineral Medium with DCMU

The material for analysis was taken from the collection of *Chlorella* pigment mutants in the Laboratory of Genetics and Cytogenetics of Microorganisms of the Biological Institute of Leningrad State University (KVITKO & BORSHCHEVSKAYA 1972). It was first essential to select a group of mutants which met all the requirements, i.e. capability for growth in autotrophic conditions, and plastid characteristics different from those of the wild strain.

For selection we used a mineral medium containing 1×10^{-4} M DCMU, to allow determination of the resistance-sensitivity of mutants to this well-known inhibitor of the non-cyclic electron flow during photosynthesis. The selected mutants differed from the wild strain in their capacity to grow on mineral medium with DCMU.

In addition to these mutants we used mutants of *Chlorella pyrenoidosa* kindly supplied by M. B. ALLEN from the U.S.A. (BENDIX & ALLEN 1962); 171 – initial strain, dark-green while growing both in the dark and in the light; g-9 – yellow-green in the dark, green in the light, phototrophic, no chlorophyll *b* in the dark; g-10 – yellow in the dark, yellow-green in the light, phototrophic, no chlorophylls in the dark, normal carotenoids in the dark and light; g-22 — photosynthesizing, yellow-green in the dark, pale-green in the light, chlorophyll amounts reduced; g-33 — light-yellow in the dark, pigment content reduced, sensitive to light.

Mutants g-9 and g-10 grew normally on the medium containing 1×10^{-4} M DCMU, whereas the wild strain 171 grew very slowly. Mutants g-22 and g-33 did not grow in similar conditions. Hence mutants g-9 and g-10 were defined as resistant to DCMU, and g-22 and g-33 as more sensitive to DCMU than the wild strain 171. The use of mutants, some resistant and some more sensitive to DCMU than the initial strain enabled us to distinguish the most DCMU sensitive unit in the photosynthetic electron transfer chain.

Effects of DCMU and Antimycin A on Photosynthetic Gas Exchange of Mutants

Another set of experiments was designed to study photosynthetic gas exchange (O_2 evolution and CO_2 absorption), cyclic and non-cyclic photophosphorylation, and effects of the inhibitors DCMU and antimycin A on these processes in vivo. O_2 evolution was estimated manometrically (SEMIKHATOVA & CHULANOVSKAYA 1965), and CO_2 absorption radiometrically (MAMUSHINA & DALETSKAYA 1968). The rate of photophosphorylation in vivo was measured indirectly as the rate of ^{14}C-incorporation into polysaccharides from phosphorylated sugars formed as a result of preceding photosynthesis in a $^{14}CO_2$ atmosphere. Using the specific inhibitors of non-cyclic (DCMU) and cyclic (antimycin A) photophosphorylation we were able to distinguish changes in photosystems 1 and 2 of the mutants and the wild strain (GLAGOLEVA et al. 1966).

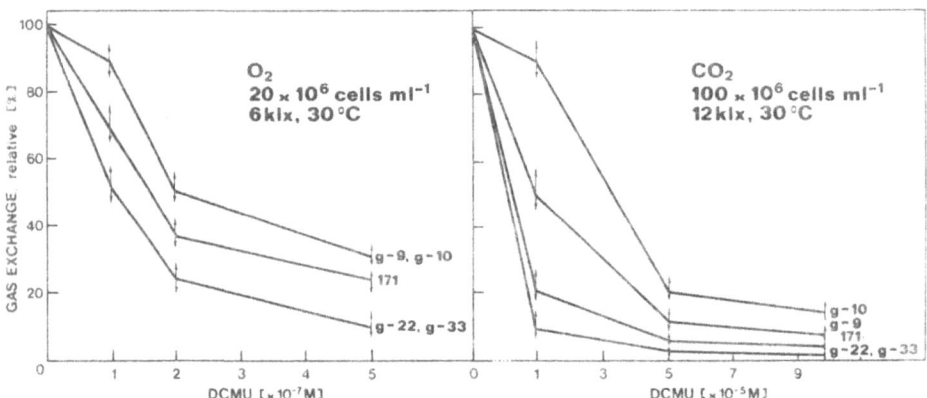

Fig. 1. Effects of DCMU on O_2 evolution (left) and CO_2 absorption (right) in *Chlorella* mutants.

The photosynthetic rates of the mutants were half that of the wild strain, i.e. 180-220 mg CO_2 $g^{-1}h^{-1}$ instead of 400 mg CO_2 $g^{-1}h^{-1}$.

The inhibiting effect of DCMU on photosynthetic oxygen and CO_2 exchange (Fig. 1) was detected in the wild and all mutant strains even at the lowest DCMU concentrations (1 x 10^{-7} M for O_2 and 1 x 10^{-5} M for CO_2). However, the level of the inhibition of O_2 evolution and CO_2 absorption in the mutants g-9 and g-10 was lower than in the wild strain; this effect is similar to the behavior of DCMU-resistant mutants with normal pigmentation (MUKHAMADIEV et al. 1971, MUKHAMADIEV & ZALENSKIĬ 1972). Mutants g-22 and g-33 were more sensitive to DCMU than the wild strain.

These DCMU effects differ from the effects of antimycin A on photosynthetic gas exchange (Fig. 2). Antimycin A caused a sharp inhibition of oxygen evolution and carbon dioxide absorption in the mutant g-10, while the wild strain was more sensitive to the antibiotics than the mutants g-9, g-22 and g-33.

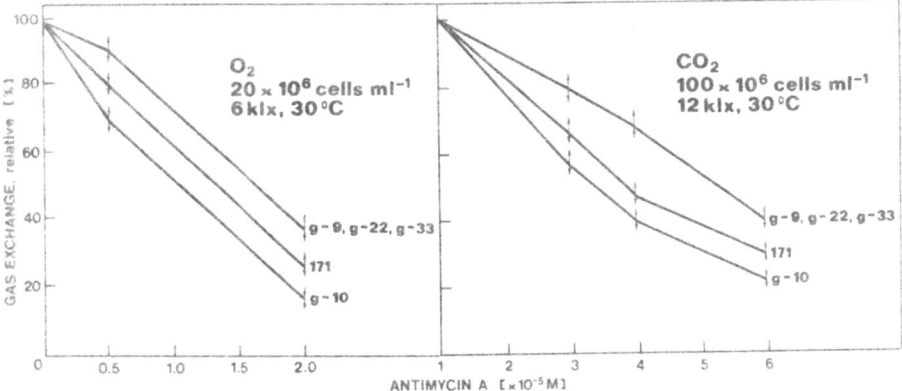

Fig. 2. Effects of antimycin A on O_2 evolution (left) and CO_2 absorption (right) in *Chlorella* mutants.

In summary, then in comparison to the wild strain 171, both sides of the photosynthetic gas exchange in the mutant g-9 were more resistant to DCMU and antimycin A. Mutant g-10 was more resistant only to DCMU, but more sensitive to antimycin A. Mutants g-22 and g-33 were resistent only to antimycin A, but more sensitive to DCMU.

Effects of DCMU and Antimycin A on Cyclic and Non-cyclic Photophosphorylation in Mutants

A similar pattern was found in experiments studying the resistance of cyclic and non-cyclic photophosphorylation in the mutants to DCMU and antimycin A. In the wild strain 171 both non-cyclic and cyclic photophosphorylation are sensitive to the inhibitors: DCMU (4×10^{-5}M) inhibited non-cyclic photophosphorylation by 60%, whereas the rate of inhibition of cyclic photophosphorylation by antimycin A (2×10^{-5} M) reached 90% (Fig. 3). Mutants g-9, g-22, and g-33 in similar conditions were resistant to antimycin A, and mutants g-9 and g-10 were resistant to DCMU. In comparison to the wild strain, mutants g-22 and g-33 were more sensitive to DCMU, while the mutant g-10 was very sensitive to antimycin A.

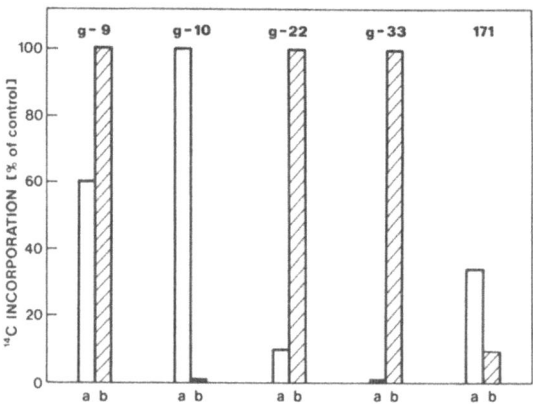

Fig. 3. Effects of inhibitors of photophosphorylation on the biosynthesis of polysaccharides (measured as ^{14}C incorporation) in *Chlorella* mutants. The control variant was not affected by inhibitors: a – DCMU (4×10^{-5} M); or b – antimycin A (2×10^{-5} M). Experimental conditions 450×10^6 cells per ml; 20 klx; 20 °C.

Chlorophyll *a* and *b* in Mutants

Among the green DCMU-resistant mutants of *Chlorella* (MUKHAMADIEV et al. 1971a, b, c) we found many forms with changed pigmentation (Table 1). As a rule, they had an increased content of chlorophyll *a*; their *a/b* ratio was either higher or lower than that of the wild strain. Mutants with changed responses to inhibitors of partial reactions of photosynthesis, selected from forms with reduced pigmentation, as well as green DCMU-resistant mutants showed a wide spectrum of changes in the ability to perform photosynthesis in the presence of inhibitors. There was no linear correlation between changes in cell pigment composition under dark (heterotrophic) cultivation and responses to inhibitors. This indicates that the two differences from the wild form, i.e. in sensitivity to inhibitors and in

Table 1. Chlorophyll a and b content [mg g^{-1} dry weight] in green mutants of *Chlorella* resistant to DCMU.

Strain		Chlorophyll			Ratio a/b
		a	b	$(a + b)$	
	82 (wild)	12.3±0.5	4.3±0.3	16.6±0.8	2.9±0.1
DCMU-resistant mutants	3$_D$	13.7±1.1	3.8±0.2	17.5±1.3	3.6±0.1
	3$_2$	17.0±0.9	6.5±0.2	23.5±1.1	2.6±0.1
	3$_9$	20.4±1.7	7.4±0.2	27.8±1.9	2.7±0.1
	B (wild)	14.8±1.2	6.1±0.3	20.9±1.5	2.4±0.1
DCMU-resistant mutants	B$_D$	33.8±3.4	11.5±1.2	45.3±4.6	3.0±0.1
	B$_1$	32.3±2.9	9.4±0.9	41.7±3.8	3.4±0.2
	B$_2$	28.2±2.2	11.4±1.0	39.6±3.2	2.5±0.1
	B$_3$	37.5±3.8	11.2±1.4	48.7±5.2	3.4±0.1
	B$_4$	24.8±2.4	8.2±0.8	33.0±3.2	3.1±0.2

the pigment system, have the same nature. Data on the frequency of these mutants (MUKHAMADIEV et al. 1971c) suggest that there are several loci responsible for DCMU-resistance. The changes in DCMU-resistant mutants are probably related to changes in the photosynthetic apparatus, because other possibilities — destruction of DCMU or changes in cell permeability — may be set aside, since DCMU evokes not a lower but rather a greater mutageneity in DCMU-resistant cells (MUKHAMADIEV et al. 1971a). Protein is the only component that can vary in the lipoproteid pigment complex which most probably changes during the pigment mutants' appearance or during mutation to DCMU-resistance. Proteins of chloroplast membranes determine both the capacity of cells to accumulate certain pigments in the corresponding complexes and changes in the responses of cells to inhibitors of photosynthesis. The predominant role of changes in the protein component of membranes is discussed in detail by GILLER et al. (1975) on the basis of experiments with synthetic complexes, and also by NASYROV et al. (1975) on mutants of green plants. Our assumption might be confirmed by analysis of proteins from chloroplast membranes in DCMU-resistant mutants of *Chlorella* (MUKHAMADIEV et al. 1971b, c).

A functional relationship between 'change in pigment composition of the plastid' and 'change in the response to inhibitors of cyclic and non-cyclic photophosphorylation' becomes evident on the basis of literature data on the participation of different chlorophylls in cyclic and non-cyclic photophosphorylation (distinguished by wavelength of irradiance). Some chlorophyll a forms in vivo, in contrast to chlorophyll b, can absorb far-red radiation (700 nm) in which non-cyclic photophosphorylation is inhibited almost completely (TAGAWA et al. 1963). Conversely, cyclic photophosphorylation is most effective in the far-red radiation absorbed only by chlorophyll a. This indicates that chlorophyll b is not required for cyclic photophosphorylation and actually may be absent in chloroplast particles where this process is localized (ARNON 1969). In fact, physical separation of the two photosystems revealed that PS-1 particles have a greatly reduced amount of chlorophyll *b* in comparison to whole chloroplasts and particles of PS-2 (HUZISIGE et al. 1969, LICHTENTHALER & TEVINI 1970, TEVINI & LICHTENTHALER 1970).

Summary

Analysis of photosynthetic gas exchange, partial reactions of photosynthesis and their relation to specific inhibitors of cyclic (antimycin A) and non-cyclic (DCMU) electron flow in a number of *Chlorella* mutants differing in pigment composition leads us to assume that different responses of mutants to the inhibitors DCMU and antimycin A may be explained by a changing ratio of photosynthetic electron transfer pathways (cyclic and non-cyclic) and changes in protein components of the pigment-lipoproteid complex.

Acknowledgement

We are indebted to Drs. O. V. ZALENSKIĬ, T. A. GLAGOLEVA and T. N. BORSHCHEVSKAYA for generous support during this study.

REFERENCES

ARNON, D. I., *Naturwissenschaften* 56: 295, 1969.
BENDIX, S. & ALLEN, M. B., *Arch. Mikrobiol.* 41: 115, 1962.
BISHOP, N. I., Annu. Rev. Biochem. 40: 197, 1971.
GILLER, Yu. E., VAKHIDOVA, L. R., ASOEVA, L. M., YUKHANANOVA, L. N., ABDULLAEVA, S. K., LIPKIND, B. I., KRASICHKOVA, G. V. & YUSUPOVA, G. A., in: NASYROV, Yu. S. & ŠESTÁK, Z. (ed.): Genetic Aspects of Photosynthesis, p. 271, Junk, The Hague 1975.
GLAGOLEVA, T. A., CHULANOVSKAYA, M. V. & ZALENSKIĬ, O. V., *Bot. Zh.* 51: 12, 1966.
HUZISIGE, H., USIYAMA, H., KIKUTI, T. & AZI, T., *Plant Cell Physiol.* 10: 441, 1969.
KVITKO, K. V. & BORSHCHEVSKAYA, T. N., in: KIRICHENKO, E. B. (ed.): Metody Issledovaniya Struktury Fotosinteticheskogo Apparata, p. 139, Pushchino-na-Oke 1972.
LEVINE, R. P., *Science* 162: 768, 1968.
LICHTENTHALER, H. K. & TEVINI, M., *Z. Pflanzenphysiol.* 62: 33, 1970.
MAMUSHINA, N. S. & DALETSKAYA, I. A., *Bot. Zh.* 53: 960, 1968.
MUKHAMADIEV, B. T., KVITKO, K. V. & ZALENSKIĬ, O. V., *Genetika* 7 (5): 36, 1971a.
MUKHAMADIEV, B. T. & ZALENSKIĬ, O. V., *Bot. Zh.* 57: 260, 1972.
MUKHAMADIEV, B. T., ZALENSKIĬ, O. V. & KVITKO, K. V., *Bot. Zh.* 56: 727, 1971b.
MUKHAMADIEV, B. T., ZALENSKIĬ, O. V. & KVITKO, K. V., *Genetika* 7 (4): 38, 1971c.
NASYROV, Yu. S., GILLER, Yu. E. & USMANOV, P. D., in: NASYROV, Yu. S. & ŠESTÁK, Z. (ed.): Genetic Aspects of Photosynthesis, p. 133, Junk, The Hague, 1975.
SAGROMSKY, H., in: NASYROV, Yu. S. & ŠESTÁK, Z. (ed.): Genetic Aspects of Photosynthesis, p. 247, Junk, The Hague 1975.
SEMIKHATOVA, O. A. & CHULANOVSKAYA, M. V.: Manometricheskie Metody Izucheniya Dykhaniya i Fotosinteza Rasteniĭ, Nauka, Moskva – Leningrad 1965.
TEVINI, M. & LICHTENTHALER, H. K., *Z. Pflanzenphysiol.* 62: 17, 1970.
TAGAWA, K., TSUJIMOTO, H. Y. & ARNON, D. I., *Nature* 199: 1247, 1963.

NATURE OF CHLOROPHYLL CHIMERAS IN THE M_1 OF ARABIDOPSIS BY MEANS OF TISSUE CULTURE. I. PRODUCTION OF REGENERANTS FROM CHLOROPHYLL-DEFICIENT TISSUES

SVETLANA I. DEMCHENKO*, V. A. AVETISOV* & RAISA G. BUTENKO**

Institute of Chemical Physics* and K.A. Timiryazev Institute of Plant Physiology**, Academy of Sciences of the U.S.S.R., Moscow, U.S.S.R.

As a result of the action of certain mutagens on seeds, plants developing from these seeds form leaves with coloured or colourless spots. The frequency of leaf spotting in the M_1 of such plants — called chlorophyll chimeras — and its correlation with the mutation frequency in the M_2 is often studied (MÜLLER 1964, ZACHARIAS & EHRENBERG 1962, GICHNER & VELEMÍNSKÝ 1966). Nevertheless, the nature and mechanism of emergence of spotting are not clear: it may be connected with various chromosomal changes (RANA 1964, ZACHARIAS & EHRENBERG 1962), plastid mutations (RÖBBELEN 1965, BELETSKIĬ et al. 1969) or morphoses (BELETSKIĬ et al. 1969, ANDROSHCHUK & MAREKHA 1968). We have suggested that chlorophyll-deficient leaf tissues are the progeny of a cell which carried a genic chlorophyll mutation and was then transferred into a homozygous state by induced somatic recombination.

In order to examine the above alternatives it is necessary to determine the genotype, or more broadly, the idiotype of the chlorophyll-deficient tissue. But this requires a genetic analysis impracticable in somatic cells. Nevertheless, it may be suggested that dedifferentiation of cells of chlorophyll-deficient mesophyll followed by callus formation and stem organogenesis could be possible employing the method of tissue culture. Gene analysis then becomes feasible as soon as the regenerant begins to flower and yield pods. The use of the technique of growing haploids from regenerant anthers permits reduction of the genotype determination, making it more rigorous without recourse to crosses.

The present work is the first step in this direction. The experiments have been carried out on controlled green regions of leaves, and our first attempts to grow regenerants from chlorophyll-deficient plant tissue in M_1 have been successful.

Material and Methods

Pieces of normal green and chlorophyll-deficient tissue of 2-5 mm in size were excised from the leaves of control green plants and from leaf spot of *Arabidopsis thaliana* HEYNH. race Enkheim in M_1 grown in a sterile test tube culture from seed treated with nitrosomethylurea solution. The pieces were transferred into test tubes with agarized media of GAMBORG & EVELEIGH (1968) and MURA-SHIGE-SKOOG (cf. BUTENKO 1964) under sterile conditions. Additions to Gamborg's medium were sucrose (30 g l^{-1}), casein hydrolysate (400 mg l^{-1}), inosite (80 mg l^{-1}) and a large set of vitamins according to STABA (0.5 mg l^{-1} Gamborg's medium I and 10 mg l^{-1} Gamborg's medium II), kinetin (1 mg/l^{-1}) and indolylacetic acid (IAA) (1 mg l^{-1}). Additions to Murashige-Skoog's medium were sucrose (30 mg l^{-1}), kinetin (1 mg l^{-1}), α-naphthylacetic acid (NAA) (1 mg l^{-1} of Murashige-Skoog's medium I and 0.6 mg l^{-1} of Murashige-Skoog's medium II), inosite (80 mg l^{-1}) and vitamin B (0.4 mg l^{-1}).

To induce callus formation tubes with the explanted tissues were kept in a thermostated room in the dark at 25°C and 70% air humidity. The explants in which callus formation was accompanied by organogenesis were put in the light. Illumination of 8.8 klx was provided by luminescent lamps LB-80. The emerging stem buds were planted on Gamborg's medium without kinetin but with a varied content of NAA or ferulic acid.

Callus formation, capability of root and stem organogenesis, the phenotype of plant-regenerants and their capability of root and pod production were studied in these cultures. The intensity of callus formation and root formation on the callus was classified visually using a 6-point scale.

Results and Discussion

The pieces of control green leaves explanted on the nutritional media developed in different ways. The following types were noted: 1. The leaf increased in size a little or displayed no signs of growth at all, became colourless and died without callus formation. 2. The leaf increased in size. Because of irregular growth of the lower and upper parts of the leaf it was somewhat twisted along the edges, gradually became colourless and after 8-14 days (depending on the composition of the medium) small parts of callus emerged, began to develop actively and in two months reached 2-8 mm of height. No organogenesis was observed. 3. Shortly after callus formation extensive root formation began; a month after leaf explantation the roots attained 1-10 mm. There were no stem buds. 4. Growth of callus and roots was insignificant while regeneration of stem buds from callus tissue was active.

Development of explants was largely dependent upon the composition of the nutritional medium on which it was grown. The first and the second types were found on both media, but very seldom. Most explants developed callus followed by root and stem organogenesis. In the control, callus grew much more actively on Murashige-Skoog's medium (Table 1). A combination of kinetin with NAA in Murashige-Skoog's medium seems to have been more stimulating for callus formation than a combination of kinetin with IAA in Gamborg's medium.

Control calluses, developed on Gamborg's and Murashige-Skoog's media, had clear-cut distinctions. On Gamborg's media callus was very dense, 'dry',

Table 1. Characteristics of callus and root formation in the 0 passage of *Arabidopsis thaliana* HEYNH. (L.)

Variant	Number of test tubes	2 weeks		Number of test tubes	5 weeks		7 weeks		10 weeks	
		Test tubes with callus [%]	Test tubes with roots [%]		Test tubes with callus [%]	Test tubes with roots [%]	Test tubes with callus [%]	Test tubes with roots [%]	Test tubes with callus [%]	Test tubes with roots [%]
Control (Gamborg's medium I)	27	74.0± ±8.4 0.85p	0	26	96.1± ±3.8 2.54p	88.0± ±6.5 1.38p	96.1± ±3.8 2.96p	88.0± ±6.5 1.45p	96.1± ±3.8 3.24p	88.0± ±6.5 1.62p
Control (Murashige-Skoog's medium I)		96.6± ±3.3 2.96p	93.1± ±4.7 2.70p	30	100.0	96.6± ±3.3 4.40p	100.0	96.6± ±3.3 4.82p	100.0	96.6± ±3.3 5.00p
'Spots' (Gamborg's medium I)	20	50.0± ±11.2 0.25p	0	17	88.2± ±7.8 2.06p	46.6± ±12.9 0.60p			88.2± ±7.8 2.78p	46.6± ±12.9 0.85p
'Spots' (Murashige-Skoog's medium I)	21	80.9± ±8.7 1.63p	82.3± ±9.1 2.12p	20	90.0± ±6.7 3.88p				90.0± ±6.7 4.50p	94.4± ±5.4 4.88p

p – point
'Spots' – chlorophyll-deficient tissue regions on M_1 plants after N-methyl-N-nitrosourea treatment.

knobby, with numerous 'centres of growth', mostly non-uniform in colour, consisting of green, yellow and white regions of different intensity of pigmentation. On Murashige-Skoog's medium callus was less dense, contained some water, was bright and had a more or less even surface, largely of yellowish-white colour.

Regeneration on Gamborg's and Murashige-Skoog's media was different. The presence of a powerful root-forming agent (NAA) in Murashige-Skoog's medium caused extensive root formation (Table 1) and blocked stem regeneration on control calluses (Fig. 1). In contrast to root organogenesis, stem regeneration was very active on Gamborg's medium (Fig. 1). During the first ten weeks, induction of stem regenerants in the control was much slower on Gamborg's medium II than on I. Seemingly, the vitamin concentration in Gamborg's II was too high. At later stages of callus growth (10-14 weeks), when the stock of vitamins was exhausted, its regenerative activity approximated that on Gamborg's I (Fig. 1).

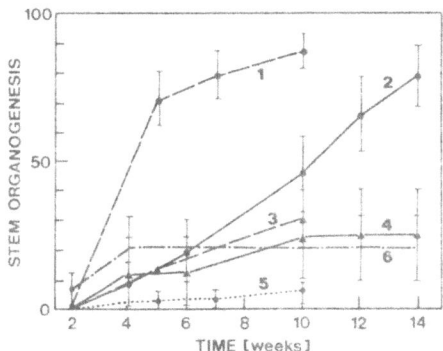

Fig. 1. Stem organogenesis in tissue cultures: control on Gamborg's medium I (1) and II (2), 'spots' on Gamborg's medium I (3) and II (4), and control on Murashige-Skoog's medium I (5) and II (6).

It should be emphasized that, in the control, regeneration of stem buds was observed only in the 0 passage, i.e. on callus in the presence of leaf tissue. There was not a single case of emergence of stem buds in the subsequent four passages. Possibly, some factor supplied by the leaf was depleted.

The root systems of control regenerants were underdeveloped as compared to seed plants. Addition of NAA to the medium (0.4 mg l^{-1}) stimulated rooting of stem buds more than observed following addition of ferulic acid (2 mg l^{-1}).

The regenerants' morphogenesis was markedly different from that of seed plants grown in test tube culture. Control plant-regenerants were much smaller than seed plants; sometimes they had no rosette leaves and their stem was appreciably thinner, bearing only one to six flowers compared to the seed plants which had up to 30 flowers.

In our experiments over 80% of the control regenerants were fertile. Morphologically, flowers of the fertile regenerants showed little, if any, difference from those of seed plants, except for their smaller size. The availability of roots was not mandatory for seed setting. Sterile control regenerant's anthers did not contain pollen. In most cases ovules were normal. At the same time there were cases of abnormal micro- and macro-sporogenesis. Regenerants had smaller pods than seed plants, or no pods at all.

During the length of a vegetative season the control plants — regenerants did not differ from the seed plants under the conditions of test tube culture.

Chlorophyll-deficient explants differed from normal green ones in the intensity of callus and root formation, as well as in the formation of regenerants and their morphogenesis (Table 2).

Table 2. Characterization of the morphogenesis of *Arabidopsis thaliana* in tissue culture and in test tube culture

Stage of the morphogenesis	Control	'Spots'	Seed plants
1. Formation of callus	++	+	
2. Induction of roots in the callus	++	+	
3. Growth of roots in the callus	++	+	
4. Induction of stem buds in the callus	++	+	
5. Growth of stem	++	+	+++
6. Induction of roots on the plant	++	+	+++
7. Growth of roots on the plant	++	+	+++
8. Formation of flowers	++	+	+++
9. Ovules	++	±	+++
10. Pollen	±	−	+++

The above mentioned distinctions of the processes of callus formation and root and stem regeneration on Gamborg's and Murashige-Skoog's media regarding the control explants held as well for chlorophyll-deficient explants with the only difference that in the latter case all the processes were inhibited to a certain extent (Table 1).

In their morphology and colour, calluses arising from defective tissues ('defective calluses') were not different from control calluses on the corresponding media. An exception was the calluses formed from the *albino* tissue; on Gamborg's medium these always remained yellowish white, even when the kinetin concentration in the medium was increased to 3 mg l^{-1}.

Chlorophyll deficiency of explants led to a sharp reduction of stem bud regeneration on the callus versus the control (Fig. 1)

The range of morphological variability of the regenerants arising from 'defective' calluses was significantly wider than in the control. In the control only green regenerants occurred with one exception where leaf spot appeared on Gamborg's medium in the 0 passage 10 weeks after explantation. The colour of regenerants arising from 'defective' calluses depended upon the colour of the explant tissue. *Albino* explants generally yielded chlorophyll-deficient regenerants. When variegated tissue was used for explantation, regenerants of both colours emerged.

Segregation for colour in the plant-regenerants was observed during stem regeneration from calluses formed on yellow-green and light green explants. The former yielded green and yellow-green, the latter green and light green regenerants. There was one case when both *albino* and green plant-regenerants occurred on the callus formed from an *albino* explant.

Rooting of regenerants induced on 'defective' callus was less successful than in the control.

The morphology of flowers emerging on the 'defective' regenerants was essentially different from the control. They had altered sepals and petals. Changes were especially marked in the stamina: apparently anthers did not contain pollen. Occasionally, apparently normal seed buds occurred, but pods seldom developed.

All attempts to produce seeds on these regenerants, by addition to the medium of sucrose (up to 4%) or by growing them on medium 199 containing all necessary vitamins, amino acids etc. with addition of sugar and mineral salts, failed.

Our experiments have demonstrated that tissue culture permits induction of all the processes necessary for the recovery of fertile plant-regenerants from *Arabidopsis* green leaf tissue (dedifferentiation of mesophyll cells, callus formation, induction of buds on the callus, rooting of buds and their development resulting in the appearance of flowering and pod-yielding regenerants). We have also proved the possibility of production, from chlorophyll-deficient leaf tissue, of plants capable of undergoing all stages of morphogenesis except normal sporogenesis.

It is unlikely that the lack of normal sporogenesis is a result of a deficiency in some specific substance. The reason for disturbance of sporogenesis may be connected with the presence in the medium of sufficiently high concentrations of biologically active substances, such as NAA, kinetin, etc. causing sterility in a part of the control regenerants. However, if the sterility of 'defective' regenerants was caused by this alone, then some fraction of the regenerants should have been fertile. Besides, the influence of these substances was considerably lessened by early transplantation of buds for rooting in medium with a very low content of active substances.

An idea of genetically-conditioned sterility of defective regenerants seems more plausible to us. Nevertheless, we do not exclude completely the possibility of induction of sporogenesis by varying the conditions of cultivation.

Attention should be paid to the fact that the green regenerants induced on both the 'defective' callus and the chlorophyll-deficient regenerants were sterile.

The emergence of green regenerants on 'defective' callus may be due to different reasons (i.e. admixture of green cells, occurrence of cells with mixed plastids, genetic reversion).

Further investigation will be aimed at explaining the reason of 'segregation' for colour in regenerants and determining the genotype of chlorophyll chimeras tissue in M_1.

Summary

The use of tissue cultures for investigating the origin of chlorophyll chimeras in M_1 after mutagenic treatment is proposed. The isolated parts of the leaf tissue with chlorophyll defects are cultivated on various nutrient media (after GAMBORG & EVELEIGH or MURASHIGE-SKOOG) in order to obtain callus and plant-regenerants originating from the defective cells.

REFERENCES

ANDROSHCHUK, A. F. & MAREKHA, L. N., *Tsitologiya Genet.* 2: 30, 1968.
BELETSKII, Yu. D., RASORITELEVA, E. K. & KUL'PINA, A. I., *Tsitologiya Genet.* 3: 421, 1969.
BUTENKO, R. G.: Kul'tura Izolirovannykh Tkaneï i Fiziologiya Morfogeneza Rastenii, Moskva 1964.
GAMBORG, O. & EVELEIGH, D., *Can. J. Biochem.* 46: 417, 1968.
GICHNER, T. & VELEMÍNSKÝ, J., *Genet. Šlechtění* 2: 81, 1966.
MÜLLER, A. J., *Züchter* 34: 102, 1964.
RANA, R., *Naturwissenschaften* 51: 642, 1964.
RÖBBELEN, G., in: RÖBBELEN, G. (ed.): Arabidopsis Research, Rep. Int. Symp. Göttingen 1965, p. 100, Arabidopsis Inform. Serv., Inst. Pflanzenbau Pflanzenzücht., Univ. Göttingen 1965.
ZACHARIAS, I. M. & EHRENBERG, L., *Hereditas* 48: 284, 1962.

GENETIC ANALYSIS OF LIGHT-SENSITIVE MUTANTS OF CHLAMYDOMONAS REINHARDI

ANNA V. STOLBOVA

The State Leningrad University, The Department of Genetics and Selection, Leningrad, U.S.S.R.

Mutants unable to grow in darkness, i.e. light-sensitive mutants, are produced in many different organisms, from the photosynthetic bacteria to the higher plants.

Investigations of light-sensitive mutants by many authors (CLAES 1954, WALLACE & SCHWARTING 1954, STANIER 1958, ANDERSON & ROBERTSON 1960) showed that bleaching, followed by chloroplast destruction, is a secondary result of biochemical disturbance of a carotenoid-forming process. These effects disappear almost completely if illumination is made in anaerobic conditions.

These facts suggest that carotenoids protect chloroplasts against photodestruction. According to KRINSKY (1966) it is highly probable that the epoxy-cycle of

Table 1. Pigments of light-sensitive mutants of *Chlamydomonas reinhardi*.

Type	Colour in darkness	Pigments						
		chlorophyll a	chlorophyll b	carotene	lutein	violaxanthin	trollein	neoxanthin
Wild type	green	++	++	++	++	++	++	++
N-13	green	++	++	++	++	++	++	++
N-174	green	++	++	++	++	++	++	tr.
N-179	green	++	++	++	++	++	++	++
N-19	orange	--	--	++	++	++	++	++
N-122	orange	--	--	++	+	+	+	--
N-172	yellow	--		++	++	++	++	--
N-135	apple-green	+	+	+	+	+	+	+
N-149	pale green	+	+	tr.	--	--	--	--
N-164	apple-green	+	tr.		--	--	--	--
N-154	pale-green	+	+	+	+	+	+	+
N-171	apple-green	+	+	--	tr.	--	--	+
N-173	apple-green	+	+	--	--	--	--	--
N-178	apple-green	+	+		--	--	--	--

tr. — traces

carotenoid pigments represents the very mechanism by which excess oxygen, contained in a chloroplast and capable of initiating photodestruction of chlorophyll and other light-absorbing porphyrins, is neutralized.

Our data obtained with light-sensitive mutants of *Chlamydomonas* are considered with respect to the ideas described above. With the help of chemical mutagens three main phenotypical classes of light-sensitive mutants were induced (STOLBOVA 1971) from the wild strain of (+) and (−) mating types (mt) of *Chlamydomonas reinhardi* (Table 1). The first class included three green forms having the same pigments as the wild type. The second class consisted of yellow chlorophyll-less mutants having some type of carotenoid disturbance. In general, these variations affected the epoxy-carotenoids trollein and neoxanthin. The third class included seven apple-green mutants with quite different pigment disturbances. Two of these had all the photosynthetic pigments of the wild type, only in a lesser quantity. The others had almost none of the carotenoids and much less chlorophylls than the wild type.

In addition to paper chromatographic studies of pigment composition we studied the ability of these mutants to grow and photosynthesize in anaerobic conditions (Table 2). For these experiments a sealed chamber with the specified atmosphere was used. In the first experiment (pressure in the chamber 0.9 bar pure nitrogen, medium with addition of acetate as a carbon source) the ability for heterotrophic growth in anaerobic conditions was studied.

In general the group of chlorophyll-less mutants alone responded to variations of the conditions. All three forms acquired the ability to grow in light in anaerobic conditions. Only one mutant of the apple-green group responded in a similar way though its response was not so distinct.

In the second experiment (pressure in the chamber 0.9 bar, 0.3% CO_2 in nitrogen, medium without acetate) only the mutants able to photosynthesize in anaerobic conditions grew. Of all forms studied, only two mutants of the first-class displayed this ability: their pigment composition did not differ from that of the wild form. Hence the ability of the green light-sensitive forms to photosynthesize in anaerobic conditions indicates that neither their pigment system nor their energy transfer chain is disturbed. Photosynthesis must stimulate the recovery of the light-protective barrier in the absence of oxygen, because these mutants cannot grow in heterotrophic conditions.

The light-sensitivity of the second group − the chlorophyll-less forms − must be due to disturbances of the carotenoid pigments, viz. to the disappearance of the epoxy-carotenoids, and thus to disturbance of the epoxy-cycle. These mutants respond to anaerobic conditions in the same way as the mutants whose light-sensitivity is interpreted as the disturbance of a carotenoid-forming process (CLAES 1954, WALLACE & SCHWARTING 1954, STANIER 1958, ANDERSON & ROBERTSON 1960).

The third group of mutants displayed almost no response to variations of growth conditions. The light-sensitivity of these forms must be due to a more severe disturbance of the photo-protective mechanism so that even exclusion of

Table 2. Growth of mutants (+ or −) in light and anaerobic conditions.

Growth conditions	Strain														
	Wild type	N-13	N-174	N-179	N-19	N-122	N-172	N-135	N-149	N-154	N-164	N-171	N-173	N-178
N_2, medium with acetate	+	−	−	−	+	+	+	−	−	+	−	−	−	−
$CO_2 + N_2$, medium without acetate	+	−	+	+	−	−	−	−	−	−	−	−	−	−

oxygen does not give protection. This disturbance may touch upon the epoxy-cycle as well, since most mutants of this group lack carotenoids and recover their ability to grow in heterotrophic conditions in the absence of oxygen. Crosses between mutants from each group were undertaken to evaluate the number of genes which determine the light-sensitivity character (Table 3). Hereditary independence of the two mutations was determined by the appearance of three types of tetrads: parental (P), nonparental ditypes (N), and tetratype (T). The appearance in crossing of only the parental ditype tetrads indicated that these mutations were allelic. Two alleles of one gene were found in the group of green mutants because there appeared no recombinations of the wild type in the crossing between mutants N-13 and N-179. In the crossing of mutants N-13 and N-174 the amount of parental ditype tetrads was much greater than those of other tetrad types, which suggested linkage of two genes. Since the mutant N-179 was capable of photosynthesis, though in anaerobic conditions, the allelic mutation N-13 probably did not disturb the photosynthetic apparatus, but rather affected the cell light-protective mechanism that functioned only during photosynthesis.

Table 3. Segregating character of crosses between the light-sensitive forms.

Parents		Tetrads			p corresponding to the ratio 1:1:4
mt(+)	mt(−)	P	N	T	
1st group − green					
N-174	N-13	34	0	17	< 0.001
N-179	N-13	112	0	0	< 0.001
2nd group − yellow					
N-172	N-19	60	72	112	< 0.001
N-172	N-122	37	22	100	> 0.95
3rd group − apple-green					
N-149	N-164	1140	0	0	< 0.001
N-154	N-135	309	0	0	< 0.001
N-149	N-135	1120	0	0	< 0.001
N-171	N-135	1314	0	0	< 0.001
N-173	N-135	156	0	0	< 0.001
N-178	N-135	509	0	0	< 0.001

In crosses between chlorophyll-less mutants neither alleles of one gene nor linkage between two genes were found. If the epoxy-cycle of these mutants is really disturbed, these disturbances may obviously occur at the level of synthesis of epoxy-carotenoids, at the level of regulation of their transformation, at the level of their structural organization, etc., and hence the epoxy-cycle is determined by more than one gene. The crosses between apple-green mutants revealed multiple allelism within this mutation group. The mutations that had different phenotypical expression, i.e. pigment mutations, affected one and the same, apparently complicated gene. The qualitative and quantitative variations of pigment composition must have resulted from some disturbance, general to the whole series of mutants, the origin of which is unknown at present. Consequently, the character of light-sensitivity has a complicated genetic determination.

In some crosses between light-sensitive mutants, a small number of tetrads appeared that did not segregate according to Mendelian inheritance. A hybridological analysis of separate clones from these tetrads showed they were heterozygotes for the parental factors, but for the nuclear genes a normal Mendelian segregation took place in the same crossings. Consequently some pigment mutations might have passed through meiosis without segregation. Of all the *Chlamydomonas* genes studied, only those genes considered extranuclear (resistance to some antibiotics) possessed a similar property (GILLHAM 1969).

A more detailed investigation of this effect was carried out on crosses between two apple-green mutants of the third group.

The crossing scheme is as follows (the light-sensitivity mutations were designated by the symbol *lts*):

$$\frac{lts_{1-2} \times +}{\downarrow}$$

$$\frac{lts_{1-2} \times lts_{1-1}}{+ \quad \downarrow}$$

$$\text{zygote} \quad \frac{lts_{1-1}}{lts_{1-2}}$$
$$+$$
$$\swarrow \searrow$$

(1) $\dfrac{lts_{1-2}}{+} \quad \dfrac{lts_{1-2}}{+} \quad \dfrac{lts_{1-1}}{} \quad \dfrac{lts_{1-1}}{}$ (5) $+ \quad lts_{1-1} \quad \dfrac{lts_{1-2}}{+} \quad \dfrac{lts_{1-1}}{lts_{1-2}}$

(2) $\dfrac{lts_{1-1}}{} \quad \dfrac{lts_{1-1}}{} \quad \dfrac{lts_{1-2}}{+} \quad \dfrac{lts_{1-2}}{+}$ (6) $+ \quad lts_{1-2} \quad \dfrac{lts_{1-1}}{+} \quad \dfrac{lts_{1-1}}{lts_{1-2}}$

(3) $\dfrac{lts_{1-2}}{} \quad \dfrac{lts_{1-1}}{} \quad \dfrac{lts_{1-1}}{+} \quad \dfrac{lts_{1-2}}{+}$ (7) $+ \quad + \quad \dfrac{lts_{1-1}}{lts_{1-1}} \quad \dfrac{lts_{1-2}}{lts_{1-2}}$

(4) $+ \quad + \quad \dfrac{lts_{1-1}}{lts_{1-2}} \quad \dfrac{lts_{1-1}}{lts_{1-2}}$ (8) $lts_{1-2} \quad lts_{1-2} \quad \dfrac{lts_{1-1}}{lts_{1-1}} \quad \dfrac{+}{+}$

(9) $lts_{1-1} \quad lts_{1-1} \quad \dfrac{lts_{1-2}}{lts_{1-2}} \quad \dfrac{+}{+}$

The cross between mutant N-149 that carried mutation lts_{1-2} and the wild type that had a normal allele yielded both tetrads that contained only green clones and tetrads that segregated in a light-sensitivity character. The assumed heterozygote was crossed with mutant N-135 that carried the allelic mutation lts_{1-1}. Both mutants were easily discerned visually, since N-149 was pale-green while N-135 was apple-green. Nine genotypical classes that distributed into five types of tetrads were expected of such a heterozygote, taking into consideration all probable cases of marker combinations.

The frequencies of the appearances of these five types of tetrads coincide with those calculated theoretically with the probability $p > 0.80$ (Table 4). Hence the data are in good agreement with the scheme based on the behavior of the unlinked mutations in meiosis.

Table 4. Comparison of the theoretically calculated segregation of a heterozygote with the data obtained experimentally.

Segregation by a phenotype	Theoretical frequencies of the tetrads appearance	Experimental frequencies	p (comparison by χ^2 method)
2 pg : 2 g	2	4	
2 ag : 2 g	2	5	$0.95 > p > 0.80$
2 ag : 1 pg : 1 g	1	1	
2 pg : 1 ag : 1 g	1	2	
2 g : 1 ag : 1 pg	3	9	

g — (as the wild type) green
ag — (as N-135) apple-green
pg — (as N-149) pale-green

Thus, the following peculiarities of the behavior of light-sensitive mutations in crossings can be pointed out: 1. They display a Mendelian type of heredity. 2. Some of them are capable of partial preservation of the heterozygote state in their passage through meiosis and following mitosis. It is this peculiarity of the pigment mutation's behaviour that enables to ascribe them to the class of extranuclear, and probably plastid mutations.

It may be supposed that the distribution of genetic factors available in the plastids must be realized by a process similar to nuclear meiosis. However, this process is not very regular, therefore the inadequate distribution of the determinants is observed rather frequently.

Summary

Using various chemical mutagens three phenotypically distinct groups of *Chlamydomonas* mutants, with lethal effects at illuminance of 2 klx, were obtained. The first group of mutants (green forms) was characterized by a pigment pattern similar to the wild type; the second group represented mutants lacking chlorophyll (yellow forms); the third group included mutants with changes in carotenoid pigments (apple-green forms).

Analysis of hybrids of these light-sensitive mutants revealed a series of multiple alleles in the group of apple-green forms. Analysis of tetrads in crossings of the light-sensitive forms indicated that, in some cases, the mutation of light-sensitivity may go through meiosis without segregation.

Acknowledgement

The author is deeply thankful to H. P. STOLBOVA for her kind help in carrying out the experiments.

REFERENCES

ANDERSON, I. C. & ROBERTSON, D. S., *Plant Physiol.* 35: 531, 1960.
CLAES, H., *Z. Naturforsch.* 9b: 461, 1954.
GILLHAM, N. W., *Amer. Naturalist* 103: 932, 1969.
KRINSKY, N. J., in: GOODWIN, T. W. (ed.): Biochemistry of Chloroplasts, Vol.1, p.423, Academic Press, London-New York 1966.
STANIER, R. Y., in: The Photochemical Apparatus, its Structure and Function, *Brookhaven Symp. Biol.* 11: 43, 1958.
STOLBOVA, A. V., *Genetika* 7: 90, 1971.
WALLACE, R. H. & SCHWARTING, A. E., *Plant Physiol.* 29: 431, 1954.

MUTATION ANALYSIS AS A METHOD OF STUDYING GENOTYPE STRUCTURE OF GREEN ALGAE

K. V. KVITKO, V. V. TUGARINOV, PHAM THAN HO, A. S. CHUNAEV, E. E. TEMPER* & B. T. MUKHAMADIEV**

Biological Institute and Genetics Department of Leningrad State University, Leningrad, U.S.S.R.

Introduction

Green unicellular algae studied in our laboratory (*Chlorella, Scenedesmus* and *Chlamydomonas*) are capable of growing in heterotrophic, autotrophic and mixotrophic conditions. Their advantages for experimental study are the same as those of *Procaryota;* in addition these algae have a typical eucaryotic cell structure analogous to higher plants cells. This enables us to consider the unicellular green algae as model objects for experimental studies of genetic, biological and molecular aspects of photosynthesis.

The main task of our study is analysis of the genotype structure. There can be two approaches: genotype structure of eucaryotic cells as a functionally indivisible, unique system, or as a morphologically and physiologically divided one. The latter case gives priority to the study of nuclear-organelle relationships, the constituents of the system being: genome (nucleus), subdivided into chromosomes, loci, sites and so on up to nucleotides, plastome, in which one can distinguish the groups of linkage, loci, etc., and chondriome with similar subunits. It is probable that some other genetically functional organelles are incorporated into the genotype of green algae, and hence it is difficult to decide which unit is the elementary one.

The approach to analysis green algae genotype structure as a functional system was based on the postulation that its units are genes irrespective of their localization. For a geneticist it would be ideal to combine these two aspects, but the logics of genetic analysis methods gives priority to the study of genotype as a functionally united system of genes. A convenient scheme of realization of genetic information in the cell and the role of replication, transcription and translation processes has been suggested by INGE-VECHTOMOV (1969).

The paths from genotype to phenotype are controlled by two groups of genes (Fig. 1): first, that which controls the structure of proteins involved in the biosynthesis of the low molecular weight moiety of cell components, and the second

---

* Present address: *Dept. of Microbiol., Medical Institute, Blagoveshchensk, U.S.S.R.*
**Present address: *Inst. of Plant Physiology and Biophysics, Dushanbe, U.S.S.R.*

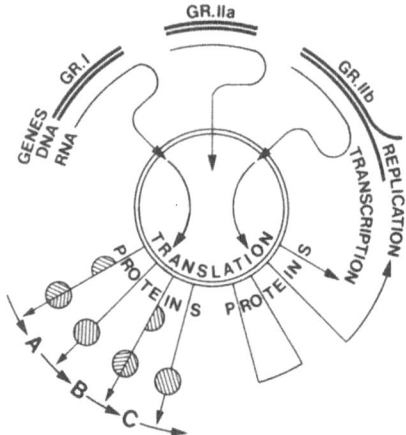

Fig. 1. Scheme of genetic control of step-wise processes (synthesis of metabolites A, B, C) and the matrix processes (DNA replication, transcription, translation) (according to INGE-VECHTOMOV 1969).

group which controls the matrix processes; group IIa exerts control over the structure of RNA-components of the translation apparatus, group IIb determines the structure of proteins which perform the replication, transcription and translation.

Mutations of genes of the first group usually block one definite function, and under particular conditions this block may be compensated by addition of a metabolite, the synthesis of which was reduced by the mutation. Hence one can maintain cells containing a mutation of the same gene to get progeny called the mutants. The viability of the studied mutants depends upon the experimenter's art to control the low molecular weight moieties of cells.

Mutations of genes of the second group as a rule change many functions because of the pleiotropic action of these genes. The phenotypic expression of such mutations is indefinable and the harmful effect of such mutations leads to degeneration or dying of the progeny. Consequently, genes of the first group are more available for experimental study, and on the basis of the collection of forms with such allelic mutations (so-called 'gene-character' mutation system) one can develop the 'gene-enzyme' system. This system enables genetic analysis of a particular gene and of the genes of Group II connected with transcription and translation processes, and capable of repairing the first (direct) mutation in genes of group I. In such situations the deleterious effect of a mutation in a gene of group II which controls, for example, some translation step, could be balanced by suppression of the metabolism disturbances after translation repair of the first mutation. Thus one may distinguish phenotypes of alleles of a gene from group II and maintain the mutants with two separately harmful mutations, because the first, conditionally lethal, mutation in a gene of group II transforms fitness of the potentially deleterious mutation of a gene with high pleiotropy to a positive one.

The logics of analysis of functional structure of green algae genotypes requires: (a) construction of mutational systems 'gene-character' and later on 'gene-polypeptide', (b) revelation of mutations in genes of group II according to their effects on mutations in genes of group I studied earlier. All this usually takes place in the course of genetic analysis, its elementary but essential step being mutation analysis. We do insist on the potential equality of the mutation and hybridological methods of genotype dissection, especially if one takes into consideration the importance of mutation analysis for the agamic species (ZAKHAROV & KVITKO 1967; KVITKO 1972).

Several characters are advantageous for studying the genetics of photosynthesis: colour of colonies and pigment contents; cell reactions to light and dark, to inhibitors of partial photosynthetic reactions and to inhibitors of protein synthesis in chloroplasts. We began to study all these characteristics, attempting to find those most convenient for the first step of analysis − the construction of a mutation system 'gene-polypeptide'.

Pigment Losses in Algae Mutants

Plastid pigments have been studied in our laboratory with mutants of *Chlorella* and *Chlamydomonas* (STOLBOVA 1971a, b; STOLBOVA & KHROPOVA 1971) and *Scenedesmus* (TEMPER & KVITKO 1971). 75 mutant forms of *Chlorella vulgaris* strain B were classified according to the combination of six pigments (Table 1). Some groups contain many mutants, especially the group without chlorophylls and carotene. Eight other pigment combinations occurred quite frequently, 14 combinations only rarely, and 41 possible groups were not represented by any mutation. The comparative study of pigment combinations in *Chlamydomonas reinhardi* and *Scenedesmus obliquus* mutants (Table 2) provided additional evidence of uneven distribution of the frequencies of pigment-loss types and of correlations between the components of normal chloroplast structure.

Mutants without chlorophylls and carotene are quite valuable for mutation analysis because their colonies have bright-yellow colour and it is easy to check the pigment contents by spectroscopic analysis of cell absorption in vivo (KVITKO et al. 1973).

The reaction of all strains to light varied: some mutants synthesized the missing pigments and thus looked like the wild type, some died, and others remained viable but manifested minor changes to the wild type or to a more bleached form. The light-sensitive (LTS) mutants were the subjects of more extensive study. We found among *Chlorella*, *Scenedesmus* and *Chlamydomonas* mutants three phenotype groups: the green, the pale-green, and the yellow forms. The green LTS-mutants did not demonstrate any change of pigment content as shown by chromatography (STOLBOVA 1975). The colonies of yellow LTS-forms usually had dark-

Table 1. The frequency of pigment losses combinations among *Chlorella vulgaris* strain B mutants as registered by paper chromatography.

Pigments			Chlorophyll a								Sum		
			+				−						
			Chlorophyll b										
			+		−		+		−				
			Carotene										
Lutein	Violaxanthin	Neoxanthin	+	−	+	−	+	−	+	−			
+	+	+	6	7	2	−	9	−	2	−	47	60	69
+	+	−	1	2	1	−	3	−	−	−	13		
+	−	+	−	−	−	−	1	−	−	−	2		
+	−	−	2	1	−	−	2	−	3	3	7		
−	+	+	−	−	−	−	1	−	1	−	1	9	6
−	+	−	−	−	−	−	−	−	−	−	−		
−	−	+	−	−	−	−	−	−	−	3		1	
−	−	−	9	−	1	−	17	−	8	25	5	5	
Sum			19	10	20	3							
			39				36				75		

228

Table 2. The frequency of pigment losses combinations among *Scenedesmus obliquus* and *Chlamydomonas reinhardi* mutants (Sc/Ch) as registered by paper chromatography.

Pigments				Chlorophyll a						Sum		
Lutein	Violaxanthin	Neoxanthin		Chlorophyll b		+		−				
				Carotene								
				+	−	+	−	+	−			
+	+	+		0/14								
+	+	−		0/3		0/2					0/20	33/36
+	−	+					0/1	0/5			33/26	
+	−	−						33/21				
−	+	+		0/1					1/2		1/2	1/2
−	+	−		0/3			0/1					
−	−	+		0/4		0/2	0/1		1/2		0/1	7/5
−	−	−						7/0			7/4	
Sum				0/17	0/21	0/24	0/3	40/26	1/3	1/2	40/26	41/53
								41/29				34/48
												7/5

red inclusions of unidentified pigments, probably porphyrins. Microspectrophotometric study of these crystal-like inclusions by BOYADZHIEV et al. (1973) showed five absorption bands in the region of 400-650 nm from small crystals found in the cells of a yellow LTS-mutant of *Chlamydomonas reinhardi*. The absence of both chlorophylls and sometimes of neoxanthin was typical of yellow LTS-mutants of *Scenedesmus* (TEMPER & KVITKO 1971) and *Chlamydomonas* (STOLBOVA 1975). The absence of β-carotene was an additional feature of yellow LTS-mutants of *Chlorella*. Two types of revertants were found for yellow LTS-mutants of *Chlamydomonas*: towards the wild type (green colour), and without any definable change in the absorption spectrum (CHUNAEV 1973).

The pale-green mutants have been identified as forms with a sharp decrease in the contents of all pigments; one *Chlorella* mutant had only 2% of the original chlorophyll *a* content and no other detectable pigments (N.46-7, see GILLER et al. 1971). The phenotype of these mutants was easily defined by the colour of colonies, which was similar in *Chlorella, Scenedesmus* and *Chlamydomonas* LTS-mutants. BENDIX & ALLEN (1962) called such colour 'apple-green'. Paper chromatography revealed a great variability in pigment content (STOLBOVA 1975).

Light-resistant Revertants

The light-resistant revertants arose spontaneously or after nitrosoethylurea or nitrosoguanidine treatments. They differed in colour and in levels of light resistance. Normalization of absorbance in bands typical of chlorophyll b and carotene indicated reversion to the wild type (CHUNAEV & KVITKO 1972; CHUNAEV 1973; KVITKO et al. 1973). The yellow light-resistant revertants were able to synthesize chlorophylls in the dark, but in the light these pigments disappeared (CHUNAEV 1973). The rate of spontaneous reversions to light-resistance was determin-

Table 3. The frequency of light-resistant revertants per 10^9 light-sensitive cells of pale-green mutants of *Chlorella* and *Chlamydomonas*.

Original culture	Mutants	Revertants	
		dark-green	'bleached'
Chlorella vulgaris			
str. B	45–4	0.9 –3.0	130 –480
str. B–15	B–15–19	0.03	4.0– 11.0
str. C–1–1–12	W 5 g *	0.01	2.7– 21.7
str. 157	157–8	0.5 –1.9	3.0– 21.0
Chlamydomonas reinhardi			
str. 137 c(–)	N–135	1.4	240.0

* Mutant from M.B. ALLEN collection (U.S.A.)

ed (Table 3). The frequency of emergence of dark-green revertants was 10-100 times lower than the frequency of revertants of other colonies. This agrees with our speculation about two types of suppression of the original LTS-mutation: through mutations of other genes or due to rare reversions of the same locus (CHUNAEV & KVITKO 1972). As shown by STOLBOVA (1975) in the recombination test, all seven independently occurring pale-green mutants of *Chlamydomonas reinhardi* have the allelic mutations in the lts_1 locus. The complementation test for allelism of these mutations is still to be done. The variability of pigment content of pale-green mutants makes it very improbable that the lts_1 locus is directly involved in the control of a pigment biosynthetic pathway. It is more probable that pigment accumulation is the function which is blocked by such a mutation.

Effects of DCMU, Antibiotics, and Mutagens

Selection of DCMU-resistant clones of *Chlorella* on solid mineral media with 10^{-4} DCMU produced different mutations (MUKHAMADIEV et al. 1971; MUKHAMADIEV & ZALENSKIĬ 1972); the frequency of spontaneous mutations was close to 10^{-7} as expected for a case where a smaller number of loci is involved. A new effect of DCMU has been discovered: the addition of non-lethal concentrations to media with a carbon source (glucose for *Scenedesmus,* acetate for *Chlamydomonas*) enabled the LTS-mutants to grow in the light. Among these were mutants sensitive to light even in the absence of oxygen (the sensitizing effect of external oxygen was shown by STOLBOVA 1975). One explanation for this phenotypic 'reversion' of the LTS-mutants could be inhibition of intracellular oxygen evolution by DCMU. If this were so, then we have criteria for identifying mutants in which oxygen, produced by photosynthetic reactions, appears to be the main cause of lethality in the light. This allows selection of not only DCMU-resistant but also conditionally DCMU-dependent mutants.

An explanation of the mutagenic effect of DCMU (MUKHAMADIEV et al. 1971) is very difficult; nevertheless chloroplast membrane changes, as one of the consequences of DCMU action, may lead to mutations in chloroplast DNA. The supposed mutageneity of sublethal doses of visible light may also be related to photooxidative disturbance of the chloroplast membrane.

With respect to the resistance of green algae to some organelle-specific antibiotics, TUGARINOV et al. (1970) found a high degree of similarity between types of resistance to streptomycin in *Chlorella* and *Chlamydomonas.* The same yellow and green mutants on streptomycin in heterotrophic conditions (such as sr-1 and sr-2 mutants of *Chlamydomonas*) were observed in *Scenedesmus obliquus.* The similarity of the mutants enables an extensive mutation analysis of genes concerned with chloroplast protein synthesis in algae of different systematic positions. The first attempt at mutational mapping was undertaken on *Chlamydomonas reinhardi* (TUGARINOV 1972). The principle of asynchronous DNA replication in organelles and nucleus (CHIANG 1971) was used in studies (Fig. 2) which suggest the preferential (96.4%) mutability of sr-2 factor during the first part of the cell cycle and the very high mutation rate of sr-1 (85.9%) at the beginning of the dark period of cultivation, when the nuclear DNA was replicated (CHIANG 1971). This fact was ascertained by LEE & JONES (1973) and LEE et al. (1973).

As was shown by CHIANG (1971), gametes have a different proportion of organelle DNA per cell than the vegetative haploid cells, and it is possible to predict the different mutability of these two haploid types of cells. The gametes have a higher mutation rate and are less viable after nitrosoethylurea treatment

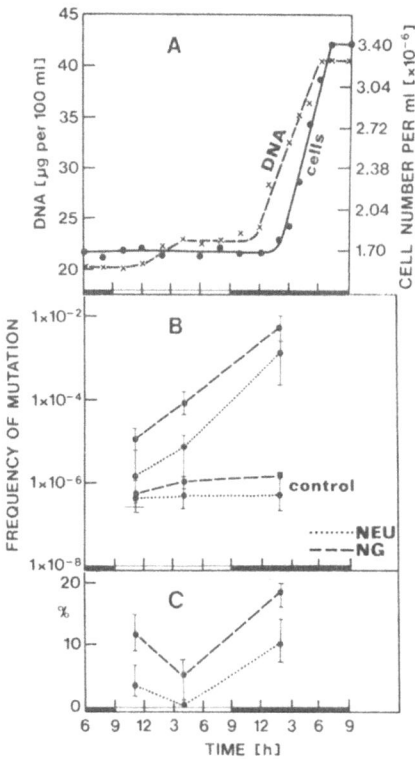

Fig. 2. Localization of genetic factors in a mutation experiment. (A) DNA synthesis in synchronized cells of *Chlamydomonas reinhardi* according to CHIANG (1971). Organelle DNA synthesis was performed during the 3rd hour of light. Nuclear DNA was synthesized at the beginning of the dark period. (B) Mutability of streptomycin resistance factors in a synchronous culture of *Chlamydomonas reinhardi,* treated with nitrosoguanidine (NG) and nitrosoethylurea (NEU). (C) Survival of cells on media without streptomycin after NG and NEU treatment in the same experiment.

Table 4. Comparison of mutability of vegetative cells and gametes of *Chlamydomonas reinhardi*, strain 137 C (−) after nitrosoethylurea (NEU) treatment (from TUGARINOV 1972).

Conditions and character	Vegetative cells	Gametes
Geometrical means and their limits of frequency for streptomycin-resistant mutants per 10^6 cells		
control	0.4 (0.2−0.9)	3.6 (1.7− 7.7)
NEU treatment	2.7 (0.8−9.9)	120.0 (68 −236)
Frequency of pigment mutants after NEU treatment [%]	0.91±0.12	0.92±0.14
Viability [%]	29.28	2.52

(Table 4), which indicates mitochondrial localization of sr-2 factor due to the excess of their DNA in gametes (CHIANG 1971).

The Yellow Mutant Without Chlorophyll

The next type of mutation we were interested in was a yellow type without chlorophyll synthesis in the dark. Such mutants look green in the light, but under preparative microscopy one could see that they were very unstable and their colonies were mosaic by colour (green sectors and spots on a yellow background under heterotrophic conditions). Pedigree analysis revealed unreciprocal segregation of yellow and mosaic, mosaic and green cells in most of the vegetative cell sporulations. Frequency of sporulations with 3 and 5 green (or yellow) segregants among eight descendants of one 'mosaic' cell was the lowest in comparison to any of the others (6:2, 4:8, 2:6, 1:7). This is an indication of not more than one (per each cell division) act of unreciprocal segregation as a cause of mosaic patterns of colonies. The general rate of emergence of green clones varied from 1.3 to 35.6×10^{-2} per sporulation. The type of inheritance of mutants with a mosaic pattern of pigmentation (MOM) was studied by tetrad analysis. Most of the tetrads were 2 : 2 green to yellow when the MOM-mutants were crossed with the wild type (FAM TKHAN KHO 1972). The three MOM-factors were linked one to another and with markers lts_4 and chl_2 localized in a fragment described by STOLBOVA (1973). We think that MOM-factors do satisfy the criteria of hypothetic plastome genes (CHIANG 1971): biparental, analogous to a Mendelian type of inheritance.

In this short communication no reference was made to the splendid works of R. P. Levine's laboratory on a series of acetate-dependent mutants (Table 5; see LEVINE & GOODENOUGH 1970; LEVINE 1971).

As reported earlier in this paper, there are at our disposal mutation systems of several genes which control light-sensitivity, probably of plastid localization, as well as factors of resistance to streptomycin, neomycin and erythromycin, localized either in the nucleus or mitochondria and, probably, in the chloroplast. There are

Table 5. Localization of the transcription and the translation processes during the synthesis of chloroplast components in *Chlamydomonas* (from LEVINE 1971).

Chloroplast component	Sensitivity to inhibitors	Transcription	Translation
DNA-polymerase		N	Cyt
DNA-dependent RNA$_2$ polymerase		N	Cyt
Ribosomal RNA (5S, 16S, 23S)	R	P	–
Ribosomal proteins		?	Cyt
Phosphoribulokinase	C	N	Cyt
RuDP-carboxylase	S,C	N	PCyt
Ferredoxin	C	N	Cyt
Ferredoxin-NADP-reductase	C	N	Cyt
Cytochrome 559		N	P
Cytochrome 553	R,S	NP	P
Cytochrome 563	R,S,C	P	PCyt
Chlorophyll	C	N	Cyt
β-carotene	C	N	Cyt
Lutein		N	?
Violaxanthin		N	?
Neoxanthin	C	N	Cyt
Trollein	C	N	Cyt

C – cycloheximide, R – rifampicin, S – spectinomycin; N – nuclear, Cyt – cytoplasmic, P – plastid localization of the processes.

premises for creating more than one mutation system of 'gene-character' type and, in the course of time – 'gene-polypeptide' systems, and it is quite probable that one of these might be of plastid origin. The final choice of the system will depend upon the experimental opportunities of finding and eliminating errors in the methods of gene localization. Parallel study of gene mutations localized in different organelle genophores would be ideal.

Green algae appear the most efficient organisms for analysis of organelle inheritance, because if speculation about the bi-parental type of chloroplast gene transmission is confirmed, then the map of mitochondrial genophores constructed by SAGER (1972) is a rather detailed and useful basis for further studies. Work with three groups of green algae of the genera *Chlorella*, *Scenedesmus* and *Chlamydomonas* has convinced us of the parallel nature of their variability (Table 6), and thus we believe that a full mutation analysis of agamic algae will be possible and then, according to N. I. Vavilov's law, one will be able to determine the homology of genotype for *Chlamydomonas*, *Chlorella* and *Scenedesmus*.

Summary

Mutation analysis of green algae (*Chlorella, Scenedesmus, Chlamydomonas*) shows that the genotype of the *Eucaryota* is to be considered either as a functionally

Table 6. Homology of induced and spontaneous mutations found among *Chlamydomonas reinhardi*, *Chlorella vulgaris* and *Scenedesmus obliquus* clones

Group	Characters	Mutants and hypothetical localization of mutated factors		
		Chlamydomonas reinhardi	*Chlorella vulgaris*	*Scenedesmus obliquus*
1	Change of the flagellar apparatus	n *	–	?
2	Auxotrophy:			
	(a) vitamins,	n	+	+
	(b) amino acids,	n	+	+
	(c) nucleic acid bases.	?	?	+
3	Resistance to the analogs:			
	(a) amino acids,	n	+	+
	(b) vitamins.	n	+	+
4	Resistance to the cytoplasmic protein synthesis inhibitors	n	+	+
5	Changes of pigment composition:			
	(a) loss of chlorophylls,	n,p **	+	+
	(b) loss of carotenoids,	n,p	+	+
	(c) loss of both,	n,p	+	+
	(d) occurrence of new pigments,	?	+	+
	(e) leading to light-sensitivity.	p	+	+
6	Mosaic colonies:			
	(a) yellow subclones greening in the light,	n	+	+
	(b) yellow subclones unable to green in the light,	?	+	+
	(c) yellow subclones dying in the light.	?	+	+
7	Inability to photosynthesise			
	(a) related to pigment deficiency,	n,p	+	+
	(b) without pigment deficiency.	n,m ***	+	+
8	Auxotrophy only in heterotrophic conditions:			
	(a) amino acids,	+	+	+
	(b) vitamins	+	+	+
	(c) nucleic acid bases,	?	?	+
	(d) concerned with pigment deficiency.	?	?	+
9	Resistance to organelle protein synthesis inhibitors	m,p	+	+
10	Light-sensitivity without pigment deficiency	p	+	+

* – localization in the nucleus (n); ** – bi-parental organelle mutations regarded as plastid ones (p); *** – uniparental inherited organelle mutations regarded as mitochondrial ones (m).

united (indivisible) system, or more probably as a morphologically and physiologically divisible system.

REFERENCES

BENDIX, S. & ALLEN, M. B., *Arch. Mikrobiol.* 41: 115, 1962.
BOYADZHIEV, P. H., SMIRNOV, A. F. & KVITKO, K. V., in: Upravlyaemyĭ Biosintez i Biofizika Populyatsiĭ, p. 97-99, Sibir. Otd. Akad. Nauk SSSR, Krasnoyarsk 1973.
CHIANG, K.-S., in: BOARDMAN, N. K., LINNANE, A. W. & SMILLIE, R. M. (ed.): Autonomy and Biogenesis of Mitochondria and Chloroplasts, p. 235, North-Holland Publ. Co., Amsterdam-London 1971.
CHUNAEV, A. S., in: Upravlyaemyĭ Biosintez i Biofizika Populyatsiĭ, p. 175-176, Sibir. Otd. Akad. Nauk SSSR, Krasnoyarsk 1973.
CHUNAEV, A. S. & KVITKO, K. V., in: Geneticheskie Aspekty Fotosinteza, Tezisy Dokladov, p. 77, Dushanbe 1972.
FAM TKHAN KHO, *Genetika* 8: 146, 1972.
GILLER, Yu. E., STOLBOVA, A. V., VAKHIDOVA, L. R. & KVITKO, K. V., *Biofizika* 16: 67, 1971.
INGE-VECHTOMOV, S. G., *Vestn. Akad. Nauk SSSR* 1969 (8): 25, 1969.
KVITKO, K. V., in: KIRICHENKO, E. B. (ed.): Metody Issledovaniya Struktury Fotosinteticheskogo Apparata, p. 119, Pushchino-na-Oke 1972.
KVITKO, K. V., BARANOV, A. A., CHUNAEV, A. S. & SAAKOV, V. S., in: Upravlyaemyĭ Biosintez i Biofizika Populyatsiĭ, p. 123-124, Sibir. Otd. Akad. Nauk SSSR, Krasnoyarsk 1973.
LEE, R. W., GILLHAM, N. W., WINKLE, R. P. & van BOYTON, J. E., *Mol. gen. Genet.* 121: 109, 1973.
LEE, R. W. & JONES, R. F., *Mol. gen. Genet.* 121: 99, 1973.
LEVINE, R. P., in: Evolution of Genetic Systems, *Brookhaven Symp. Biol.* 23: 503, 1971.
LEVINE, R. P. & GOODENOUGH, U. W., *Annu. Rev. Genet.* 4: 397, 1970.
MUKHAMADIEV, B. T., KVITKO, K. V. & ZALENSKIĬ, O. V., *Genetika* 7 (5): 36, 1971.
MUKHAMADIEV, B. T. & ZALENSKIĬ, O. V., *Bot. Zh.* 57: 260, 1972.
MUKHAMADIEV, B. T., ZALENSKIĬ, O. V. & KVITKO, K. V., *Genetika* 7 (4): 38, 1971.
SAGER, R., Cytoplasmic Genes and Organelles, Academic Press, New York-London 1972.
STOLBOVA, A. V., *Genetika* 7 (9): 90, 1971a.
STOLBOVA, A. V., *Genetika* 7 (11): 124, 1971b.
STOLBOVA, A. V., *Genetika* 8(4):123, 1972.
STOLBOVA, A. V., in: NASYROV, Yu. S. & ŠESTÁK, Z. (ed.): Genetic Aspects of Photosynthesis, p. 217, Junk, The Hague 1975.
STOLBOVA, A. V. & KHROPOVA, V. I., in: ZALENSKIĬ, O. V. (ed.): Fotosintez i Ispol'zovanie Solnechnoĭ Energii, p. 261, Nauka, Leningrad 1971.
TEMPER, E. E. & KVITKO, K. V., *Nauch. Dokl. vyssh. Shkoly, biol. Nauki* 14 (4): 106, 1971.
TUGARINOV, V. V., Tezisy II S"ezda VOGIS, I-B-293, Moskva 1972.
TUGARINOV, V. V., THRUNI, F. N. & KVITKO, K. V., *Issled. Genet.* 4: 142, 1970.
ZAKHAROV, I. A. & KVITKO, K. V., Genetika Mikroorganizmov, Izd. lening. gos. Univ., Leningrad 1967.

MODELS FOR STUDYING THE MECHANISM OF PHOTOSYNTHESIS

PIGMENT SYNTHESIS AND PHOTOSYNTHETIC ACTIVITY IN CAROTENOID DEFICIENT MUTANTS OF MAIZE

ÁGNES FALUDI-DÁNIEL

Institute of Plant Physiology, Biological Research Center, Hungarian Academy of Sciences, Szeged, Hungary

Chloroplasts without a normal set of carotenoids possess, as a rule, the ability for chlorophyll synthesis, but these chlorophylls become photooxidized when seedlings are grown under normal light conditions (ANDERSON & ROBERTSON 1960; FALUDI et al. 1960). Chloroplasts with a low carotenoid content (WALLES 1966) or with a high concentration of carotenoid precursors (GYURJÁN et al. 1969) are not able to form or maintain normal grana and the leaves develop an *albina* or *xantha* phenotype. The increased photosensitivity of chloroplasts may be a consequence of the impaired transfer of energy from chlorophyll to carotenoid triplets (MATHIS 1971) and/or to an intensive photodeaggregation of chlorophylls (KRASNOVSKY 1972). Despite extensive study of these phenomena, details of molecular characteristics of carotenoids important in preventing chlorophyll destruction, critical concentrations and molecular spacing have not been elucidated. Studies were conducted in vitro (SINESHCHEKOV et al. 1972) and also in artificially modified chloroplasts (KONISHI et al. 1972) but a comparative investigation of the carotenoid mutants should also contribute to the solution of these problems.

This paper is a review of our studies comparing some structural and functional characteristics of carotenoid mutant chloroplasts of maize leaves, which accumulate intermediate compounds of carotenoid biosynthesis. Data are reported on the carotenoid and chlorophyll contents, on the electron microscopic structure of chloroplasts, on the binding state and aggregation of chlorophylls, on the activity of individual photosystems and for some photosynthetic dark reactions.

Plant Material

The seed material was a segregation population of two inbred lines of maize, heterozygous for the lycopenic (ly) and ζ-carotenic (ζ) genes. The strain containing the ly-gene was a gift of prof. A. BIANCHI (Italy), the strain containing the ζ gene was obtained from dr. E. PAP (Hungary). Due to the increased photosensitivity of chloroplasts, homozygote recessive seedlings are lethal, thus ly and ζ genes can be propagated only in a heterozygous condition. Ly and ζ genes follow a Mendelian inheritance, and the ζ-gene is in a recessive epistatic relation to the gene responsible for the lycopenic character (FALUDI-DANIEL et al. 1967).

Seedlings were grown under continuous illuminance (25 lx), favourable for the mutants. Leaves of 7-9-d-old seedlings were used in the experiments.

Pigment Contents

Carotenoid analysis of the leaves has shown that the ζ-leaves contain phytoene, phytofluene and trans and cis-ζ-carotene indicating impaired desaturation of the double bonds and deficiency of ring closure of the C_{40} precursor. The biosynthetic block due to the ly-gene was 'leaky' permitting the synthesis of several percent of the normal level of α + β-carotenes. The main component of this mutant was δ-carotene. However, phytoene, phytofluene, γ-carotene, lycopene, antheraxanthin and traces of other xanthophylls could also be detected. This indicated that the ly gene reduced cyclization of the carotenoid molecule (HORVÁTH et al. 1972).

Deficiencies of carotenoid synthesis were manifested in etiolated leaves. Under illumination carotenoid synthesis was enhanced, as in normal leaves, but to a smaller extent. Under favourable conditions the total carotenoid content of normal and mutant leaves was about the same.

In the ly chloroplasts δ-carotene and lycopene were localized in System 2 particles obtained by digitonin fragmentation. β-carotene was found mostly in System 1 fragments. System 2 particles of ζ-plastids contained mainly ζ-carotene, while System 1 particles were enriched in phytoene (HORVÁTH & FALUDI-DÁNIEL 1972).

Chlorophyll synthesis of both mutants could be regarded as normal in a qualitative sense. Phytolization of mutants differed only slightly from that of normal leaves (SHLYK et al. 1967). Under high illuminance the chlorophyll content of the mutants decreased, and this was followed by a depression of protochlorophyllide synthesis (FALUDI-DÁNIEL et al. 1969).

Continuous breakdown of chlorophylls, which occured even at low illuminances, resulted in very high carotenoid to chlorophyll ratios (FALUDI-DÁNIEL et al. 1968).

Binding State and Aggregation of Pigments

By extracting leaves with petroleum ether it has been demonstrated that the pigments in mutant leaves are bound to proteins less firmly than in normal leaves (FALUDI-DÁNIEL et al. 1969).

Aggregation of chlorophylls was reduced in the carotenoid mutants. This has been shown by low temperature emission spectra (FALUDI-DÁNIEL et al. 1966), by circular dichroism spectra (GARAY et al. 1972), and by decomposing the in

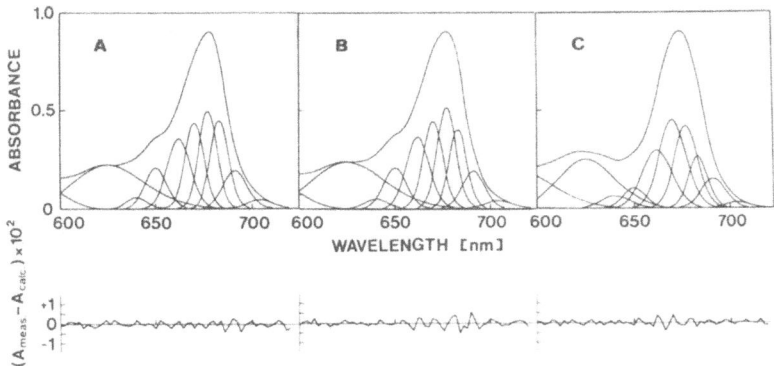

Fig. 1. The absorption spectra of chloroplasts from normal (A), lycopenic (B) and ʓ-carotenic (C) leaves of maize, as fitted by the sum of Gaussian components. Fitting is indicated by the error.

Table 1. Approximate distribution of different forms of chlorophylls in normal, lycopenic and ʓ-carotenic chloroplasts.

Material		normal [%]	lycopenic [% per normal %]	ʓ-carotenic [% per normal %]
chlorophyll *b*				
form	640*	21.0	0.93	2.08
	650	79.0	1.01	0.71
chlorophyll *a*				
form	662	21.1	1.00	1.05
	670	20.8	1.09	1.47
	677	22.7	1.09	1.03
	683	21.1	0.84	0.61
	692	11.6	0.95	0.81
	704	2.7	1.00	0.67

* Peak positions fixed at the respective wavelengths.

vivo spectra of chloroplasts. Curve decomposition was based on the 'constant component concept' of FRENCH (1971). The computer program was constructed by S. DEMETER from our laboratory according to the principles of MARQUARDT (1963). The curve decomposition (Fig. 1 and Table 1) suggested that, in chloroplasts containing large amounts of intermediate compounds of carotenoid synthesis, the chlorophyll-protein complexes are weak and/or chlorophyll aggregation is inferior to the normal level.

Ultrastructure of Carotenoid-Deficient Chloroplasts

In mesophyll chloroplasts of carotenoid mutant leaves normal grana were absent. In the ly mutant, some chloroplasts contained macrograna, with long tightly appressed lamellae, some others had concentrically arranged lamellae characteristic of chromoplasts. ʓ-carotenic chloroplasts developed only small amounts of lamellae. In most cases they contained only vesicles or degenerated prolamellar bodies, but occasionally macrograna and chromoplast-like bodies could also be observed (FALUDI-DÁNIEL et al. 1968; GYURJAN et al. 1969).

Despite the monogenic determination of carotenoid-deficiency, the chloroplast population of mutants was very heterogeneous (Fig. 2).

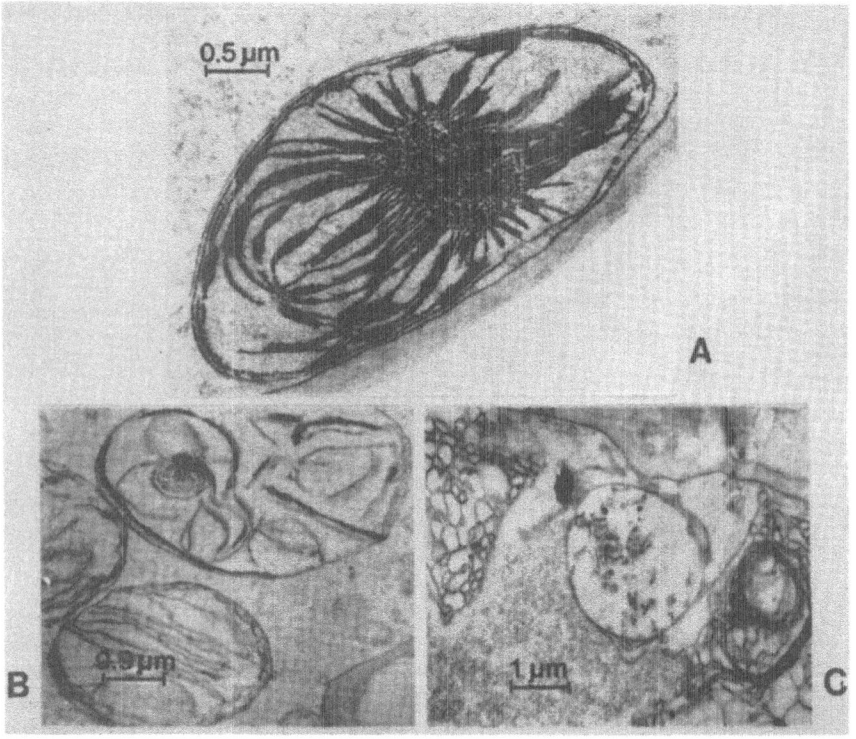

Fig. 2. Electron micrographs of chloroplasts prepared from normal (A), lycopenic (B) and ʓ-carotenic (C) leaves of maize.

Structural Stability and Activity of Photosystems

The structural stability of photosystems was studied in chloroplasts subjected to digitonin fragmentation. The distribution of chlorophyll between the particulate fractions demonstrated that System-2-chlorophyll of the mutants had been solubilized to a much higher degree than that of normal chloroplasts. This indicated a decreased stability of System 2 of mutant chloroplasts (NAGY et al. 1971).

System 1 activity was tested by measuring the P_{700} content of chloroplasts. In this respect, ζ-carotenic chloroplasts were superior to normal ones. The P_{700} content of ly plastids was the same as in normal chloroplasts. These data show that in carotenoid deficient chloroplasts System 1 activity was only slightly affected. System 2 activity was measured by the rate of ferricyanide photoreduction. In ly chloroplasts System 2 activity was only half of the normal value. In the ζ-plastids, however, the rate of ferricyanide reduction was about the same as in normal samples, but the activity of mutant chloroplasts decayed very fast. In ly chloroplasts the activity of the photosystems was not affected by a strong illuminance in contrast to the rapid decrease of ferricyanide reduction observed in ζ-carotenic chloroplasts (NAGY et al. 1972).

CO_2 Fixation by Mutants

In the mutants CO_2 fixation was depressed to a much larger extent than could be expected from the chlorophyll contents. Low CO_2 fixation of leaves was also in contrast with the high in vitro capacity of carboxylating enzymes (NAGY et al. 1971). In mutant leaves a strong illuminance induced an oxidative inactivation of ribulose 1,5-diphosphate carboxylase, resulting in an auto-inhibition of the photosynthetic apparatus (NAGY et al. 1972).

Discussion

It has been generally accepted (GOODWIN 1965) that the number of conjugated double bonds is important in enabling carotenoids to prevent photooxidation of chlorophylls. However, as shown by the photosensitivity of ly chloroplasts, cyclization of carotenoid molecules may also have some importance.

The concentration of carotenoids has been shown to influence the degree of chlorophyll aggregation (KONISHI et al. 1972). By adding β-carotene to chloroplast suspensions the amount of monomeric forms of chlorophyll was increased at the expense of aggregated forms. This deaggregation was accompanied by the inactivation of System 2 activity, whereas System 1 reactions were only slightly affected. These experimental conditions may be similar to the in vivo conditions in carotenoid mutant chloroplasts. The approximate content of different chlorophyll forms, as calculated from the absorption spectra in vivo, showed smaller amounts of chlorophyll forms absorbing at longer wavelengths. A further analogy is that in the carotenoid mutants System 2 is mainly affected.

Conclusions

Comparison between normal and carotenoid-deficient chloroplasts suggested that ionon rings are necessary for enabling carotenoids to prevent chloroplasts from photodestruction.

Incomplete desaturation of double bonds and/or deficiency of carotenoid cyclization decreases the structural and functional stability of System 2. As a secondary consequence ribulose 1,5-diphosphate will also be inhibited.

Regarding the concentration of carotenoids, the importance of an optimum level can be suggested. Low carotenoid to chlorophyll ratios are disadvantageous because of limiting conditions for the energy transfer between chlorophyll and carotenoid triplets. Very high carotenoid to chlorophyll ratios are also not favourable because excess carotenoids decrease the binding and aggregation of chlorophylls.

REFERENCES

ANDERSON, J. M. & ROBERTSON, D. S., *Plant Physiol.* 35: 531, 1960.
FALUDI, B., FALUDI-DÁNIEL, Á. & KELEMEN, G., *Physiol. Plant.* 13: 227, 1960.
FALUDI-DÁNIEL, Á., FRIDVALSZKY, L. & GYURJÁN, I., *Planta* 78: 184, 1968.
FALUDI-DÁNIEL, Á., LÁNG, F. & FRADKIN, L. I., in: GOODWIN, T. W. (ed.): Biochemistry of Chloroplasts, Vol. 1, p. 269, Academic Press, London-New York 1966.
FALUDI-DÁNIEL, Á., LÁNG, F., NAGY, Á. & FALUDI, B., *Acta agron. Acad.Sci.hung.* 16: 1, 1967.
FALUDI-DÁNIEL, Á., NAGY, Á. H. & NAGY, Á., in: METZNER, H. (ed.): Progress in Photosynthesis Research, Vol. 2, p. 592, Tübingen 1969.
FRENCH, C. S., *Proc.nat.Acad.Sci. U.S.A.* 11: 2893, 1971.
GARAY, A., DEMETER, S., KOVÁCS, K., HORVÁTH, G. & FALUDI-DÁNIEL, Á., *Photochem. Photobiol.* 16: 139, 1972.
GOODWIN, T. W., in: PRIDHAM, J. B., SWAIN, T. (ed.): Biosynthetic Pathways in Higher Plants, p. 57, Academic Press, London-New York 1965.
GYURJÁN, I., RAKOVÁN, J. N. & FALUDI-DÁNIEL, Á., in: METZNER, H. (ed.): Progress in Photosynthesis Research, Vol. 1, p. 63, Tübingen 1969.
HORVÁTH, G., KISSIMON, J. & FALUDI-DÁNIEL, Á., *Phytochemistry* 11: 183, 1972.
HORVÁTH, G. I. & FALUDI-DÁNIEL, Á., in: FORTI, G., AVRON, M. & MELANDRI, A. (ed.): Photosynthesis, Two Centuries after Its Discovery by Joseph Priestley, Vol. 3, p. 2443, Junk, The Hague 1972.
KONISHI, K., YAGINUMA, N. & SHIBATA, K., *Plant Cell Physiol.* 13: 531, 1972.
KRASNOVSKY, A. A., *Biophys. J.* 12: 749, 1972.
MARQUARDT, D. W., *J.Soc.ind.appl.Math.* 11: 431, 1963.
MATHIS, P., Étude de Formes Transitoires des Carotenoïdes, Thesis, Saclay 1971.
NAGY, A. H., BOKÁNY, A., DOMAN, N. G. & FALUDI-DÁNIEL, Á., in: FORTI, G., AVRON, M. & MELANDRI, A. (ed.): Photosynthesis, Two Centuries after Its Discovery

by Joseph Priestley, Vol. 3, p. 1861, Junk, The Hague 1972.
NAGY, A. H., PACSÉRY, M. & FALUDI-DANIEL, Á., *Physiol. Plant.* 24: 301, 1971.
SHLYK, A. A., FRADKIN, L. I., SAVCHENKO, G.E. & FALUDI-DÁNIEL, Á., *Photosynthetica* 1: 241, 1967.
SINESHCHEKOV, V. A., LITVIN, F. F. & DAS, M., *Photochem. Photobiol.* 15: 187, 1972.
WALLES, B., *Hereditas* 56: 131, 1966.

MUTANTS AS OBJECTS FOR INVESTIGATIONS ON THE FUNCTION OF CHLOROPHYLLS

HERTA SAGROMSKY

Zentralinstitut für Genetik und Kulturpflanzenforschung der Akademie der Wissenschaften der DDR, 4325 Gatersleben, D.D.R.

Introduction

Many physiological reactions overlap or interfere with each other to such an extent that it is difficult to study the course and significance of each process independently. In order to have an idea on the intensity and mechanism of action of each reaction, the plant material is often treated with specific inhibitors to eliminate some of the overlapping reactions and to study the activity of the remaining ones.

Another possibility, often used, to examine characteristic reaction pecularities is to observe plants during their ontogenetic development, especially during their earliest stages where single organelles are built up and functions therefore appear successively. With respect to photosynthesis, a similar approach is also possible when dark grown plants are exposed to light.

The third method for observing specific functions is to select mutants that lack special components and to compare their reactions with those of normal plants.

The main aim of this study was to examine the role of chlorophyll *b* in photosynthesis and photosynthetic productivity.

The Role of Chlorophyll *b* in Photosynthesis and Photosynthetic Productivity

Chlorophyll *a* and chlorophyll *b* have similar absorption spectra. According to DUYSENS (1952) & RABINOWITCH (1956) and many others, there exists an effective energy transfer from chlorophyll *b* to chlorophyll *a*, hence one hypothesis for the function of chlorophyll *b* is to gather quanta of radiation.

WITT et al. (1965) & WITT (1966) suppose that chlorophyll *b* is involved in primary reactions of photosynthesis, is bound to the plastoquinone reaction in the electron transport chain, and is coupled with the formation of ATP. According to the hypothesis of SEYBOLD (1941) which is also supported by MÜLLER (1964) and others, chlorophyll *b* is necessary for the polymerisation of sugar to starch. Another theory supposes that chlorophyll *b* is needed for the differentiation of plastid structure, especially for thylakoid membranes.

In order to test some of these hypotheses we compared the form *Hordeum*

Fig. 1. Vegetative phase of the barley mutant 3613 and the parent form Donaria.

vulgare cv. Donaria with three of its non-lethal mutants that lack chlorophyll *b*. The mutants were somewhat smaller and less dark green than the parent form (Fig. 1). When they were grown under the same illuminance and temperature, their production was lower than that of the normal plant (Table 1). The mutants formed lesser amounts of fresh and dry matter of shoots and roots than the parent form Donaria. Their production was lowered to 75.5%, 58.7%, or 56.6% of the normal cv. Donaria.

The three mutants were light green. Their chlorophyll *a* content was nearly one half that of the form Donaria (Table 2). The carotenoids were present in nearly the same amounts as in the original form Donaria: the amounts of lutein and neoxanthin were somewhat diminished, but violaxanthin was a little augmented. All three mutants produced seeds (Fig. 2). The grain yield of the form Donaria was higher than that of the mutants (Table 3). The mutants had a lower number of ears per plant and a diminished weight of seeds from one plant (47%, 66%, 45.8% of the original form Donaria respectively). The weight of 1000 grains was only slightly changed.

MÜLLER (1964) studied chloroplasts of leaves from a pea mutant that

Table 1. Production [g] of the original form Donaria and the three mutants without chlorophyll b cultivated for 7 weeks in nutrient solution.

	Shoot		Root		Whole plant		Whole plant
	Fresh weight	Dry weight	Fresh weight	Dry weight	Fresh weight	Dry weight	Relative
Donaria	39.6±1.5	3.7±0.3	14.0±0.8	1.2±0.1	53.0	4.9	100
Mutant 3613	30.8±1.2	2.6±0.2	9.3±0.9	0.8±0.1	40.1	3.4	75.5
Mutant 2807	23.7±1.4	1.9±0.3	7.2±0.5	0.6±0.1	30.9	2.5	58.7
Mutant 2800	23.8±1.0	1.9±0.2	6.2±0.4	0.5±0.1	30.0	2.4	56.6

Table 2. Amounts of pigments [mg g^{-1} fresh weight] in barley cv. Donaria and its three mutants without chlorophyll b.

	Chlorophyll a	Chlorophyll b	β-Carotene	Lutein	Violaxanthin	Neoxanthin
Donaria	2.34	0.86	0.074	0.106	0.034	0.036
Mutant 3613	1.12	—	0.078	0.078	0.034	0.016
Mutant 2807	1.36	—	0.078	0.090	0.038	0.014
Mutant 2800	1.25	—	0.074	0.084	0.046	0.014

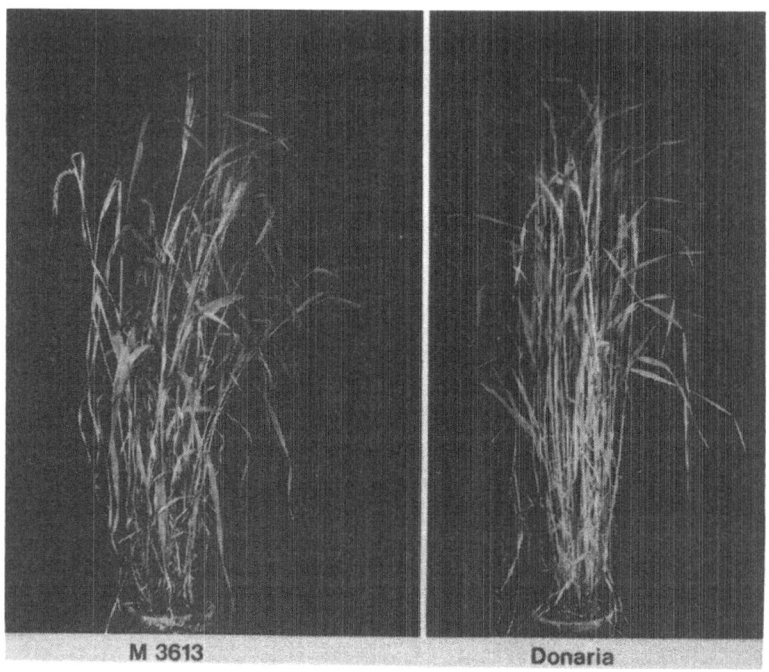

Fig. 2. Fertile phase of the barley mutant 3613 and the parent form Donaria.

Table 3. Grain yield of the original form Donaria and the three mutants without chlorophyll b.

	Number of ears	Grain yield per plant [g]	Weight of 1000 grains [g]	Grain yield, relative
Donaria	12.4±1.0	11.7±1.2	44.7±0.6	100.0
Mutant 3613	7.6±1.0	5.5±1.3	42.3±0.6	47.0
Mutant 2807	9.4±1.2	7.7±1.3	40.6±0.8	66.0
Mutant 2800	7.8±1.4	5.0±1.5	39.2±0.7	45.8

lacked chlorophyll b. He found starch only in chloroplasts of the guard cells of the stomata, but never in chloroplasts of the parenchyma cells. Therefore he supported the hypothesis of SEYBOLD (1941) according to which chlorophyll b is required for the polymerisation of sugar into starch. In our mutants, especially in the mutant 3613, we did find starch in the chloroplasts of the parenchyma cells. Hence chlorophyll b is not always necessary for the formation of starch.

HIGHKIN & FRENKEL (1962) observed a mutant of *Hordeum vulgare* that lacked chlorophyll b. This mutant was nearly half the size of the normal plant, but possessed equal rates of photosynthesis and respiration.

Also WILD & EGLE (1967) found a lowered cell production in a mutant of *Chlorella* nearly without chlorophyll b, but the rates of photosynthesis and respiration were the same as in the normal form, in spite of the smaller amount of chlorophyll. They therefore concluded that chlorophyll b cannot be necessary for the primary reactions. Our mutants, for example the mutant 3613 (Fig. 3), had, according to the measurement of APEL (1967), a lowered rate of CO_2 uptake. This diminished photosynthetic rate was not caused by a higher photorespiration rate: the decline of CO_2-uptake in the presence of 21% O_2 (compared with the atmosphere of 1% O_2) was similar in the mutant 3613 as in the parent plant (Table 4).

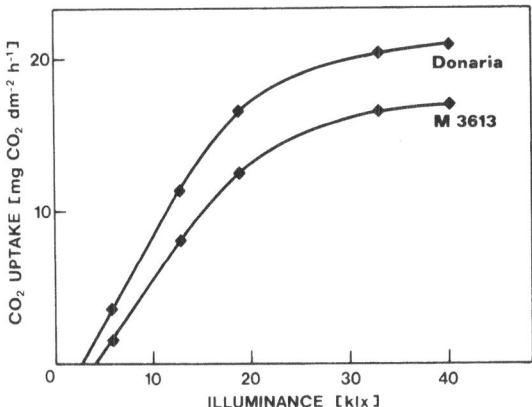

Fig. 3. CO_2 uptake of the barley mutant 3613 and its parent form Donaria at different illuminances.

Table 4. Decline of CO_2-uptake in the presence of 21% O_2 compared with 1% O_2 [%]. Measurements in white light of 40 klx (APEL 1967).

Plant age [d]	Donaria	Mutant 3613
8	28.5	31.5
20	31.5	34.5

In contrast to these results, the activity of the Hill reaction with ferricyanide in chloroplasts of the mutants at light saturation was higher than that of the parent form Donaria (Fig. 4). If the Hill reaction rate is related to the chlorophyll (a + b) content (and not to the chlorophyll a content as in Fig. 4), the activity of the plastids of the mutant is nearly twice as high as the activity of the chloroplasts

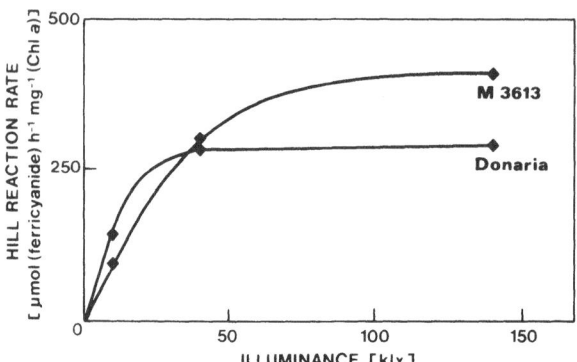

Fig. 4. Hill reaction rate with ferricyanide in chloroplasts from the barley mutant 3613 and the respective parent form Donaria, measured at various illuminances.

from cv. Donaria. The cells of Donaria had almost the same number of chloroplasts as adequate cells of the mutant 3613, however the chlorophyll content of the mutant was only half that of the original form of Donaria. Hence the activity of one chloroplast of the mutant 3613 is nearly equal to that of one chloroplast of the cv. Donaria.

The mutant 3613 had, according to the electron microscopic studies of DOEBEL (unpublished), a well developed thylakoid system, and therefore chlorophyll *b* is not necessary for the formation of chloroplast structure.

The mutant 3613 showed non-cyclic photophosphorylation. This result contradicts the assumption of WITT et al. (1965) & WITT (1966) according to which chlorophyll *b* is necessary for non-cyclic photophosphorylation.

Summary

For the investigation of the function of chlorophyll *b*, three mutants without chlorophyll *b* were compared with the original form of *Hordeum vulgare* cv. Donaria. They were lower in dry matter production, chlorophyll *a* content, and grain yield. The mutants were able to form starch in their chloroplasts. Their rate of photosynthesis was a little lower but not as a consequence of higher photorespiration. Their Hill-reaction with ferricyanide at light saturation was higher than that of the parent form Donaria when related to chlorophyll *a*, but equal when related to one chloroplast. The mutants were able to form a well developed thylakoid system and were active in non-cyclic photophosphorylation.

REFERENCES

APEL, P., *Stud. biophys.* 5: 105, 1967.
DUYSENS, L. N. M.: Transfer of Excitation in Photosynthesis, Thesis, Utrecht 1952.
HIGHKIN, H. R. & FRENKEL, A. W., *Plant Physiol.* 37: 814, 1962.
MÜLLER, F., *Planta* 63: 65, 1964.
RABINOWITCH, E. I.: Photosynthesis and Related Processes, Vol. II, Part 2, Interscience Publishers Inc., New York 1956.
SEYBOLD, A., *Bot. Arch.* 42: 254, 1941.
WILD, A. & EGLE, K., *Biol. Zentralbl.* 86 (Suppl.): 495, 1967.
WITT, H. T., *Umschau Wiss. Tech.* 1966: 596, 1966.
WITT, H. T., RUMBERG, B., SCHMIDT-MENDE, P., SIGGEL, U., SKERRA, B., VATER, J. & WEIKARD, J., *Angew. Chem.* 77: 821, 1965.

FLUORESCENCE INDUCTION OF NORMAL AND MUTANT MAIZE SEEDLINGS

N. V. KARAPETYAN, V. V. KLIMOV, F. LANG & A. A. KRASNOVSKIÏ

A.N. Bakh Institute of Biochemistry, Academy of Sciences of the U.S.S.R., Moscow, U.S.S.R.

Introduction

At the beginning of illumination of green plants and algae complex kinetics of fluorescence changes are observed. These patterns are known as fluorescence induction (KAUTSKY & HIRSCH 1931). According to DUYSENS & SWEERS (1963) fluorescence (F) yield is determined by the redox state of the primary electron acceptor of Photosystem 2, Q, which in its oxidized form quenches F. Different stages in the complex kinetics of photoinduced changes of F (ΔF) reflect the interaction of photosystems mediated by the electron transport chain of photosynthesis (DUYSENS & SWEERS 1963; MUNDAY & GOVINDJEE 1969; KARAPETYAN et al. 1971). If the scheme of photosynthetic electron transport is

$$H_2O \rightarrow Z \rightarrow Chl\,a_2 \rightarrow Q \rightarrow A \rightarrow P700 \rightarrow X \rightarrow NADP,$$

then the photoreduction of Q by Photosystem 2 causes an increase in F, and the oxidation of QH by Photosystem 1 results in a decrease in F.

Induction of F is one of the important tests for the presence and functioning of Photosystem 2. Therefore it is often used to characterize the photosystems of mutants. In turn, the investigation of pigment mutants helps in understanding the mechanisms of photosynthetic electron transport and the role of different forms of chlorophylls in that process. For this reason we investigated lycopenic and ζ-carotenic mutants of maize having different pigment contents (SHLYK et al. 1967; LANG et al. 1969, 1971) and low rates of CO_2 uptake (GYURJÁN et al. 1969). We attempted to explain the low photosynthetic activity of mutants by studying changes in ΔF and oxygen evolution during the greening of their leaves in comparison to leaves of normal seedlings.

Methods

The kinetics of ΔF of maize leaves with $\lambda > 710$ nm was recorded oscilloscopically by means of a double-beam apparatus described by KARAPETYAN & KLIMOV

(1971) and KARAPETYAN (1972). A low intensity monochromatic beam (650 nm, 0.05 W m^{-2}) excited the F of the leaves; this light was exchanged for a perpendicular actinic beam (λ>710 nm, 1.7 x 10^3 W m^{-2}, absorbed mainly by Photosystem 1 or 600-700 nm, 1.5 x 10^2 W m^{-2}, more effectively absorbed by Photosystem 2). The initial level of F (F_i) excited by the monochromatic beam and its changes (ΔF) under actinic irradiance were successively recorded. The photosynthetic activity of the leaves was measured amperometrically as the rate of oxygen evolution (KLIMOV et al. 1972). The rate of oxygen evolution, ΔF and spectra of F were measured successively in the same leaf. Sometimes leaves attached to seedlings were used for ΔF measurements; this enabled utilization of the same leaf during the whole greening period.

Results

Green leaves

The values of ΔF of the leaves of normal and mutant seedlings are essentially different. The relationship $\Delta F/F_i$ is 2.5-3.0 for normal leaves, 1.0-1.5 for lyco-

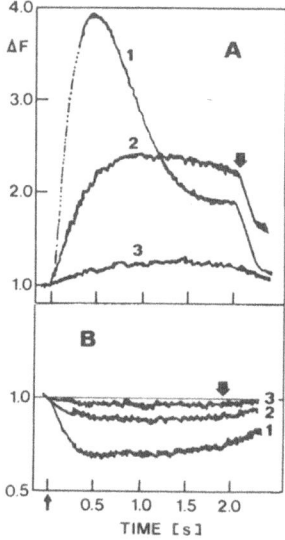

Fig. 1. The kinetics of ΔF of green maize leaves irradiated with λ = 600-700 nm (A) and λ>710 nm (B) in anaerobic conditions: 1 — normal seedlings, 2 — lycopenic, 3 — ξ-carotenic mutants. $\Delta F = \Delta F/F_i$, when $F_i = 1$.

penic, and 0.1-0.15 for ζ-carotenic leaves (Fig. 1A). The kinetics of ΔF are different in mutant and normal leaves:

(1) Although the values of the fast component of ΔF for normal and lycopenic leaves are equal, the value of the slow component of the mutant is 1.5-2.0 times lower (Fig. 1A); the slow increase in F is observed only with the ζ-carotenic mutant.

(2) The increase in F to a maximal level and its decrease were slower in mutant leaves than in normal leaves.

Also, the rates of oxygen evolution in normal and mutant leaves are essentially different (Fig. 2): in normal and lycopenic leaves the maximum level is achieved in 6-8 min and a slow decrease follows. A constant rate is achieved more rapidly in normal leaves (curve 1) than in leaves of the lycopenic mutant which require more than 30 min (curve 2). The rate of oxygen evolution of the ζ-carotenic mutant is only 3-5% of that of normal leaves (curve 3).

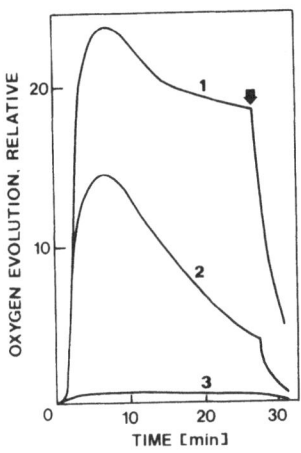

Fig. 2. The kinetics of oxygen evolution by green maize leaves of normal seedlings (1), and lycopenic (2) and ζ-carotenic (3) mutants under saturating irradiance (λ>640 nm, 100 W m^{-2}).

Illuminance of green normal and mutant leaves in anaerobic conditions with λ > 700 nm induces a decrease in F which is due to oxidation of QH by Photosystem 1. This effect is less apparent in mutant leaves. Illuminance of mutant leaves in anaerobic conditions with λ = 600-700 nm results in complex kinetics for ΔF with an intermediate dip characteristic of normal leaves. The addition of DCMU does not enhance ΔF during illumination with λ = 600-700 nm, because the reaction is saturated with radiant energy. However, the intermediate dip disappears in the presence of DCMU and only a fast rise of F to a miximal level is observed. The decrease in F caused by $\lambda > 700$ nm disappears in the presence of DCMU in both normal and mutant leaves.

When the leaf structure is mechanically damaged by a glass stick, the complex kinetics for ΔF disappear and only an increase in F is observed, similar to that of chloroplasts. The addition of dithionite to buffer containing the damaged leaf results in a slow increase in F_i (KARAPETYAN et al. 1971; KARAPETYAN 1972). Wavelengths >700 nm cause only a decrease in F, and $\lambda>680$ nm result in an increase of F with an intermediate dip. The addition of hydroxylamine to damaged normal and mutant leaves does not increase the value of ΔF.

Etiolated and Greening Leaves

Irradiance of etiolated normal and mutant leaves with $\lambda = 600\text{-}700$ nm induces an increase in F which is irreversible and disappears only when the etiolated leaf is irradiated beforehand for several minutes. The observed ΔF is not related to the functioning of the electron transport chain; it is caused by a phototransformation of protochlorophyllide to chlorophyllide which is accompanied by a long-wave spectral shift of the maximum of F (KLIMOV et al. 1972).

A reversible ΔF was observed in normal maize leaves only after 2.5 h of illumination (Fig. 3A) (see also GOEDHEER 1961; BUTLER 1965). The ΔF increased during a further 24 h of greening and its kinetics was similar to the ΔF

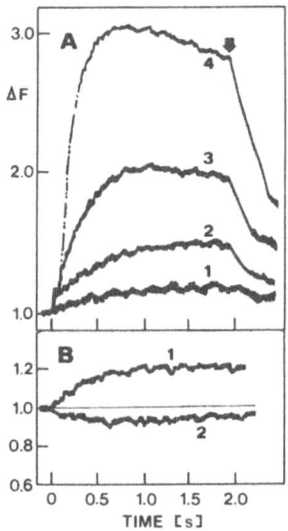

Fig. 3. The kinetics of ΔF in normal maize leaves: A – after 2.5 (1), 6 (2), 12 (3) and 24 (4) h of illumination of etiolated leaves; B – illumination of normal leaves with $\lambda>700$ nm (1) and $\lambda = 600\text{-}700$ nm (2) after 2.5 h of greening.

kinetics characteristic for a fully green leaf (curves 2-4). The increase of ΔF during greening was due to an increase in the slow component of ΔF. The reversible ΔF was observed in lycopenic leaves only after 4.5 h and in ζ-carotenic leaves only after 12 h of greening. The effect of wavelengths >700 nm on F is apparent at the same time as the effect of wavelengths 600-700 nm in both greening normal and greening mutant leaves (Fig. 3B). This shows that Photosystem 1 is already active when the activity of Photosystem 2 (ΔF) first appears.

The ability of leaves to evolve oxygen coincides with the appearance of ΔF and formation of a long-wave maximum of F. During greening the oxygen evolution rate increased in parallel with ΔF (Fig. 4A). Photoinhibition of greening leaves caused a parallel decrease and simultaneous disappearance of ΔF and oxygen evolution (Fig. 4B). These data present additional evidence for the relation of $\overline{\Delta F}$ to the activity of the photosynthetic apparatus.

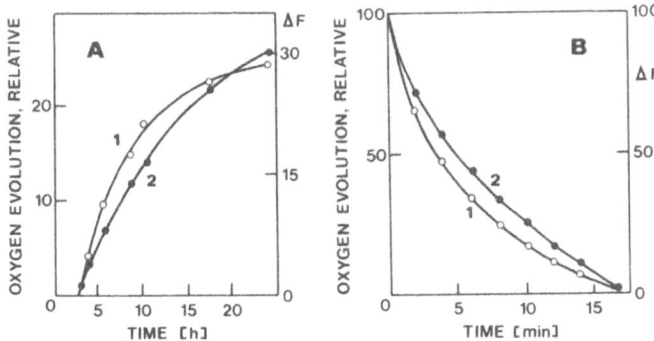

Fig. 4. The dependence of oxygen evolution rate (1) and the value of ΔF (2) on the time of greening of etiolated leaves (A) and on the time of photoinhibition of normal green leaves (B).

The ΔF kinetics of green mutant leaves and greening normal leaves are similar at different stages of greening. Mutants could be considered as greening normal leaves whose development of photosynthetic apparatus is blocked at a definite stage of greening. In the ζ-carotenic mutant this blocking takes place earlier than in the lycopenic mutant.

Discussion

The comparative study of the photosynthetic activity and ΔF of maize mutants allowed an estimation of their ability for the primary processes of photosynthesis. Both mutants have a functioning Photosystem 2. The decreasing ΔF under illuminance with λ>700 nm shows the presence of Photosystem 1 in mutants and a functioning electron transport chain between the photosystems.

However, in comparison with normal leaves, the mutants possess a lower rate of oxygen evolution. The low CO_2 incorporation of lycopenic and ζ-carotenic mutants found by GYURJÁN et al. (1969) was not only caused by a low chlorophyll content in mutant leaves. The chlorophyll content of the ζ-carotenic mutant is 20-25% of the normal value (LANG et al. 1971) but its photosynthetic activity is less then 5%.

The data presented above indicate that the low photosynthetic activity of both mutants is caused by a blocking of the light-induced reactions of photosynthesis, mainly of Photosystem 2. The low value of ΔF for leaves of these mutants under saturating illuminance without or with DCMU (which blocks electron transport between the photosystems) could be caused by a low content of active centres of Photosystem 2 or by a damaging of the structure of these centres and other sites of the photosynthetic electron transport chain.

If the low ΔF of mutants was caused by a low ratio of active centres to total chlorophyll content, then the kinetics of ΔF would be the same as for normal leaves. In fact, ΔF's of mutants and normal leaves differ not only in values but also in kinetics. This indicates some damage of the photosynthetic electron transport chain of mutants, possibly localized in either donor or acceptor parts of Photosystem 2.

Hence the results of our experiments suggest that the low photosynthetic activity of the studied mutants is related mainly to blocking of the activity of Photosystem 2.

Summary

The reversible light-induced fluorescence changes of greening and green leaves of normal and mutant (lycopene and ζ-carotene) maize seedlings are observed simultaneously with the appearance of oxygen evolution and long-wave fluorescence maximum. The activity of Photosystem 2 of normal leaves is observed after 2.5-3 h, that of the lycopene mutant after 5 h and that of the ζ-carotene mutant after 12 h of greening. The rate of oxygen evolution of the green mutant leaves is lower than that of normal leaves. The time course of photo-induced changes of fluorescence of mutant leaves is different from that of normal leaves. The interaction of Photosystem 1 and Photosystem 2 of normal and mutant leaves, detected as a fluorescence decrease upon irradiation of leaves with $\lambda > 700$ nm, takes place immediately after the appearance of Photosystem 2 activity, i.e. at this time Photosystem 1 is already active. It seems that the low photosynthetic activity of the studied mutants is due first of all to insufficient functioning of Photosystem 2.

REFERENCES

BUTLER, W. L., *Biochim. biophys. Acta* 102:1, 1965.
DUYSENS, L. N. M. & SWEERS, H. E., in: ASHIDA, J. (ed.): Studies on Microalgae and Photosynthetic Bacteria, p. 353, Univ. Tokyo Press, Tokyo 1963.
FALUDI-DÁNIEL, Á., AMESZ, J. & NAGY, A. H., *Biochim. biophys. Acta* 197: 60, 1970.
GOEDHEER, J. C., *Biochim. biophys. Acta* 51: 494, 1961.
GYURJÁN, I., RAKOVÁN, J.N. & FALUDI-DÁNIEL, Á., in: METZNER, H. (ed.): Progress in Photosynthesis Research, Vol. I, p. 63, Tübingen 1969.
KARAPETYAN, N.V., in: FORTI, G., AVRON, M. & MELANDRI, A. (ed.): Photosynthesis, Two Centuries after Its Discovery by Joseph Priestley, Vol. I, p. 180, Junk, The Hague 1972.
KARAPETYAN, N.V. & KLIMOV, V.V., *Fiziol. Rast.* 18: 223, 1971.
KARAPETYAN, N.V. & KLIMOV, V.V., *Fiziol. Rast.* 20: 545, 1973.
KARAPETYAN, N.V., KLIMOV, V.V., KRAKHMALEVA, I.N. & KRASNOVSKIĬ, A.A., *Dokl. Akad. Nauk SSSR* 201: 1244, 1971.
KARAPETYAN, N.V., KLIMOV, V.V., LANG, F. & KRASNOVSKIĬ, A.A., *Fiziol. Rast.* 18: 507, 1971.
KAUTSKY, H. & HIRSCH, A., *Naturwissenschaften* 19: 964, 1931.
KLIMOV, V.V., LANG, F., KARAPETYAN, N.V. & KRASNOVSKIĬ, A.A., *Fiziol. Rast.* 19: 151, 1972.
LANG, F., VOROB'EVA, L.M. & KRASNOVSKIĬ, A.A., *Biokhimiya* 34: 257, 1969.
LANG, F., VOROB'EVA, L.M. & KRASNOVSKIĬ, A.A., *Mol. Biol.* (Moskva) 5: 366, 1971.
MUNDAY, J.C. Jr. & GOVINDJEE, *Biophys. J.* 9: 22, 1969.
SHLYK, A.A., FRADKIN, L.I., SAVCHENKO, G.E. & FALUDI-DÁNIEL, A., *Photosynthetica* 1: 241, 1967.

HILL REACTION AND DELAYED FLUORESCENCE IN MUTANTS OF GOSSYPIUM HIRSUTUM

MUKHIBA M. YAKUBOVA*, A. B. RUBIN**, GALINA A. KHRAMOVA* & D. N. MATORIN**

V.I. Lenin Tajik State University, Dushanbe, U.S.S.R., and *M.V. Lomonosov State University, Moscow, U.S.S.R.***

Introduction

The electron-transport chain (ETC) in mutants with altered photosynthetic activity and defects in the biosynthesis of pigments has recently become a very interesting topic of research. The use of lethal and viable mutants allows identification of the components which function in the ETC of photosynthesis, allows the coupling of the photosystems to be blocked and also permits determination of the regulating effect of genetic factors of the nucleus and chloroplasts upon these processes (HERRMANN 1971; NASYROV et al. 1971; VOSKRESENSKAYA et al. 1971). Variations in the pigment content in chloroplasts of mutants may be related not only to a genetic block in the biosynthetic chains of chlorophyll and carotenoids, but also to disturbances in other biochemical reactions (VOSKRESENSKAYA et al. 1968; GILLER et al. 1971). Consequently, a detailed study of chlorophyll mutants may clarify the mechanism of action of genetic factors upon the whole chloroplast.

In this connection we studied photochemical processes of photosynthesis (Hill reaction and delayed fluorescence) in viable pigment mutants of *Gossypium hirsutum*.

Delayed fluorescence discovered by STREHLER & ARNOLD (1951), is a universal peculiarity of all photosynthesizing plants. Fluorescence is produced by the fully formed photosynthetic apparatus and depends on a functioning reaction centre. Delayed fluorescence is closely related to the primary photochemical reactions of photosynthesis (LITVIN et al. 1966).

Plants and Methods

The quantitative and qualitative composition of plastid pigments in initial cultivars and mutants of cotton were evaluated during plant ontogenesis and the most interesting mutant forms were chosen (YAKUBOVA et al. 1972). Three cotton cultivars – C-460, 108-F, C-4727 – and their mutants produced by γ-irradiation of seeds in the Laboratory of Physiological Genetics of the Institute of Plant

Physiology and Biophysics of the Academy of Sciences of the Tajik S.S.R. by P. USMANOV and G. BIKASYAN were grown in soil culture in a chamber with artificial climate (25-28°C; 10 klx; day-length 16 h).

Chloroplasts were isolated from plants in the 4-5 leaf stage using a modified method of ARNON & WHATLEY (1963). The medium contained 0.3 M NaCl, 0.5 M sucrose, 0.05 Tris-HCl buffer (pH 7.8) and 3% bovine serum albumin. The rate of the Hill reaction was estimated spectrophotometrically by reduction of ferricyanide in a medium containing in 3 ml: 104 μmol NaCl, 45 μmol Tris-HCl buffer (pH 7.8), 3 μmol $K_3Fe(CN)_6$ and chloroplast suspension (50-100 μg of chlorophyll). The samples were illuminated 10 min at 40 klx, temperature 18-20°C. After illumination the samples were fixed with 1 ml of 10% TCA. Chlorophyll content was measured according to ARNON (1949).

Some specific biochemical products accumulating in cells (fatty acids, phenolic compounds, organic acids, etc.) may inhibit the activity of isolated chloroplasts. Gossipol, the prevailing polyphenol of cotton plants, is formed already during the first days of growth and is present in all vegetative and generative organs (SADYKOV 1971). This compound and other endogenous inhibitors extracted into the medium during chloroplast isolation were bound with added bovine albumin (FRY 1970). Delayed fluorescence was recorded with an apparatus using the phosphoroscope developed in the Laboratory of Cosmic Biology of M.V. Lomonosov University of Moscow (MATORIN 1971); this apparatus separated the moments of excitation and fluorescence. The period between excitation and the time of fluorescence measurement was 2.5 ms. The source of radiation was an incandescent lamp (750 W) with a water filter.

The inhibitor DCMU which blocks the photosynthetic ETC before plastoquinone in between Photosystems 1 and 2 (TREBST et al. 1970) was used at a concentration of 10^{-5} M. The block deprives Photosystem 1 of the normal supply of electrons generated in Photosystem 2, and inhibits the millisecond component of fluorescence which is an expression of non-cyclic electron transport (see LITVIN & KRASNOVSKIĬ 1967).

Results

The induction curve of changes in the intensity of delayed fluorescence of chloroplasts from cotton cv. C-460 consists of two distinct components — a long-living and a millisecond component (Fig. 1, top). In the mutant of this variety, 50/8, the transition from the millisecond to the long-living component is smooth.

Uncoupling of Photosystem 1 from Photosystem 2 with DCMU changes the induction curve of delayed fluorescence in cv. C-460, namely its millisecond component. Very little change was induced by DCMU in the mutant 50/8 (Fig. 1, top) perhaps indicating defects in non-cyclic electron transport in this mutant. Experiments with leaves confirmed these results (Fig. 2).

The rate of the Hill reaction expressed as ferricyanide reduction per unit

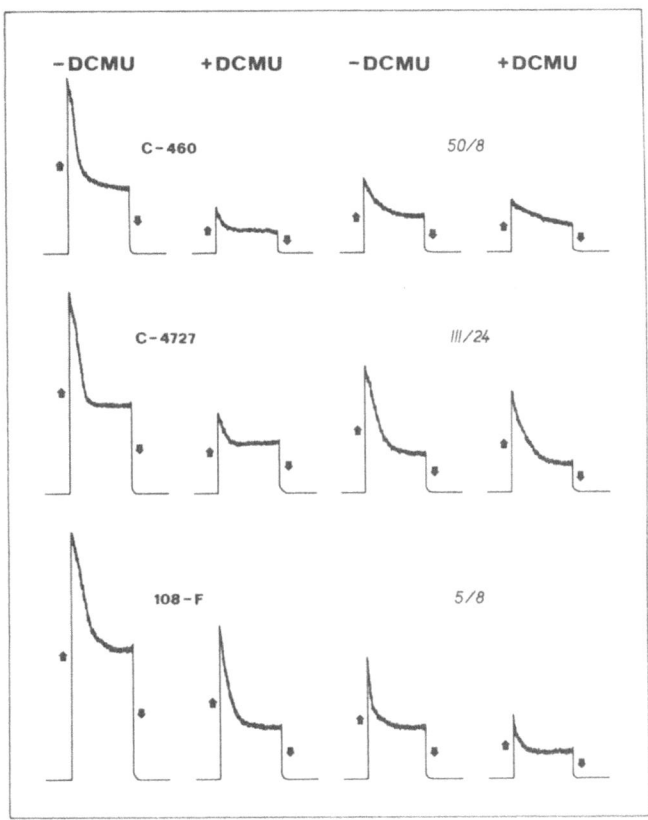

Fig. 1. Intensity of delayed fluorescence of chloroplasts in isolation medium and after DCMU treatment. Cotton cv. C-460, C-4727, and 108-F; mutants 50/8, III/24 and 5/8.

amount of chlorophyll (Table 1) was in cv. C-460 twice as high as in the mutant 50/8. Hence the mutant 50/8 truly has a defect in the electron flow between Photosystems 1 and 2 in addition to altered biosynthesis of pigments and may be regarded as both a pigment and a photosynthetic mutant.

The induction curves of delayed fluorescence for chloroplasts from the cotton mutant III/24 (Fig. 1, middle) in the isolation medium and after DCMU treatment do not differ in shape from those for the initial cv. C-4727. The difference in intensity of delayed fluorescence might be explained by a lower chlorophyll content in the sample from the mutant (ca. 50% of that of cv. C-4727). As similar differences have been observed in the Hill reaction (Table 1), both the lower intensity of delayed fluorescence and the impaired capacity for ferricyanide reduc-

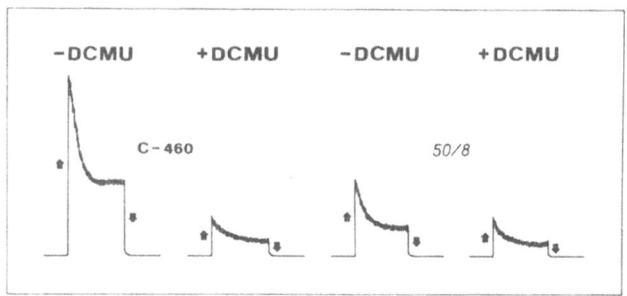

Fig. 2. Intensity of delayed fluorescence in leaves of cotton cv. C-460 and mutant 50/8 before and after DCMU treatment.

Table 1. Rate of Hill reaction in chloroplasts isolated from cotton plants

Variant	Rate of $K_3Fe(CN)_6$ reduction [μM mg^{-1} (Chl) h^{-1}]	Chlorophyll ($a+b$) [mg g^{-1} (fresh w.)]
Wild form C-460	94.8	0.980
Mutant 50/8	37.8	1.051
Wild form C-4727	66.0	0.790
Mutant III/24	30.0	1.198
Wild form 108-F	114.0	1.240
Mutant 5/8	81.6	1.150

tion in the mutant III/24 might be due only to the decreased content of chlorophyll.

The intensity of delayed fluorescence per chlorophyll unit in the cotton cv. 108-F was twice as high as in its mutant 5/8 (Fig. 1, bottom). DCMU halved the intensity of delayed fluorescence in both the initial cultivar and the mutant. These data together with the Hill reaction rates (Table 1) constitute evidence that in chloroplasts of the mutant 5/8 electron transport is only somewhat slower than in the initial cultivar.

The leaves of a variegated mutant of cotton cultivar 108-F have green, light-green and yellow sections (Fig. 3). The chlorophyll content in the green section corresponded to that of cv. 108-F, while the light-green portion contained half as much chlorophyll as in the green part, the yellow section did not contain chlorophyll at all.

The induction curves of delayed fluorescence of the green portions of leaf from the variegated mutant were of normal shape, with both components (Fig. 4). The light-green leaf part had a decreased delayed fluorescence, probably due to the reduced content of chlorophylls. The yellow section of the leaf did not emit any delayed fluorescence. These data agree well with existing concepts on the

Fig. 3. Variegated mutant of cotton cv. 108-F. (Photo by P. USMANOV & G. BIKASYAN.)

relationship between the intensity of delayed fluorescence and both the chlorophyll content and formation of the photosynthetic apparatus. This process is a photo-induced chemiluminescence of chlorophyll which arises as a result of reversible reactions of the primary products of photosynthesis (FERRARI et al. 1957; LITVIN & KRASNOVSKIĬ 1967).

Summary

The relationship between the values of delayed fluorescence and Hill reaction indicates a close connection between delayed fluorescence and the primary reac-

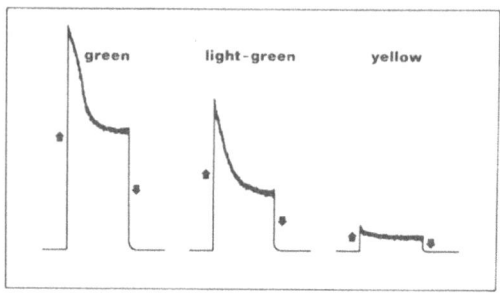

Fig. 4. Intensity of delayed fluorescence of green, light-green, and yellow sections of a leaf from the variegated-leaf mutant of cotton cv. 108/F.

tions of photosynthesis. Both the intensity of delayed fluorescence and the Hill reaction depend on the concentration of chlorophyll, as illustrated by experiments with the mutant III/24 of cotton cv. C-4727 and the variegated mutant 5/8 of cv. 108-F. In the mutant 50/8 of the cv. C-460 a nuclear mutation blocked not only the biosynthesis of pigments but also some photochemical component of the electron transport chain between Photosystems 1 and 2. These results suggest that formation of electron transport chain components in Photosystem 1 is controlled by nuclear genes.

REFERENCES

ARNON, D. I., *Plant Physiol.* 24: 1, 1949.
ARNON, D. I. & WHATLEY, F. R., in: COLOWICK, S. P. & KAPLAN, N. O. (ed.): Methods in Enzymology, Vol. 6, p. 308, Academic Press, New York-London 1963.
FERRARI, A. G., STREHLER, B. L. & ARTHUR, W. E., in: GAFFRON, H. et al. (ed.): Research in Photosynthesis, p. 45, Interscience Publ., New York-London 1957.
FRY, K. E., *Plant Physiol.* 45: 465, 1970.
GILLER, Yu. E., ASOEVA, L. M. & KAS'YANENKO, A. G., in: NASYROV, Yu. S. (ed.): Geneticheskie Aspekty Fotosinteza, p. 107, Donish, Dushanbe 1971.
HERRMANN, F., *Photosynthetica* 5: 258, 1971.
LITVIN, F. F. & KRASNOVSKIĬ, A. A. Jr., *Dokl. Akad. Nauk SSSR* 173: 451, 1967.
LITVIN, F. F., SHUVALOV, V. A. & KRASNOVSKIĬ, A. A., *Dokl. Akad. Nauk SSSR* 168: 1195, 1966.
MATORIN, D. N., Izuchenie Svyazi Protsessov Poslesvecheniya s Elektron-transportnymi Reaktsiyami Fotosinteza, Thesis, Mosk. gos. Univ., Moskva 1971.
NASYROV, Yu. S., KAS'YANENKO, A. G. & ABDURAKHMANOVA, Z. N., in: Biokhimiya i Biofizika Fotosinteza, p. 121, Irkutsk 1971.
SADYKOV, A. S., *Vestn. Akad. Nauk SSSR* 1971 (4): 37, 1971.
STREHLER, B. L. & ARNOLD, W. G., *J. gen. Physiol.* 34: 809, 1951.
TREBST, A., HARTH, E. & DRABER, W., *Z. Naturforsch.* 25 h: 1157, 1970.
VOSKRESENSKAYA, N. P., DROZDOVA, I. S. & GOSTIMSKIĬ, S. A., in: ZALENSKIĬ, O. V. (ed.): *Fotosintez i Ispol'zovanie Solnechoĭ Energii,* p. 236, Nauka, Leningrad 1971.

VOSKRESENSKAYA, N. P., OSHMAROVA, I. S. & GOSTIMSKIĬ, S. A., *Genetika* 4 (2): 41, 1968.
YAKUBOVA, M. M., USMANOV, P. D., BIKASIYAN, G. R., KHRAMOVA, G. A. & SHCHERBAKOVA, I. Yu., Sbornik Rabot Aspirantov (Seriya Biologicheskikh Nauk), Chast´ 4, Tadzh. Gosuniv., Dushanbe 1972.

SYNTHETIC PIGMENT-PROTEIN-LIPOID COMPLEXES — MODELS OF MOLECULAR ORGANIZATION AND FUNCTIONAL PROPERTIES OF THE PIGMENT SYSTEM OF THE PHOTOSYNTHETIC APPARATUS

YU. E. GILLER, LYUBOV' R. VAKHIDOVA, LERA M. ASOEVA, LYUBOV' N. YUKHANANOVA, SAODAT K. ABDULLAEVA, B. I. LIPKIND, GALINA V. KRASICHKOVA & GALINA A. YUSUPOVA

Institute of Plant Physiology and Biophysics, Academy of Sciences of the Tajik S.S.R., Dushanbe, U.S.S.R.

Introduction

The concept of structural and functional differences among the native forms of chlorophyll and its analogues is now universally recognized, and the analysis of factors responsible for the occurence of pigment forms differing in their molecular organization and their role in photosynthesis is highly desirable.

Pigment-pigment interactions which, together with the state of the protein or protein-lipoid matrix of the native pigment-lipoproteid complex, account for the different aggregation of chlorophyll molecules (KRASNOVSKIĬ 1962, 1965, 1970; LITVIN 1965; GULYAEV & LITVIN 1967; BYSTROVA 1972) explain also the existence of discrete spectral forms of chlorophyll in vivo. Proteins and lipids play an important role in the arrangement of different chlorophyll forms in native structures (OSTROVSKAYA 1972) and in model systems (SEMICHAEVSKIĬ & LOZOVAYA 1971; SEMICHAEVSKIĬ et al. 1971; GILLER 1972; GILLER et al. 1972; SEMICHAEVSKIĬ 1972; GILLER & YUKHANANOVA 1973) as well as for bacteriochlorophyll forms in chromatophores (EROKHIN 1969, 1972; EROKHIN & SINEGUB 1970 a, b; SINEGUB & EROKHIN 1971).

The negligible role of the concentration of chlorophyll in the formation of its aggregated forms was ascertained by the appearance of these forms during the first hours of greening of etiolated leaves, when the total pigment concentration is far from normal (LITVIN & BELYAEVA 1971). Additional evidence is the existence of mutant plants characterized either by the absence of the pigment-protein complex in Photosystem 1 (chlorophyll forms absorbing at the longest wavelengths) and a normal pigment content (GREGORY et al. 1971), or, conversely, noted for the presence of an abnormally aggregated form of chlorophyll with an absorption maximum at 745 nm (von WETTSTEIN et al. 1971).

Synthetic complexes of pigments with proteins and lipoproteids with spectral properties and photochemical characteristics similar to those of chlorophylls in vivo (GILLER & KHAITOVA 1966; SAPOZHNIKOV et al. 1966; GILLER 1968, 1969; GILLER et al. 1968, 1970; GILLER & YUKHANANOVA 1969, 1970;

GILLER & YUSUPOVA 1969, 1970; KRASICHKOVA 1972; YUSUPOVA & GILLER 1972) are good models for studying the role of the state of the protein or protein-lipoid carrier in the formation of different spectral forms of chlorophyll. This paper reviews our data on the spectral and photochemical properties and energetic interactions of photosynthetic pigments in synthetic complexes, simple models, and in vivo. They present valid proof of the important (if not predominant!) role of pigment-protein interaction in the formation of discrete native forms of pigments. A hypothesis has been suggested on the possible mode of genetic control of the molecular organization of the pigment system in the photosynthetic apparatus.

Experimental techniques for obtaining the synthetic pigment-protein or pigment-protein-lipoid complexes and analysis of their spectral and photochemical properties are described elsewhere (GILLER & KHAITOVA 1966; SAPOZHNIKOV et al. 1966; GILLER et al. 1968, 1970; TOLIBEKOV et al. 1968, 1969; GILLER & YUKHANANOVA 1969, 1970).

State of Pigments in The Synthetic Complex

Using the synthetic pigment-protein complex with caseic acid as a protein component of the system, the dependence of pigment incorporation into the complex on their concentration in the reaction mixture was studied (TOLIBEKOV et al. 1968, 1969) in order to compare the nature of our model to the analogous characteristics of the simpler models and to the quantitative composition of the pigment-containing chloroplast structures. The amount of pigment incorporation into the synthetic complex (Table 1) far exceeds that for sorption models with inorganic and simple polymer carriers (NEKRASOV et al. 1961, 1962; NEKRASOV & KAPLER 1966), and is nearly equal to that of quantasomes (PARK 1964), lamellae (RADUNZ et al. 1971), and native chlorophyll-protein complexes (THORNBER 1969; BOARDMAN 1970; VERNON et al. 1971).

At comparatively low concentrations of chlorophyll in the complex the main red absorption band was at 672 nm (Fig. 1A) and there appeared a form of pigment typical for one of the stages of formation of pigment-containing chloroplast structures (disaggregation of subunits containing Chl_{684}, reorientation) (HENNINGSEN 1970). This complex fluoresces intensively at room temperature (Fig. 1B), an unusual property of the pigment native state, and not consistent with its extremely high concentration in chloroplasts and chromatophores (FRANCK & ROSENBERG 1964; GURINOVICH et al. 1968). At liquid nitrogen temperature the intensity of long-wave fluorescence of chlorophyll rapidly increases, thus indicating the presence in the system of an aggregated pigment form typical of the native state (LITVIN et al. 1962). This chlorophyll form is not removed from the complex by petroleum ether, i.e. it is closely associated with the lipoproteid matrix (GILLER et al. 1968, 1970).

With the increase in chlorophyll concentration in the complex (Fig. 1), the

Table 1. Maximum amounts of chlorophyll incorporated into model systems in comparison to analogous characteristics of native systems.

System	Pigment content		References
	[%]	[mol chlorophyll per mol protein]	
Sorption models (chl. *a* and *b*) on carriers:			
MgO, alumogel, kapron	0.15		NEKRASOV et al. (1961, 1962)
Polyacrylonitril	1.0		NEKRASOV & KAPLER (1966)
Synthetic pigment-protein complexes:			
chlorophyll *a* and *b*	3.5	0.7	
chlorophyll *a*	17.5	3.5	
chlorophyll *b*	6.7	1.3	
Quantasomes:			
chlorophyll *a*	8.3		
chlorophyll *b*	3.7		PARK (1964)
Chloroplasts free from stroma:			
chlorophyll *a* and *b*	12.4		RADUNZ et al. (1971)
Chlorophyll-protein complexes of reaction centres of			
Photosystem 1	11	5	THORNBER (1969); BOARDMAN (1970); VERNON et al. (1971)
Photosystem 2	7-8		VERNON et al. (1971)

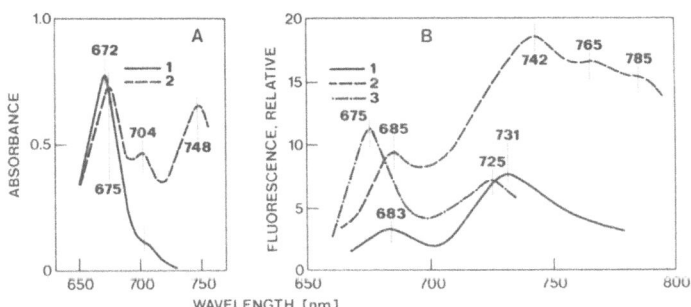

Fig. 1. Absorption (A) and fluorescence (B) spectra of chlorophyll *a* in the synthetic pigment-protein-lipoid complexes. (1) chlorophyll concentration in the complex 0.9%, fluorescence at 77K; (2) 17.5% chlorophyll, fluorescence at 77K; (3) 0.9% chlorophyll, fluorescence at room temperature.

Table 2. Molar absorption coefficients (α) of photosynthetic pigments in model systems and *in vivo*.

Pigment	Molecular solution [1 mol^{-1} cm^{-1} × 10^{-4}]		Hard layer Pigment concentration [M]	α	Synthetic complex Pigment concentration [M; %]	α	Cells Pigment concentration	α
Chlorophyll *a*	9.1	(SMITH & BENITEZ 1955)	0.46-0.50	3.1-3.4	10^{-6} 0.009%	9.0	8.3% (BRANDT & TAGEEVA 1967)	3.54
	8.51	(SEELY & JENSEN 1965)			1.2 · 10^{-4}; 1.1%	3.67		
Chlorophyll *b*	5.6	(SMITH & BENITEZ 1955)			0.5 · 10^{-5}; 0.045%	2.6		
					10^{-4}; 0.9%	2.2		
β-carotene	13.9	(SAPOZHNIKOV et al. 1964)			1.6 · 10^{-6}; 0.009%	12.3		
					1.6 · 10^{-4}; 0.9%	3.5		

maximum of the main red absorption band shifts to 675 nm, and additional long-wave maxima appear at 704 and 748 nm. In the fluorescence spectrum at 77 K the main bands shift to longer wavelengths, and additional bands are detected which belong to the aggregated forms of chlorophyll observed in vivo (LITVIN et al. 1962) and in model systems (LITVIN 1965).

The above data suggest that packing of pigment molecules into aggregates (especially into the form fluorescing at 77 K in the range of 730-740 nm) depends not only on the quantity of the pigment, but also on the specific conditions for aggregation created by the lipoproteid matrix of the complex. This assumption was also confirmed by data on the state of chlorophyll in abnormal native systems, both genetically determined (mutants) and caused by antibiotics (NASYROV et al. 1975).

Values of molar absorption coefficients (α) of chlorophylls and carotene in the complex (Table 2) are another indication of the irregular distribution of pigment molecules in our model which leads to the formation of aggregated forms even at low pigment concentrations; it is determined by a specific bond between chlorophyll and the lipoproteid or protein carrier. At 1.1% chlorophyll the value α is already comparable to the in vivo value, estimated according to BRANDT & TAGEEVA (1967). The chlorophyll molar absorption coefficient in the complex containing 1.1% of the pigment is considerably lower than α for solutions (SMITH & BENITEZ 1955; SEELY & JENSEN 1965) and almost equivalent to that for the hard pigment layer on glass where the chlorophyll concentration is 3-4 fold higher than in the complex (Table 2). Hence our model contains zones with locally increased chlorophyll concentrations, similar to the irregular variation in chlorophyll concentration in native structures during greening resulting from the conformational changes in holochrome protein (HENNINGSEN 1970). Similar changes of the molar absorption coefficients have been observed for chlorophyll b and β-carotene (Table 2).

The possible nature of the discrete spectral forms of chlorophyll in the artificial complexes is reviewed in Table 3 where the absorption peaks of chlorophyll a in the complex and simple models are compared. Hence the complex contains both the monomer forms and aggregates of different types found in model systems without any carrier and in vivo. In our model we also observed aggregates of chlorophyll b and β-carotene (Table 3) somewhat different from the native forms of these pigments (NISHIMURA & TAKAMATSU 1957; LITVIN et al. 1970).

Changes of the Chlorophyll State in the Complex after Action of the Carrier

In order to assess the degree of association of the discrete spectral forms of chlorophyll developed in the complex with lipoproteid or protein carrier we studied the acetone-induced changes in optical properties of the complexes. Acetone at concentrations lower than 50% changes the spectral properties of chlorophyll by changing the pigment-pigment and pigment-protein interaction in chloroplasts

Table 3. Maxima of absorption bands and possible nature of the forms of photosynthetic pigments in synthetic pigment-protein-lipoid complexes.

Maxima of absorption bands	Possible nature of the form responsible for the band	Presence of the form in native structures and/or references
Chl. *a* 668	Monomer form associated with lipoproteid carrier	(LITVIN *et al.* 1970)
Chl. *a* 672	Slightly aggregated form generating on water-lipid borders (CHAPMAN & FAST 1968) and in synthetic chlorophyll membranes (STEINEMANN *et al.* 1971)	(HENNINGSEN 1970)
Chl. *a* 673-675	Slightly aggregated form analogous to associations formed in colloidal solutions (GURINOVICH & STRELKOVA 1968; BYSTROVA 1972)	(LITVIN & GULYAEV 1969; GULYAEV & LITVIN 1970; MACHOLD *et al.* 1971)
Chl. *a* 704	Aggregated form analogous to that formed in the concentrated solution in pyridine (BRODY & BRODY 1966)	(LITVIN & GULYAEV 1969; GULYAEV & LITVIN 1970)
Chl. *a* 740-750	Crystallic structure (JACOBS *et al.* 1957)	(von WETTSTEIN *et al.* 1971)
Chl. *b* 655-658	Slightly aggregated form, analogous to associations formed in colloidal solutions (GURINOVICH & STRELKOVA 1968; BYSTROVA 1972)	*in vivo* 650-651 (LITVIN *et al.* 1970)
β-carotene 500-504	Aggregated form analogous to that formed in hard layers (LITVIN 1965)	*in vivo* 550 (NISHIMURA & TAKAMATSU 1957)

(TUMERMAN et al. 1961; THOMAS & FLIGHT 1964; THOMAS & VAN DER WAL 1964), leaf homogenates (VOROB'EVA & KRASNOVSKIĬ 1967), and quantasomes (SAUER & PARK 1964), and induces 're-packing' of bacteriochlorophyll molecules (probably as a result of conformational changes of the carrier) in bacterial chromatophores (EROKHIN & SINEGUB 1970 a, b; SINEGUB & EROKHIN 1971). At higher concentrations of acetone the carrier is denatured and the pigment dissociates from the carrier (THOMAS & FLIGHT 1964; THOMAS & VAN DER WAL 1964; EROKHIN & SINEGUB 1970 a, b; SINEGUB & EROKHIN 1971).

With increasing acetone concentration in the system the red maximum of chlorophyll absorption shifts to shorter wavelengths but not as far as the position typical of the acetone solution of chlorophyll, as was earlier observed for chloroplasts (THOMAS & VAN DER WAL 1964). This indicates that in 60% acetone, when the protein is denatured, pigment-pigment interaction stops completely, whereas pigment-protein interaction responsible for the observed 'red shift' (GILLER et al. 1968, 1970; GILLER 1968, 1969) still continues.

The chlorophyll a form with an absorption maximum of 704 nm disappears completely in 30% acetone (Fig. 2), demonstrating its weak association with the carrier. This association is probably only an absorption. The presence of aggregates of this type in vivo (THOMAS & VAN DER WAL 1964; BRODY & BRODY 1966; LITVIN & GULYAEV 1969; GULYAEV & LITVIN 1970) led to the conclusion that at least part of the chlorophyll in chloroplasts and chromatophores is fixed on lipoproteid carriers by adsorption (NEKRASOV 1967).

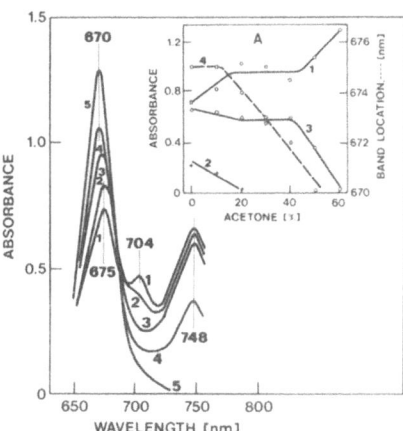

Fig. 2. Effect of acetone on absorption spectrum of chlorophyll a in the synthetic pigment-protein-lipoid complex.
(1) control; (2) 10% acetone; (3) 20, 30, 40%; (4) 50%; (5) 60%. A: Changes in absorbance of the peaks and location of the main maximum: (1) band 675-670 nm; (2) band 704 nm; (3) band 748 nm; (4) location of the main maximum.

Acetone concentrations of 10-40% have little effect on the content of chlorophyll a_{748} in the complex. This form disappears with the denaturation of protein at concentrations of 40-60%. The relationship between chlorophyll a_{748} destruction and denaturation of the carrier, and also its spasmodic character (Fig. 2A) similar to that earlier observed for one of the native forms of bacteriochlorophyll (EROKHIN & SINEGUB 1970a, b; SINEGUB & EROKHIN 1971) indicate the dependence of the structure of these aggregates on the state of the protein-carrier. This is supported by data on the possibility of induction of chlorophyll a_{748} formation in the native complex by treatment with hot organic solvents (LIPPINCOTT et al. 1962; AGHION 1963; AGHION et al. 1963) and in the synthetic complex by decreasing pH (GILLER & YUKHANANOVA 1969, 1970), i.e. treatments affecting both the pigment-pigment interaction and the carrier. Hence the abnormal structure of lamellar proteins – carriers of the chlorophyll native forms – is the reason for chlorophyll a_{745} formation in the so-called 'infra-red' barley mutants (von WETTSTEIN et al. 1971).

Fluorescence spectra at 77 K (Fig. 3) provided additional information on the

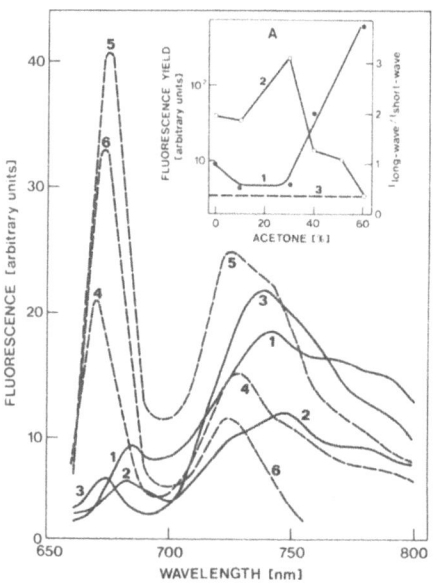

Fig. 3. Effect of acetone on fluorescence spectrum of chlorophyll in the synthetic pigment-protein complex.
(1) control; (2) 10% acetone; (3) 30%; (4) 40% (the scale reduced 2-fold); (5) 50% (the scale reduced three-fold); (6) 60% (the scale reduced 20-fold). A: Changes in the fluorescence yield (1); ratio of intensities of fluorescence in both peaks (Ilong-wave/Ishort-wave) (2); the ratio Ilong wave/Ishort-wave for chlorophyll a solution (RABINOWITCH 1951) (3).

chlorophyll state in the complex. At acetone concentrations higher than 30% disaggregation of the pigment takes place followed by disappearance of the long-wave maxima and appearance of a new band at 723 nm as well as a marked increase in the fluorescence yield in the region of the short-wave maximum. The pigment form responsible for fluorescence in the region 740-742 nm at 77 K disappears completely when the carrier is fully denatured (50-60% acetone) and the ratio $I_{\text{long-wave}}/I_{\text{short-wave}}$ reaches the value typical for chlorophyll a solutions (RABINOWITCH 1951) (Fig. 3A).

Concentrations of acetone lower than 30% probably induce 're-packing' of chlorophyll molecules with intensive aggregation, similar to that in bacterial chromatophores (EROKHIN & SINEGUB 1970a, b; SINEGUB & EROKHIN 1971); the fluorescence yield decreases with an increasing ratio of $I_{\text{long-wave}}/I_{\text{short-wave}}$ (Fig. 3).

The spasmodic (co-operative) character of changes in the pigment spectral and photochemical properties under increasing concentrations of organic solvents and detergents which break the conformation of lipoproteid carriers and disturb the hydrophobic interaction, and finally, pigment-pigment and pigment-protein interactions in the complex, also indicate the dependence of the packing pattern of chlorophyll molecules in native spectral forms upon the state of the protein or protein-lipoid matrix. In this way the spectral and photochemical parameters of chlorophyll are altered depending on its state in chloroplasts, leaf homogenates and quantasomes treated by organic solvents and detergents (KOSOBUTSKAYA & KRASNOVSKIĬ 1953; TUMERMAN et al. 1961; SAUER & PARK 1964; THOMAS & FLIGHT 1964; THOMAS & VAN DER WAL 1964; VOROB'EVA & KRASNOVSKIĬ 1966, 1967). The co-operative character of pigment (bacteriochlorophyll) 're-packing' has been particularly observed after addition of solvents

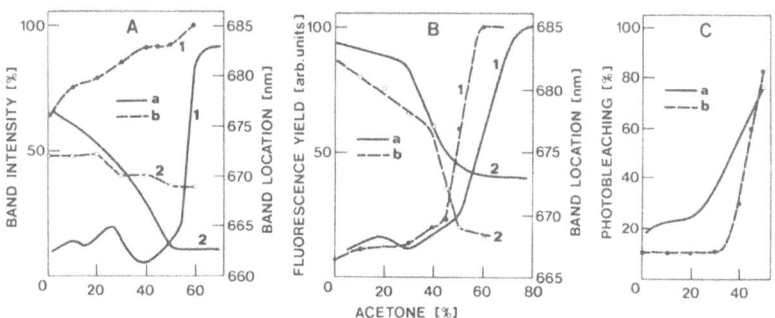

Fig. 4. Effect of acetone on the spectral and photochemical properties of chlorophyll a in the synthetic pigment-protein-lipoid complex and in native structures. A: Absorption. (1) intensity of band; (2) location of maximum; a – chloroplasts (THOMAS & VAN DER WAL 1964), b – complex. B: Fluorescence. (1) relative yield; (2) location of maximum; a – chloroplasts (THOMAS & FLIGHT 1964); b – complex. C: Photodestruction. a – leaf homogenates (VOROB'EVA & KRASNOVSKIĬ 1966); b – complex.

and detergents into suspensions of *Chromatium* chromatophores (EROKHIN & SINEGUB 1970a, b; SINEGUB & EROKHIN 1971).

The variations in absorption, fluorescence and photo-chemical properties of chlorophyll in synthetic pigment-protein-lipid complexes and in native structures (chloroplasts and leaf homogenates treated by acetone) are very similar (Fig. 4) and are observed at approximately the same concentration of solvent. Analogous results have been obtained in experiments with pyridine and the detergent sodium dodecylsulphonate (GILLER & YUKHANANOVA 1974).

Energy Interaction of Pigments in the Complex

A study of the energetic interactions of photosynthetic pigments (heterogeneous energy migration – RABINOWITCH & GOVINDJEE 1969) in this system gave some additional information about the specificity of molecular organization of the photosynthetic pigments in the artificial pigment-protein-lipid complex. The efficiency of energy transfer from chlorophyll b to chlorophyll a in the synthetic complex where the concentration of each pigment was varied from 10^{-5} to 10^{-7} M was estimated (Fig. 5) by the ratio of the intensities of chlorophyll a and chlorophyll b absorption in the excitation spectrum of chlorophyll a fluorescence (I_b/I_a). Comparison of variations of this value as affected by different concentrations of pigments in the complexes and molecular solutions (the efficiency of energy migration was estimated according to DUYSENS 1952) revealed that, in complexes, energy transfer from chlorophyll b to the short-wave forms of chlorophyll a remains at concentrations one order lower, and to the long-wave forms at least two orders lower than in chlorophyll solutions. This observation also con-

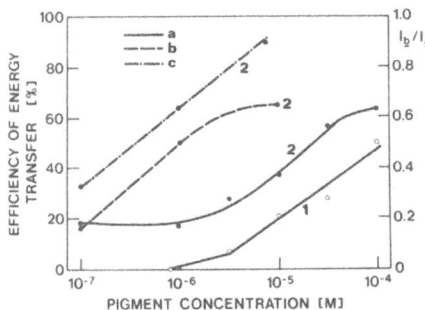

Fig. 5. Efficiency of energy migration from chlorophyll b to chlorophyll a in solution and in a synthetic pigment-protein-lipid complex. (1) efficiency of energy migration estimated according to DUYSENS (1952); (2) ratio I_b/I_a for (a) solution, (b) short-wave fluorescence band of the complex at room temperature, and (c) long-wave fluorescence band of the complex at 77 K.

firms that pigment molecules are irregularly distributed in the complex leading to the formation of zones with increased pigment concentration, probably localized in the 'slits' and 'pores' of protein layers in the membrane (RADUNZ et al. 1971). In these zones the aggregated forms of chlorophyll a which appear accept most effectively exitation energy from the molecules of chlorophyll b.

Analogous results have been obtained from a study of the energy migration from β-carotene to chlorophyll a in the complex (Fig. 6) during excitation in the region of absorption of chlorophyll a or β-carotene. Energy absorbed by β-carotene is transferred directly to the aggregated form of chlorophyll a fluorescing only at 77 K. These data agree well with information on the localization of β-carotene in the 'antenna' complex of Photosystem 1 (BOARDMAN 1970; VERNON et al. 1971).

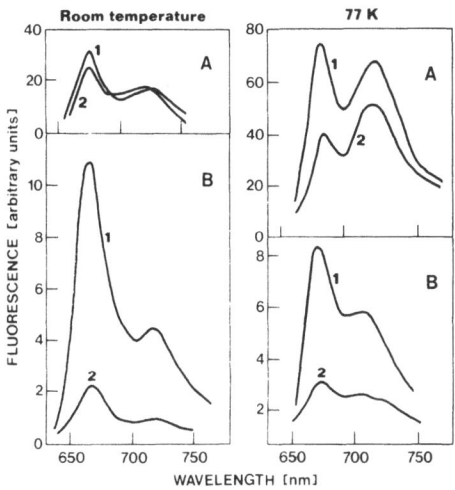

Fig. 6. Fluorescence spectra of synthetic pigment-protein-lipoid complexes containing different concentrations of chlorophyll a and β-carotene (A – chlorophyll a 2.7×10^{-5} M; β-carotene – 3.0×10^{-5} M; B – chlorophyll a 1.2×10^{-6} M; β-carotene 1.2×10^{-6}) and excited in the region of 432 nm (1) or 500 nm (2).

Conclusions

The molecular organization of photosynthetic pigments in synthetic pigment-protein and pigment-protein-lipoid complexes is rather similar to that of the native pigment-containing structures of chloroplasts and chromatophores. The artificial systems studied represent an adequate model of the function of the protein or lipoproteid carrier in the formation and activity of native forms of photosynthetic pigments.

The proposed hypothesis on the decisive role of the lipoproteid matrix in the formation of discrete forms of photosynthetic pigments which carry out specific functions in the initial act of light energy conversion during photosynthesis might be a threshold towards investigations on the genetic control of the molecular organization of the pigment system of chloroplasts. If the structure and state of the protein-lipoid components in native complexes determine the structure of pigment forms, then genetic regulation of the organization of pigment systems in the photosynthetic apparatus might be realized through control of the synthesis of the specific proteins and lipoproteins – carriers of the native pigment forms. This type of regulation controls only the composition of the pigment structures and acts to some extent independently (according to the data described above) of the now well established mechanisms of regulating pigment quantity through the control of the synthesis and activity of enzymes catalysing the biosyntheses of chlorophylls and carotenoides (NASYROV 1972).

Summary

The synthetic complex may incorporate an amount of chlorophyll similar to that in in vivo chlorophyll complexes. The bulk of chlorophyll in the complex is in a slightly aggregated state and is bound to the lipoprotein carrier. Significant fluorescence at $+20\,°C$ and the greatly increased fluorescence yield at 77 K in the region of 725-735 nm are typical of the native state of the pigment. Strongly aggregated forms of the pigment with absorption maxima at 704 and 748-750 nm and fluorescence maxima at 750 and 770-780 nm appear only under certain conditions. The action of organic solvents on the complex revealed the nature of aggregated forms of the pigment (according to the strength of binding with the carrier). Changes of the spectral and photochemical properties of chlorophyll in the complex induced by organic solvents and detergents are analogous to those for native systems.

The lowest concentration of individual pigments in the complex enabling an effective energy migration from chlorophyll *b* and carotene to chlorophyll *a* have been found. Comparison of these data with parameters for molecular solutions confirms the irregular distribution of pigments on the particles of the carrier. The excitation energy from carotene is most effectively transferred to a form of chlorophyll fluorescing at 77 K in the region of 725-735 nm.

The relation of the structure of the chlorophyll spectral forms to the state of the carrier protein is the basis of a hypothesis on the mechanism of genetic regulation of formation of native pigment forms.

REFERENCES

AGHION, J., *Biochim. biophys. Acta* 66: 212, 1963.
AGHION, J., PORCILE, E. & LIPPINCOTT, J. A., *Nature* 197: 1110, 1963.
BOARDMAN, N. K., *Annu. Rev. Plant Physiol.* 21: 115, 1970.
BRANDT, A. B. & TAGEEVA, S. V.: Opticheskie Parametry Rastitel'nykh Organizmov, Nauka, Moskva 1967.
BRODY, M. & BRODY, S. S., *Biochim. biophys. Acta* 112: 54, 1966.
BYSTROVA, M. I., in: KIRICHENKO, E. B. (ed.): Metody Issledovaniya Struktury Fotosinteticheskogo Apparata, p. 81, Pushchino-na-Oke 1972.
CHAPMAN, D. & FAST, P. G., *Science* 160: 188, 1968.
DUYSENS, L. N. M.: Transfer of Excitation Energy in Photosynthesis, Thesis, Univ. Utrecht 1952.
EROKHIN, Yu. E., in: Vtoroï Vsesoyuznyï Biokhimicheskiï S''ezd, Tezisy Dokladov, p. 278, FAN, Tashkent 1969.
EROKHIN, Yu. E., in: KIRICHENKO, E. B. (ed.): Metody Issledovaniya Struktury Fotosinteticheskogo Apparata, p. 155, Pushchino-na-Oke 1972.
EROKHIN, Yu. E. & SINEGUB, O. A., *Mol. Biol. (Moskva)* 4: 401, 1970a.
EROKHIN, Yu. E. & SINEGUB, O. A., *Mol. Biol. (Moskva)* 4: 541, 1970b.
FRANCK, J. & ROSENBERG, J. L., *J. theor. Biol.* 7: 276, 1964.
GILLER, Yu. E., *Biofizika* 13: 1006, 1968.
GILLER, Yu. E., in: Mekhanizm Fotosinteza, Tezisy Dokladov II Vsesoyuznogo Biokhimicheskogo S''ezda, Sektsiya 'Problemy Fotosinteza', p. 40, FAN, Tashkent 1969.
GILLER, Yu. E., in: IV Mezhdunarodnyï Biofizicheskiï Kongress, Tezisy Sektsionnykh Dokladov, Vol. 1, p. 368, Moskva 1972.
GILLER, Yu. E. & KHAITOVA, L. T., *Dokl. Akad. Nauk Tadzh. SSR* 9 (12): 32, 1966.
GILLER, Yu. E., KRASICHKOVA, G. V. & SAPOZHNIKOV, D. I., *Dokl. Akad. Nauk SSSR* 182: 1230, 1968.
GILLER, Yu. E., KRASICHKOVA, G. V. & SAPOZHNIKOV, D. I., *Biofizika* 15: 38, 1970.
GILLER, Yu. E. & YUKHANANOVA, L. N., *Izv. Akad. Nauk Tadzh. SSR, Otd. biol. Nauk* 1969 [2 (35)]: 47, 1969.
GILLER, Yu. E. & YUKHANANOVA, L. N., *Biokhimiya* 35: 873, 1970.
GILLER, Yu. E. & YUKHANANOVA, L. N., *Izv. Akad. Nauk Tadzh. SSR, Otd. biol. Nauk* 1974 [2(55)]: 34, 1974.
GILLER, Yu. E., YUKHANANOVA, L. N. & ABDULLAEVA, S. K., *Dokl. Akad. Nauk SSSR* 207: 1475, 1972.
GILLER, Yu. E. & YUSUPOVA, G. A., *Dokl. Akad. Nauk Tadzh. SSR* 12 (4): 63, 1969.
GILLER, Yu. E. & YUSUPOVA, G. A., *Dokl. Akad. Nauk SSSR* 190: 1470, 1970.
GREGORY, R. P. F., RAPS, S. & BERTSCH, W., *Biochim. biophys. Acta* 234: 330, 1971.
GULYAEV, B. A. & LITVIN, F. F., *Biofizika* 12: 845, 1967.
GULYAEV, B. A. & LITVIN, F. F., *Biofizika* 15: 670, 1970.
GURINOVICH, G. P., SEVCHENKO, A. N. & SOLOV'EV, K. N., Spektroskopiya Khlorofilla i Rodstvennykh Soedineniï, Nauka i Tekhnika, Minsk 1968.
GURINOVICH, G. P. & STRELKOVA, T. I., *Biofizika* 13: 782, 1968.
HENNINGSEN, K. W., *J. Cell Sci.* 7: 587, 1970.
JACOBS, E. E., HOLT, A. S., KROMHOUT, R. & RABINOWITCH, E., *Arch. Biochem. Biophys.* 72: 495, 1957.
KOSOBUTSKAYA, L. M. & KRASNOVSKIÏ, A. A., *Biokhimiya* 18: 340, 1953.
KRASICHKOVA, G. V.: Issledovanie Spektral'nykh i Fotokhimicheskikh Svoïstv Vodorastvorimykh Pigmentnobelkovolipoidnykh Kompleksov, Thesis, Dushanbe 1972.
KRASNOVSKIÏ, A. A., in: Mekhanizm Fotosinteza, Trudy V Mezhdunarodnogo Biokhimicheskogo Kongressa, Simp. VII, p. 196, Izd. Akad. Nauk SSSR, Moskva 1962.
KRASNOVSKIÏ, A. A., in: KRASNOVSKIÏ, A. A. et al. (ed.): Biokhimiya i Biofizika Fotosinteza, p. 26, Nauka, Moskva 1965.

KRASNOVSKIĬ, A. A., in: Funktsional'naya Biokhimiya Kletochnykh Struktur, p. 15, Nauka, Moskva 1970.
LIPPINCOTT, J. A., AGHION, J., PORCILE, E. & BERTSCH, W. F., *Arch. Biochem. Biophys.* 98: 17, 1962.
LITVIN, F. F., in: KRASNOVSKIĬ, A. A. et al. (ed.): Biokhimiya i Biofizika Fotosinteza, p. 96, Nauka, Moskva 1965.
LITVIN, F. F. & BELYAEVA, O. B., *Biokhimiya* 36: 615, 1971.
LITVIN, F. F. & GULYAEV, B. A., *Nauch. Dokl. vyssh. Shkoly, biol. Nauki* 12 (2): 118, 1969.
LITVIN, F. F., GULYAEV, B. A. & KARNEEVA, N. V., *Nauch. Dokl. vyssh. Shkoly, biol. Nauki* 13 (4): 95, 1970.
LITVIN, F. F., RIKHIREVA, G. T. & KRASNOVSKIĬ, A. A., *Biofizika* 7: 578, 1962.
MACHOLD, O., MEISTER, A. & ADLER, K., *Photosynthetica* 5: 160, 1971.
NASYROV, Yu. S., *Zh. obshch. Biol.* 33: 683, 1972.
NASYROV, Yu. S., GILLER, Yu. E. & USMANOV, P. D., in: NASYROV, Yu. S. & ŠESTÁK, Z. (ed.): Genetic Aspects of Photosynthesis, p. 133, Junk, The Hague 1975.
NEKRASOV, L. I., *Biofizika* 12: 215, 1967.
NEKRASOV, L. I. & KAPLER, R., *Biofizika* 11: 48, 1966.
NEKRASOV, L. I., KOBOZEV, N. I. & KOMISSAROV, G. G., *Vestn. mosk. gos. Univ., Ser. II. Khim.* 17 (6): 36, 1962.
NEKRASOV, L. I., KOBOZEV, N. I., PICHUGINA, N. G. & PROKOF'EVA, N. A., *Vestn. mosk. gos. Univ., Ser. II. Khim.* 16 (2): 9, 1961.
NISHIMURA, M. & TAKAMATSU, K., *Nature* 180: 699, 1957.
OSTROVSKAYA, L. K., in: KIRICHENKO, E. B. (ed.): Metody Issledovaniya Struktury Fotosinteticheskogo Apparata, p. 8, Pushchino-na-Oke 1972.
PARK, R. B., in: BONNER, J. & VARNER, J. E. (ed.): Plant Biochemistry, 2nd Ed., Academic Press, New York-London 1964.
RABINOWITCH, E., *Annu. Rev. Plant Physiol.* 2: 361, 1951.
RABINOWITCH, E. & GOVINDJEE: Photosynthesis, John Wiley and Sons, Inc., New York-London-Sydney-Toronto 1969.
RADUNZ, A., SCHMID, G. H. & MENKE, W., *Z. Naturforsch.* 26 b: 435, 1971.
SAPOZHNIKOV, D. I. (ed.): Pigmenty Plastid Zelenykh Rasteniĭ i Metodika ikh Issledovaniya, Nauka, Moskva-Leningrad 1964.
SAPOZHNIKOV, D. I., TOLIBEKOV, D. & GILLER, Yu. E., *Izv. Akad. Nauk Tadzh. SSR, Otd. biol. Nauk* 1966 [2 (23)]: 48, 1966.
SAUER, K., PARK, R. B., *Biochim. biophys. Acta* 79: 476, 1964.
SEELY, G. R. & JENSEN, R. G., *Spectrochim. Acta* 21: 1835, 1965.
SEMICHAEVSKIĬ, V. D., *Biofizika* 17: 530, 1972.
SEMICHAEVSKIĬ, V. D., LOS´, S. I. & LOZOVAYA, G. I., *Biofizika* 16: 1117, 1971.
SEMICHAEVSKIĬ, V. D. & LOZOVAYA, G. I., *Dokl. Akad. Nauk SSSR* 199: 965, 1971.
SINEGUB, O. A. & EROKHIN, Yu. E., *Mol. Biol. (Moskva)* 5: 472, 1971.
SMITH, J. H. C. & BENITEZ, A., in: PAECH, K. & TRACEY, M. V. (ed.): Modern Methods of Plant Analysis, Vol. 4, p. 142, Springer-Verlag, Berlin-Göttingen-Heidelberg, 1955.
STEINEMANN, A., ALAMUTI, N., BRODMANN, W., MARSCHALL, O. & LÄUGER, P., *J. Membrane Biol.* 4: 284, 1971.
THOMAS, J. B. & FLIGHT, W. F. G., *Biochim. biophys. Acta* 79: 500, 1964.
THOMAS, J. B. & van der WAL, U. P., *Biochim. biophys. Acta* 79: 490, 1964.
THORNBER, J. P., *Biochim. biophys. Acta* 172: 230, 1969.
TOLIBEKOV, D., ABDULLAEVA, S. K. & GILLER, Yu. E., *Dokl. Akad. Nauk Tadzh. SSR* 11 (2): 62, 1968.
TOLIBEKOV, D., ABDULLAEVA, S. K. & GILLER, Yu. E., *Dokl. Akad. Nauk Tadzh. SSR* 12 (12): 55, 1969.
TUMERMAN, L. A., BORISOVA, O. F. & RUBIN, A. B., *Biofizika* 6: 645, 1961.

VERNON, L. P., SHAW, E. R., OGAWA, T. & RAVEED, D., *Photochem. Photobiol.* 14: 343, 1971.
VOROB'EVA, L. M. & KRASNOVSKIĬ, A. A., *Biokhimiya* 31: 573, 1966.
VOROB'EVA, L. M. & KRASNOVSKIĬ, A. A., *Biofizika* 12: 240, 1967.
von WETTSTEIN, D., HENNINGSEN, K. W., BOYNTON, J. E., KANNANGARA, G. C. & NIELSEN, O. F., in: BOARDMAN, N. K., LINNANE, A. W. & SMILLIE, R. M. (ed.): Autonomy and Biogenesis of Mitochondria and Chloroplasts, p. 205, North-Holland Publ. Comp., Amsterdam-London 1971.
YUSUPOVA, G. A. & GILLER, Yu. E., *Dokl. Akad. Nauk Tadzh. SSR* 15 (10): 54, 1972.

REGULATION BY PYRUVATE KINASE AND PHOSPHATASE OF INORGANIC PHOSPHATE INCORPORATION DURING PHOTOPHOSPHORYLATION

A. A. YASNIKOV, BETYA I. BERSHTEÏN, NINA V. VOLKOVA, LYUDMILA I. VASILENOK, OLGA I. VOLOVIK, NINA A. ZAÏTSEVA, N. P. KANIVETS, LINA S. MUSHKETIK, A. S. OKANENKO, LYUDMILA K. OSTROVSKAYA, SVETLANA S. PETRENKO, ANNA I. POLISHCHUK, TAMARA A. REÏNGARD & IRINA I. SEMENYUK

Institute of Organic Chemistry and Institute of Plant Physiology of the Academy of Sciences of the Ukrainian S.S.R., Kiev, U.S.S.R.

Introduction

A study of the effect on the photophosphorylating system of those enzymes, which transform phosphoric esters, may be of great importance for discovering a mechanism in the regulation of some stage of the bioenergetic processes. Enzymes added to intact chloroplasts may interact with intermediates of photophosphorylation or with enzymes of this system, and thus evoke changes in the rate of inorganic phosphate incorporation during photophosphorylation. In this connection the effects of three enzymes — acid and alkaline phosphatases and pyruvate kinase — on the rate of photophosphorylation were studied. The enzymes were introduced into suspensions of chloroplasts at the beginning of or after definite periods of illumination. In the course of the experiment changes in inorganic phosphate and nucleotides were determined.

Methods

Chloroplasts were isolated from leaves of 14 d-old seedlings of pea cv. Ramonskiï 77. The methods of isolation and other analytical procedures are described in BERSHTEÏN et al. (1969). Nucleotides were separated by horizontal electrophoresis, and chromatographic eluates were analysed spectrophotometrically (STRÁNSKÝ 1963). The enzyme preparations were: Acid phosphatase from wheat bran (Reanal), 0.3 unit per 1 mg of enzyme; alkaline phosphatase from chicken intestine (Reanal), 0.14 unit per 1 mg of enzyme; pyruvate kinase (Sigma), type 1.

The Effects of Phosphatases

Control experiments showed that acid phosphatase splits ATP and ADP at pH 7.8. AMP is not hydrolyzed, and relatively high concentrations of AMP inhibit the hydrolysis of ADP. In the dark, chloroplasts do not prevent hydrolysis of ADP and ATP by added phosphatase (Fig. 1). Illuminated photophosphorylating chloroplasts affect the course of ADP and ATP hydrolysis by phosphatase. When phosphatase is added 2 min after the beginning of illumination no noticable changes are found in the concentration of inorganic phosphate. During alternation of 'light-dark' (Fig. 1, curve 5) ADP and ATP are hydrolysed in the dark, while in the light the hydrolysis ceases and there is some incorporation of inorganic phosphate during photophosphorylation.

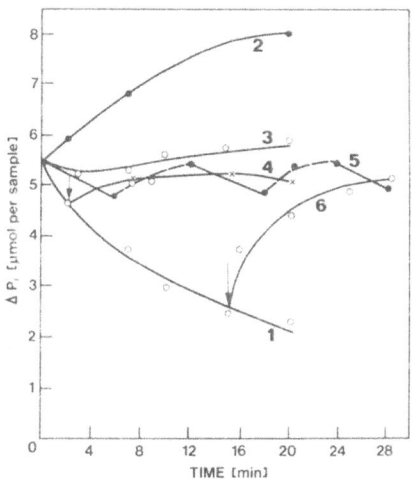

Fig. 1. Change in the content of inorganic phosphate in the presence of acid phosphatase. The reaction mixture (volume 1.5 ml) contained (μmol): ADP, 3.3; KH_2PO_4, 5; Na-ascorbate, 5; $MgCl_2$, 2.5; Tris-HCl (pH 7.8), 30; NaCl, 20; FMN, 0.03. Also added were chloroplasts (containing 300 μmol chlorophyll) and, where indicated, 0.8 units phosphatase. (1) Control (without phosphatase); (2) dark variant; (3) phosphatase added at beginning of illuminance; (4) phosphatase added 2 min after beginning illuminance; (5) alternate light and dark (- - - -).

On the other hand, addition of phosphatase after 15 minutes of illumination, i.e. when photophosphorylation is over, results in intensive hydrolysis of formed ATP (Fig. 2). In this case the illuminated chloroplasts do not inhibit the hydrolytic effect of phosphatase. The addition of phosphatase at the beginning of illumination does not change the course of ADP usage. ATP is formed more slowly than in the control, and AMP appears simultaneously with ATP. With the introduction of phosphatase during the first minute of illumination ADP continues to decrease at the same rate as in the experiment without phosphatase, but the rate of ATP formation is approximately half the control value.

Formation of ATP is accompanied by accumulation of AMP. But when phosphatase is introduced in the 15th minute (Fig. 2) hydrolysis of formed ATP and

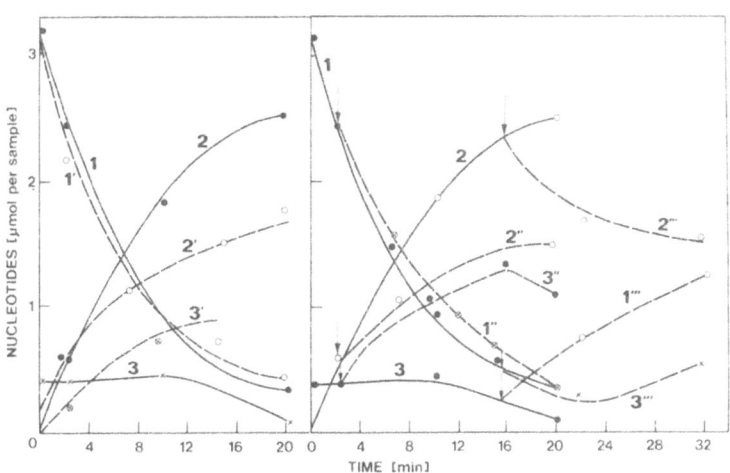

Fig. 2. Changes in nucleotides during phosphorylation. 1, 1', 1" and 1"'− ADP; 2, 2', 2" and 2"'− ATP; 3, 3', 3" and 3"'− AMP; 1, 2, 3 − control. 1', 2', 3'− acid phosphatase added at beginning of illumination; 1". 2" and 3" − acid phosphatase added 2 min after beginning of illumination; 1"', 2"' and 3"' acid phosphatase added after 15 min illumination.

ADP occurs, and hence ADP and AMP begin to accumulate. In some experiments the presence of phosphatase in the photophosphorylating system induced a more rapid fall in ADP concentration, probably because the ATP hydrolysis is not completely inhibited by the photophosphorylating system.

Phosphatase does not affect the light-dependent transport of protons inside the chloroplasts, i.e. the functioning of 'the proton pump'. The blocking of ATP and ADP hydrolysis by the photophosphorylating system indicates the presence of a phosphorylated product which is hydrolyzed at a higher rate than ATP and ADP. There is a peculiar competitive relationship between hydrolysis of this substance, on the one hand, and hydrolysis of ATP and ADP, on the other. Since the inhibition of ATP and ADP hydrolysis occurs during the whole experiment, we suspect that the phosphorylated product is formed constantly. This is possible only if the system of non-phosphorylated precursors X and X_E (in the terms of JAGENDORF & URIBE 1967) continues to function in the presence of phosphatase (see Fig. 3).

Thus, the experimental data provide evidence for the formation of a macroergic phosphorylated product preceding the formation of ATP. According to our hypothesis (BERSHTEÏN et al. 1969) the intermediate product has an enolphosphate structure.

The results of the experiments on the changes of nucleotides may be interpreted on the basis of the hypothesis of ROY & MOUDRIANAKIS (1971 a, b) according to which AMP, and not ADP, is the acceptor of inorganic phosphate; the

AMP transforms firstly into ADP linked with the chloroplasts membrane, then two molecules of ADP form ATP and AMP by transphosphorylation. This pathway is evidenced by the coïncidence of the rates of ADP transformation into ATP during photophosphorylation and into AMP and ATP in the chloroplasts where incorporation of P_i is inhibited by phosphatase. Acid phosphatase does not inhibit the process of transphosphorylation. Control experiments showed that phosphatase in the absence of chloroplasts does not produce transformation of ADP into ATP and AMP.

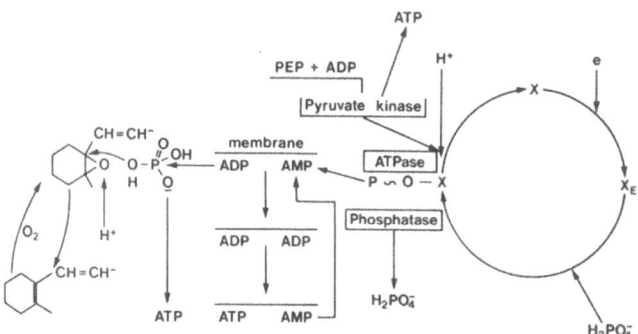

Fig. 3. Mechanism of interaction of phosphatase and pyruvate kinase with the photophosphorylating system.

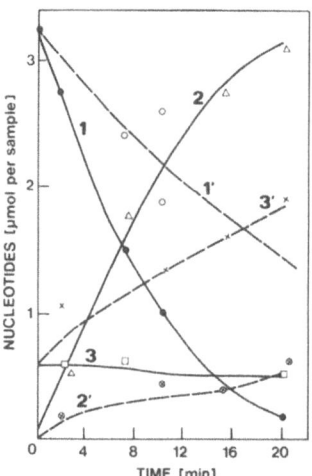

Fig. 4. Changes in nucleotides during photophosphorylation. 1 and 1' – ADP; 2, 2' – ATP; 3, 3' – AMP. 1, 2, 3 – control; 1', 2', 3'' – alkaline phosphatase added at beginning of illumination.

The mechanism of the inhibition by phosphatase of inorganic phosphate incorporation during photophosphorylation and the process of transphosphorylation are given in Fig. 3. Contrary to acid phosphatase, alkaline phosphatase not only inhibits the incorporation of inorganic phosphate but also eliminates transphosphorylation; ATP is not formed and ADP hydrolyses into AMP (Fig. 4). Alkaline phosphatase, as with the acid enzyme, does not block the light-dependent transport of protons inside the chloroplast.

The Effect of Pyruvate Kinase

Pyruvate kinase synthesises ATP during substrate phosphorylation. Low concentrations of this enzyme block photophosphorylation (YASNIKOV et al. 1971; REÏNGARD et al. 1972): one molecule of pyruvate kinase eliminates the work of 4 000 molecules of chlorophyll during photophosphorylation. The photophosphorylating system does not inhibit the functioning of pyruvate kinase itself; the synthesis of ATP from ADP and PEP occurs at approximately the same rate as with the enzyme alone.

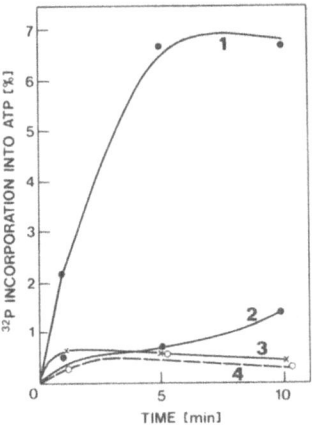

Fig. 5. Effect of pyruvate kinase on incorporation of ^{32}P from $H^{32}PO_4^{2-}$ into ATP during phosphorylation. Experimental conditions as in Fig. 1. 1 – without pyruvate kinase; 2 – with 1.5 mg pyruvate kinase; 3 – with 2.5 mg pyruvate kinase; 4 – with 1.5 mg pyruvate kinase and 5 μmol phosphoenolpyruvate.

Using the isotope it was shown that pyruvate kinase inhibits this process, although not completely (Fig. 5): 10% of the total flow is not blocked. The phosphorylation which is insensitive to pyruvate kinase is concentrated in small digitonin fragments (REÏNGARD et al. 1972). Photophosphorylation by these fragments is almost unaffected by pyruvate kinase whereas phosphorylation in large fragments is inhibited by this enzyme.

Firstly we believed that pyruvate kinase interacts with the intermediate phosphorylated product. However, it was later established that pyruvate kinase blocks not only inorganic phosphate incorporation during photophosphorylation but also inhibits the functioning of the proton pump. The latter probably shows that pyruvate kinase interacts with one of the proteins of the photophosphorylating system, or a complex forms between this protein and one of the intermediate compounds (see Fig. 3). Evidently, phosphorylation independent of pyruvate kinase is linked with the transformation of epoxy-carotenoids (YASNIKOV et al. 1972).

Regulation of Photophosphorylation by Phosphatase and Pyruvate Kinase

The enzymatic regulation is probably conditioned first of all by induced in vivo synthesis of the enzymes. Accumulation of phosphorylated products causes intensive synthesis of phosphatase. The acid phosphatase formed blocks only the transport of phosphate into the transphosphorylation system without affecting the pathway of non-phosphorylated intermediates. Pyruvate kinase is synthesized in a cell in response to the appearance of phosphoenolpyruvate which can itself support ATP synthesis without photophosphorylation.

Pyruvate kinase blocks the enzymatic system of photophosphorylation which is responsible for proton transport and the transfer of phosphate from the intermediate phosphorylated product into ATP. But this linkage with the photophosphorylating system does not influence the activity of pyruvate kinase which continues to metabolise the surplus of formed PEP. After a fall in the concentration of PEP in the system the amount of pyruvate kinase decreases, and the phosphorylation system begins to operate again.

Hence the study of the mutual effects of the photophosphorylating system and exogenous added enzymes may give information on the regulatory mechanisms in chloroplasts associated with the photophosphorylation process.

Summary

The addition of pyruvate kinase in a molar ratio with chlorophyll of 1 : 3 600 inhibited photophosphorylation in chloroplasts by 90-95%. The photophosphorylation not inhibited by pyruvate kinase represented 5-10% of the total photophosphorylation, as determined in experiments with ^{32}P-labeled phosphate. One of the paths of photophosphorylation involves the formation of an intermediate phosphate ester; this step was uncoupled by acid phosphatase added to the suspension. This suggests that this enzyme blocks the transport of phosphate to ADP. Addition of acetolphosphate and pyridoxal phosphate to intact chloroplasts stimulated the incorporation of inorganic phosphate into ATP and pyrophosphate. The phosphorylated intermediate is formed from acetolphosphate and pyridoxal phosphate and takes part in the enolphosphate path of inorganic phosphate incorporation during photophosphorylation. The path-insensitive to pyruvate kinase is probably associated with the transformation of epoxy-carotenoids, and appears to be a 'starting mechanism' of photophosphorylation.

REFERENCES

BERSHTEĬN, B.I., VOLKOVA, N.V., VOLOVIK, O.I., IVANISHCHEVA, S.Yu., OKANENKO, A.S., OSTROVSKAYA, L.K., PETRENKO, S.G., POLISHCHUK, A.I., PSHENICHNAYA, A.K., REĬNGARD, T.A., SEMENYUK, I.I. & YASNIKOV, A.A., *Fiziol. Biokhim. kul't. Rast.* 1: 21, 1969.
JAGENDORF, A.T. & URIBE, E., in: Energy Conversion by the Photosynthetic Apparatus, Brookhaven Symp. Biol. 19: 215, 1967.
REĬNGARD, T.A., VOLOVIK, O.I., ZAĬTSEVA, N.A., POLISHCHUK, A.I., OSTROVSKAYA, L.K., VOLKOVA, N.V., OKANENKO, A.S., PETRENKO, S.G., SEMENYUK, I.I. & YASNIKOV, A.A., *Fiziol. Biokhim. kul't. Rast.* 4: 345, 1972.
ROY, H. & MOUDRIANAKIS, E.N., *Proc. nat. Acad. Sci. U.S.A.* 68: 464, 1971a.
ROY, H. & MOUDRIANAKIS, E.N., *Proc. nat. Acad. Sci. U.S.A.* 68: 2720, 1971b.
STRÁNSKÝ, Z., *J. Chromatogr.* 10: 456, 1963.
YASNIKOV, A.A., BERSHTEĬN, B.I., VOLKOVA, N.V., OKANENKO, A.S., OSTROVSKAYA, L.K., PETRENKO, S.G., REĬNGARD, T.A. & SEMENYUK, I.I., in: Teoreticheskie Osnovy Fotosinteticheskoĭ Produktivnosti, p.71, Nauka, Moskva 1972.
YASNIKOV, A.A., BERSHTEĬN, B.I., VOLKOVA, N.V., VOLOVIK, O.I., ZAĬTSEVA, N.A., OKANENKO, A.S., POLISHCHUK, A.I., PSHENICHNAYA, A.K., REĬNGARD, T.A. & SEMENYUK, I.I., *Fiziol. Biokhim. kul't. Rast.* 3: 468, 1971.

THE METABOLIC AND EPIGENETIC CONTROL OF CHLOROPHYLL BIOSYNTHESIS

V.L. KALER

Institute of Experimental Botany, Academy of Sciences of the B.S.S.R., Minsk, U.S.S.R.

The chlorophyll forming system may well consist of complex dynamic systems, one important peculiarity of which – the hierarchy of their control mechanism – is to be ascribed to this system as well. The genetic control of chlorophyll biosynthesis is considered to belong to a rather high level of control mechanisms, its commands being fulfilled through the functioning of control mechanisms at lower levels; i.e. the epigenetic and metabolic mechanisms (RATNER 1965). Hence the investigation of the regulation of chlorophyll biosynthesis at the epigenetic and metabolic levels provides valuable information about the mechanisms of realization of the genetic code.

The many papers concerning the regulation of chlorophyll biosynthesis have been surveyed in reviews (LASCELLES 1964, 1965; BOGORAD 1967; KIRK & TILNEY-BASSETT 1967), but as yet nobody has emphasized the level of regulation. Extreme points of view on the control of chlorophyll biosynthesis can be found in the literature. Some authors, for instance, consider chlorophyll formation to be controlled only via repression of key-enzyme synthesis (GASSMAN & BOGORAD 1967 a, b). The lifetime of the control enzyme is thought to be rather short (NADLER & GRANICK 1970), and the lag of chlorophyll formation during the very first hours of greening of dark-grown leaves is considered to reflect the time for synthesis of the control enzyme. In these theories only the epigenetic level of regulation is taken into account.

Other authors support metabolic regulation and believe the lag is caused by either the exhaustion of some of the metabolites important for chlorophyll formation (NOACK 1934; SMITH & YOUNG 1956; SISLER & KLEIN 1963), or allosteric inhibition of the key-enzyme by the chlorophyll precursor – protochlorophyllide (GODNEV 1963), which accumulates in the absence of light (SCHIFF & EPSTEIN 1967).

Experimental results from different authors may be explained by taking into account the hierarchy of control mechanisms in the chlorophyll forming system.

The Metabolic Control of Chlorophyll Biosynthesis

Dark-grown barley leaves were illuminated in a large series of experiments in which both the duration and level of illuminance with 'white' and 'red' light were varied. The mean rate of dark re-synthesis of protochlorophyllide during the first 5-6 h after the end of illumination was proportional to the fraction of protochlorophyllide transformed (Fig. 1). The highest rate was achieved when practically all the protochlorophyllide was transformed. The rate of protochlorophyllide resynthesis asymptotically approximates some steady level (Fig. 2), corresponding to the rate of its formation from the exogenic specific precursor of porphyrins – δ-aminolevulinic acid (ALA) (GODNEV 1963). Hence, the light activation of protochlorophyllide resynthesis is accomplished at the stage of ALA formation. The ALA synthesis proves to be allosterically inhibited by protochlorophyllide, and the protochlorophyllide seems to be formed by an integrated enzyme system (KALER 1971).

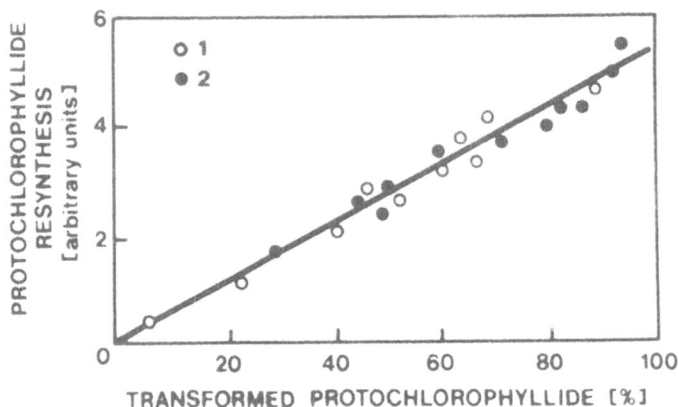

Fig. 1. The ratio between the rate of protochlorophyllide resynthesis and the amount of protochlorophyllide transformed during illumination of etiolated barley seedlings: (1) experiments with different levels but constant duration of illumination; (2) experiments with different durations but a constant level of illumination.

The rate of protochlorophyllide resynthesis may be determined by the activity of the integrated enzyme system obtained after protochlorophyllide phototransformation.

We could not detect any inhibition of protochlorophyllide resynthesis either by the specific inhibitor of chloroplast protein synthesis – chloramphenicol (SMILLIE et al. 1968), or by 'far-red' illumination, which inactivates phytochrome, during the 5-6 h following protochlorophyllide phototransformation

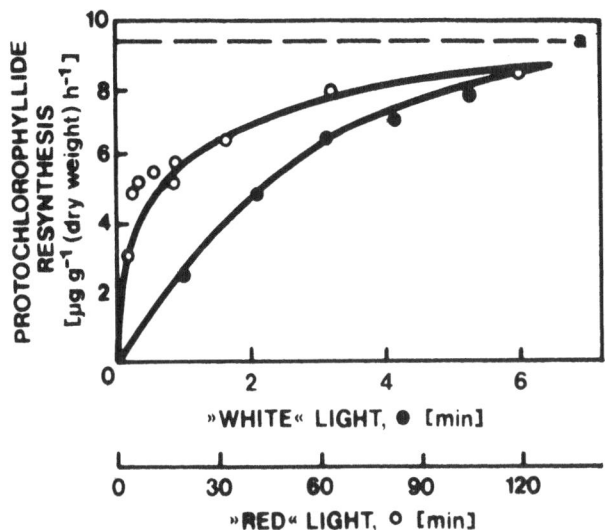

Fig. 2. The relationship between the rate of protochlorophyllide resynthesis in etiolated barley seedlings and the length of illuminance with 'red' or 'white' light. a: The steady level (to which the protochlorophyllide resynthesis rate approximates in both cases) is the rate of synthesis observed in barley seedlings exposed to 0.01 M ALA solution.

(Fig. 3 and 4). The absence of chloramphenicol inhibition can not be explained by a slow penetration of the inhibitor into plant tissue (SCHLENDER et al. 1969). This indicates that protochlorophyllide resynthesis in the first stage of greening of dark-grown leaves is neither regulated through chloroplast protein formation, nor by the phytochrome system. Hence, protochlorophyllide synthesis at this stage is controlled only at the level of metabolic regulation.

The Epigenetic Level of Chlorophyll Biosynthesis Control

The chloramphenicol inhibition of protochlorophyllide formation could be observed only during 6 h following illumination of dark-grown barley seedlings exposed to 0.001 M ALA (Fig. 3), and only after more prolonged dark intervals in plants exposed to water (KALER 1971). The 'red' light therefore activates not only formation of the key-enzyme (ALA-synthase) but also the formation of other enzymes of protochlorophyllide synthesis. Such an activation appears at the third stage of greening, and may be depressed by inhibitors of chloroplast protein synthesis. Far-red light acts in a similar way (Fig. 5 and 6), indicating the participation of the phytochrome system in the activation of protochlorophyllide resynthesis at the third stage of greening of dark-grown seedlings.

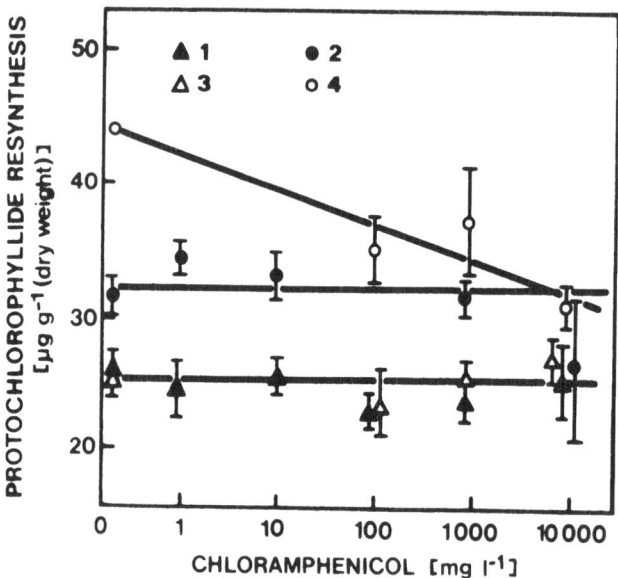

Fig. 3. The effect of chloramphenicol on protochlorophyllide resynthesis in the first 3 h (1 and 3) or 6 hours (2 and 4) after illumination of etiolated barley leaves. The plants were exposed to water solutions of the inhibitor (1 and 2) or to 0.001 M ALA + inhibitor solution (3 and 4) for 2 h and further illuminated until complete transformation of protochlorophyllide occurred; plants were then returned to the dark on the same solutions.

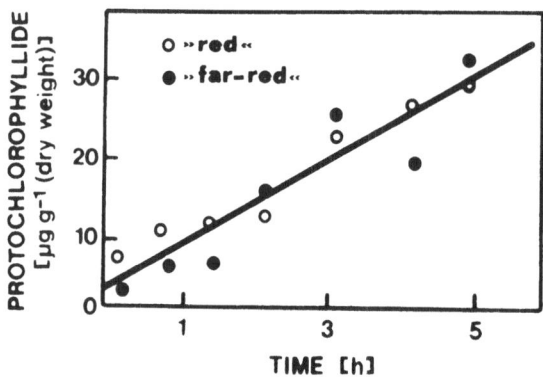

Fig. 4. Accumulation of protochlorophyllide after illumination of 7 d old etiolated barley seedlings by 'red' light or 'far-red' light. Doses: 'red' light – 1.1×10^{-2} J cm^{-2} ($\lambda=659$ nm), 'far-red' light -4.5×10^{-7} J cm^{-2} ($\lambda > 700$ nm). 'Red' light was obtained by passing 'white' light through a 5 cm thick water layer and a $\overline{659}$ nm interference filter, and 'far-red' light using a 5 cm water filter and a KC-19 filter transmitting $\lambda > 700$ nm.

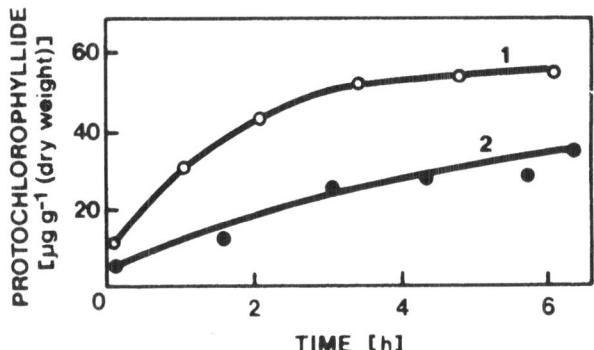

Fig. 5. Accumulation of protochlorophyllide in 7 day-old etiolated barley seedlings. Illumination conditions: (1) 'red' light 1.1×10^{-2} J cm^{-2} + 18 h in the dark + 'red' light 1.1×10^{-2} J cm^{-2} + 6 h in the dark; (2) 'red' light 1.1×10^{-2} J cm^{-2} + 'far-red' light 4.5×10^{-2} J cm^{-2} + 18 h in the dark + 'red' light 1.1×10^{-2} J cm^{-2} + 6 h in the dark.

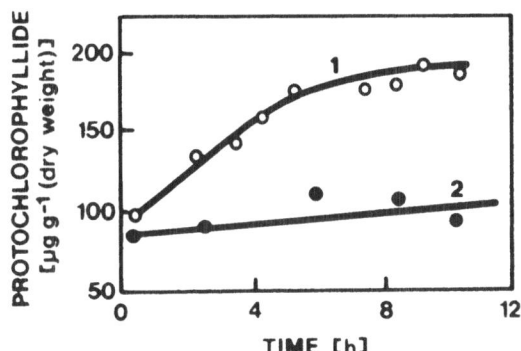

Fig. 6. Accumulation of protochlorophyllide in etiolated barley seedlings exposed to a 0.005 M solution of ALA for 2 h. Illumination conditions (see Fig. 5): (1) 'red' light + 18 h in the dark + 'red' light + 10 h in the dark; (2) 'red' light + 'far-red' light + 18 h in the dark + 'red' light + 10 h in the dark.

The specific inhibitor of DNA-dependent RNA synthesis — aurantin inhibited the 'red' light activation of protochlorophyllide resynthesis in an analogous manner (Fig. 7). Hence phytochrome may activate chlorophyll synthesis at the transcription level (KALER 1971). Phytochrome may well be considered as the light acceptor of a system regulating the transcription of the whole cistron chain, which controls the differentiation of etioplasts into chloroplasts, including the formation of the oligomers of the enzyme system of chlorophyll synthesis. These processes form the epigenetic level of the regulation of chlorophyll biosynthesis.

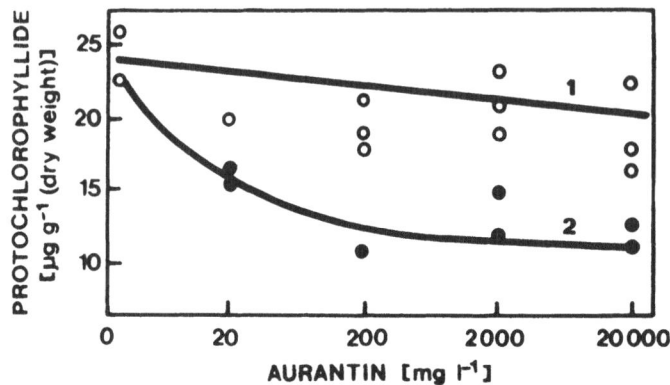

Fig. 7. The effect of aurantin on the accumulation of protochlorophyllide in etiolated barley seedlings exposed to solutions of different concentrations of the antibiotic for 2 h. Illumination conditions (see Fig. 5): (1) 2 h in the dark + 'red' light + 6 h in the dark; (2) 2 hours in the dark + 'red' light + 18 h in the dark + 'red' light + 6 h in the dark.

Conclusions

The ideas discussed about the control mechanisms of chlorophyll formation were drawn from experimental data. These concepts could be summarised in a single scheme allowing comparison with the object studied on a model level. The mathematical model of control mechanisms of chlorophyll biosynthesis was constructed, taking into account the screening of protochlorophyllide and phytochrome by accumulating chlorophyll, as well as the photooxidative destruction of chlorophyll. This model reproduced practically all the details of the behaviour of the system studied, up to the light adaptation of the chlorophyll apparatus (KALER 1971, 1972).

REFERENCES

BOGORAD, L., *Developm. Biol., Suppl.* 1: 1, 1967.
GASSMAN, M. & BOGORAD, L., *Plant Physiol.* 42: 774, 1967a.
GASSMAN, M. & BOGORAD, L., *Plant Physiol.* 42: 781, 1967b.
GODNEV, T.N., Khlorofill, ego Stroenie i Obrazovanie v Rastenii, Izd. Akad. Nauk BSSR, Minsk 1963.
KALER, V.L., Mekhanizmy Avtoregulyatsii v Sisteme Biosinteza Khlorofilla, Thesis, Akad. Nauk BSSR, Minsk 1971.
KALER, V.L., in: IV Mezhdunarodnyĭ Biofizicheskiĭ Kongress, Tezisy Sektsionnykh Dokladov, E IV 6 215, Vol. 1, p. 370, Moskva 1972.

KIRK, J.T.O., *Annu.Rev.Plant Physiol.* 21: 11, 1970.
KIRK, J.T.O. & TILNEY-BASSETT, R.A.E.: The Plastids. Their Chemistry, Structure, Growth and Inheritance, W.H. Freeman and Co., London – San Francisco 1967.
LASCELLES, J.: Tetrapyrrole Biosynthesis and Its Regulation. N.A. Benjamin Inc., New York – Amsterdam 1964.
LASCELLES, J., in: PRIDHAM, J.B. & SWAIN, T. (ed.): Biosynthetic Pathways in Higher Plants, p. 163, Academic Press, London – New York 1965.
NADLER, K.D. & GRANICK, S., *Plant Physiol.* 46: 240, 1970.
NOACK, K., *Deut. Forsch.* 23: 68, 1934.
RATNER, V.A.: Geneticheskie Upravlyayushchie Sistemy. Kibernetika v Monografiyakh, Vol. 3, Nauka – sibir. Otd., Novosibirsk 1965.
SCHIFF, J.A. & EPSTEIN, H.T., in: GOODWIN, T.W. (ed.): Biochemistry of Chloroplasts, Vol. 1, p. 341, Academic Press, London – New York 1967.
SCHLENDER, K.K., SELL, H.M. & BUKOVAC, M.J., *Phytochemistry* 8: 957, 1969.
SISLER, E.C., & KLEIN, W.H., *Physiol. Plant.* 16: 315, 1963.
SMILLIE, R.M., SCOTT, N.S. & GRAHAM, D., in: SHIBATA, K., TAKAMIYA, A., JAGENDORF, A.T. & FULLER, R.C. (ed.): Comparative Biochemistry and Biophysics of Photosynthesis, p. 332, Univ. Tokyo Press, Tokyo, and Univ. Park Press, State College, Pa. 1968.
SMITH, J.H.C. & YOUNG, V.M.K., in: HOLLAENDER, A. (ed.): Radiation Biology, Vol. 3, p. 393, McGraw-Hill, New York – Toronto – London 1956.

GENETIC BASIS OF OPTIMIZATION OF PLANT PHOTOSYNTHETIC ACTIVITY

THE IMPLICATIONS OF GENETIC, PHYSIOLOGICAL AND ENVIRONMENTAL CONTROL OF CHLOROPLAST ULTRASTRUCTURE TO THE OPTIMIZATION OF PHOTOSYNTHETIC ACTIVITY

R. M. SMILLIE, D. G. BISHOP & J. CONROY

Plant Physiology Unit, C.S.I.R.O. Division of Food Research and School of Biological Sciences, Macquarie University, North Ryde 2113, Sydney, Australia

Introduction

Previous research on chloroplasts in our laboratory has been concentrated on early biochemical synthetic processes associated with chloroplast development such as the synthesis of chloroplast nucleic acids and proteins (SMAÏLI* et al. 1972). In this paper another aspect of chloroplast development will be considered in which the most apparent changes are not synthetic, but ultrastructural, although some synthesis, certainly lipid synthesis, takes place. In this stage of chloroplast differentiation individual chloroplast lamellae become appressed to form grana. The genetic, physiological and especially the environmental control of this structural elaboration and the implications of these ultrastructural changes to photosynthetic activity will be discussed.

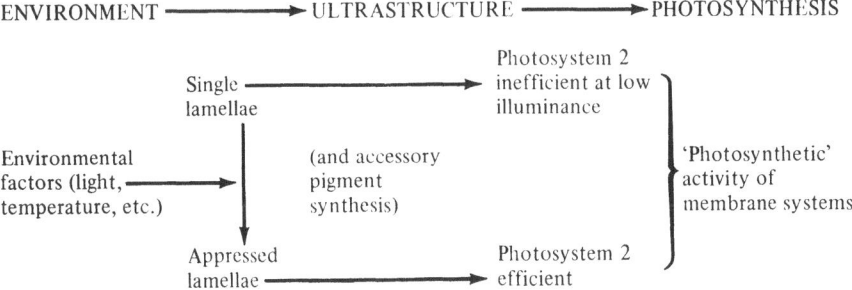

Fig. 1. Scheme relating environmental factors, chloroplast ultrastructure and photosynthetic activity.

Based on our own experiments and those of others we envisage environmental factors directly influencing ultrastructural development of chloroplasts and in turn this affects photosynthetic activity (see Fig. 1). Specifically, we propose that factors such as light, temperature and nutrition can affect the extent of the

* Paper of R.M. SMILLIE published in Russian.

formation of appressed lamellae. Individual non-appressed lamellae contain both photosystems 1 and 2 but photosystem 2 in particular is inefficient at low illuminances. In contrast, appressed lamellae, which in higher plants are usually referred to as grana lamellae, are more efficient in utilizing harvested light depending on the degree of the structural elaboration. The resultant photosynthetic activity of chloroplast membrane systems at different illuminances will then depend, among other things, upon the relative proportions of the two types of lamellae. This scheme suggest a possible mechanism by which plants might adjust photosynthetic requirements to changing environmental conditions. Genetic control of grana formation would provide a basis for long-term adaption of plants to different environments and regulation of the expression of genes for grana formation by environmental and other factors would provide a mechanism for acclimation of plants to a changing environmental condition within the lifetime of an individual plant.

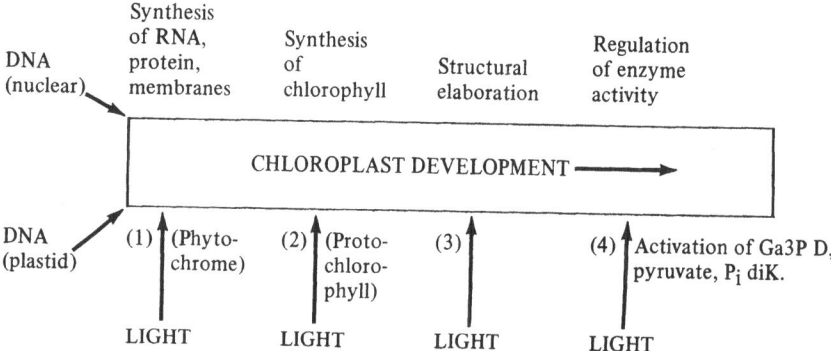

Fig. 2. Some steps in chloroplast development that can be regulated by light.

These proposals are seen in perspective with our earlier work and with studies by others in Fig. 2, in which stages of chloroplast development that can be regulated by light are indicated. Light can influence the synthesis of chloroplast RNA and chloroplast protein through the phytochrome system (GRAHAM et al. 1971, SCOTT et al. 1971, SMAÏLI et al. 1972). Light is also required for the conversion of protochlorophyll to chlorophyll *a*. The effect of light on chloroplast ultrastructure, designated (3) in the diagram, is the subject of this paper. Light is also known to activate certain enzymes of CO_2 fixation (HATCH & SLACK 1969, ZIEGLER et al. 1969).

Results and Discussion

Genetic, Physiological and Environmental Factors Controlling the Formation of Grana

The formation of grana is under strict control of genes and it is not unexpected that many mutants have been found in which pigment synthesis has been blocked or lamellar structure altered (HOMANN & SCHMID 1967, VON WETTSTEIN 1967, GOODENOUGH et al. 1969, NASYROV et al. 1971, WALLES 1971). Mutations of both nuclear genes and chloroplast genes are known to affect grana formation. In addition expression of genes involved in grana formation can be modified by a variety of physiological and environmental factors. The wide diversity of structure in chloroplasts found in plant tissues of different physiological age or type has been illustrated by several studies such as those carried out by TAGEEVA et al. (1971). The variation in ultrastructure found in the maize chloroplast depending upon its location within the leaf and the age of the leaf (ANDERSEN et al. 1972) also illustrates this point. The structure of chloroplasts in the mesophyll cell of an expanded leaf of maize is fairly typical of chloroplasts of higher plants. Granal stacks are prominent but numerous single lamellae (stroma lamellae) are also present. Grana formation in chloroplasts of the mesophyll cell of a young leaf is not very extensive and where appressed lamellae are seen they occur mostly in stacks of two or three (ANDERSEN et al. 1972). Chloroplasts from the bundle sheath cells of young maize leaves also contain grana

Table 1. Environmental factors affecting grana formation.

Environmental factor inhibiting grana formation	Plant	Reference
Light		
High illuminance	Soybean	BALLANTINE & FORDE (1970)
High illuminance	*Amaranthus*	LYTTLETON et al. (1971)
Red light	*Elodea*	TAGEEVA et al. (1969); PUNNETT (1971)
Far-red light	Bean	DE GREEF et al. (1971)
Flashing white light	Bean	SIRONVAL et al. (1969); ARGYROUDI-AKOYUNOGLOU et al. (1971)
Temperature		
Low temperature (10 °C)	*Sorghum* and other plants	TAYLOR & CRAIG (1971); TAYLOR & ROWLEY (1971)
Nutrition		
Growth on acetate	*Chlamydobotrys stellata*	WEISSNER & AMELUNXEN (1969)

consisting of two or three appressed lamellae but individual lamellae are even more evident. The bundle sheath chloroplasts of the mature leaf are almost completely agranal.

Table 1 lists some environmental factors which are known specifically to affect grana formation and while not an exhaustive list it is sufficient to indicate that the formation of appressed lamellae in a number of plants and algae is subject to environmental control by factors which include temperature, nutrition and the intensity, spectral quality and duration of light.

The relationship between these structural changes and the photochemical activities of the chloroplast has been investigated using three experimental systems, (1) a comparison of granal and agranal chloroplasts found in C_4 plants, (2) the agranal chloroplasts of red algae and (3) chloroplasts from C_3 plants grown in a flashing light regime.

Photochemical Activities of Granal and Agranal Chloroplasts from Maize

The existence of two distinct types of chloroplast, one containing grana, the other almost devoid of grana, in mature leaves of certain plants containing the C_4 pathway of photosynthesis such as maize allows a comparison to be made of the photochemical activities of the two types of chloroplast. Mesophyll chloroplasts which contain grana and bundle sheath chloroplasts which are essentially agranal were separated using the procedure of WOO et al. (1970) and ANDERSON et al. (1971).

Both mesophyll and bundle sheath chloroplasts isolated from maize leaves contain photosystem 1 and photosystem 2 and catalyze the photoreduction of NADP from water (ANDERSEN et al. 1972, SMILLIE et al. 1972a). However a striking difference between the two types of chloroplasts can be seen if the activities are plotted as a function of increasing red illuminance (SMILLIE et al. 1972b). Photosystem 1 activity, measured by the photoreduction of NADP in the presence of ascorbate and DCIP as electron donors, in both mesophyll and bundle sheath chloroplasts and photosystem 2 activity, measured by the photoreduction of DCIP or ferricyanide from water, in mesophyll chloroplasts all approach light saturation at fairly low illuminances. In contrast, photosystem 2 activity in bundle sheath chloroplasts, also measured by either ferricyanide or DCIP reduction, does not approach light saturation even at very high illuminances (SMILLIE et al. 1972b). Thus photosystem 2 in bundle sheath chloroplasts appears to be relatively inefficient in light energy capture and utilization compared with the other photosystem.

The different responses of photosystem 2 to illuminance in mesophyll and bundle sheath chloroplasts are illustrated in Fig. 3. Both electron acceptors for photosystem 2 – ferricyanide and the dye tetranitro blue tetrazolium – yield similar results. The use of the tetrazolium dye allows us to see if the photosystem 2 activity of *intact* bundle sheath cells behaves in the same way as a function of the irradiance as in the case of the isolated chloroplasts, since the tetrazolium

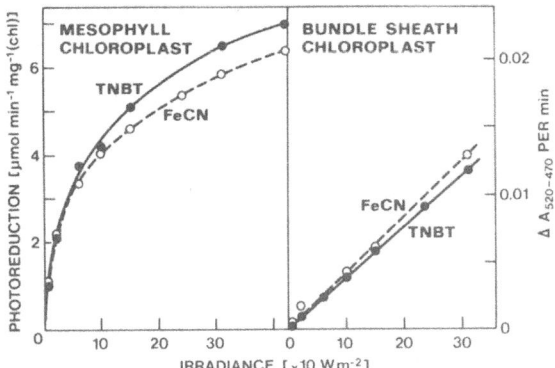

Fig. 3. The photoreduction of ferricyanide (FeCN) and tetranitro blue tetrazolium (TNBT) at different irradiances by red beams by mesophyll and bundle sheath chloroplasts isolated from maize leaves. The reaction mixture (0.75 ml) for mesophyll chloroplasts contained chloroplasts (3.5 µg chlorophyll), 100 mM NaCl, 2.5 mM $KH_2PO'_4$, 5 mM $MgCl_2$, 0.5% (w/v) bovine serum albumin, 50 mM tricine buffer pH 8.5 and 66 µM potassium ferricyanide or 133 µg ml^{-1} tetranitro blue tetrazolium. Reaction mixtures for bundle sheath chloroplasts were the same except that the buffer used was 50 mM phosphate pH 7.0. (Data from SMILLIE et al. 1973.)

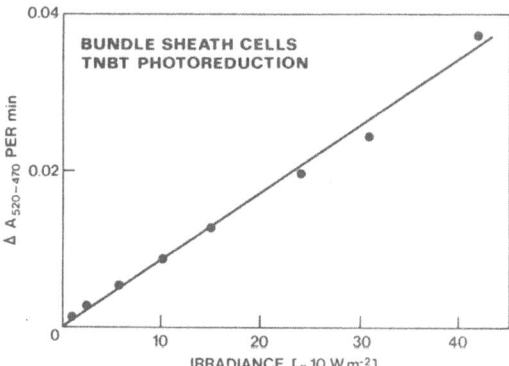

Fig. 4. Photoreduction of tetranitro blue tetrazolium by isolated strands of bundle sheath cells. The reaction mixture was as in Fig. 3 except bundle sheath cells (7.0 µg chlorophyll) replaced the bundle sheath chloroplasts and *Ficoll* (0.26 g) was also added. (Data from SMILLIE *et al.* 1973.)

readily penetrates cells whereas electron acceptors such as ferricyanide and DCIP do not. Preparations of maize bundle sheath strands containing intact bundle sheath cells may be prepared from maize leaves by first removing the epidermal and mesophyll cells by low-speed blending (SMILLIE et al. 1973). The photosystem 2 activity of these cells as a function of irradiance, measured using tetranitro blue tetrazolium as the electron acceptor, is shown in Fig. 4. The same dependence upon irradiance as was shown by the isolated chloroplasts is also shown by the intact cells. Hence it can be concluded that the results obtained with isolated chloroplasts reflect the conditions existing in the intact cell.

Photosystem 2 Activity and Illuminance in Red Algae

The structure of the lamellae in maize bundle sheath chloroplasts and red algal chloroplasts is similar in that neither contain appressed lamellae. Unless the result obtained with the bundle sheath chloroplasts is peculiar to these agranal chloroplasts of C_4 plants it might be expected that photosystem 2 in chloroplasts of red algae would show a similar relationship with illuminance if the accessory pigment, phycoerythrin, were first removed. When cells of the red alga *Griffithsia monile* are broken, phycoerythrin leaks from the released chloroplasts, and after centrifugation a pellet of green chloroplasts is obtained. Photosystem 2 activity of these chloroplasts, measured by the reduction of ferricyanide, shows a dependence upon illuminance similar to that of the bundle sheath chloroplast of maize (SMILLIE et al. 1973).

Hence this phenomenon of low efficiency of light capture and utilization for photosystem 2 activity by nonappressed lamellar systems is by no means restricted to C_4 plants.

The same phenomenon has been demonstrated in another way using partially fixed but otherwise intact cells of the red alga *Porphyridium cruentum* (unpublished experiments). By exposing a culture of this organism to high illuminances the ratio of phycoerythrin to chlorophyll is drastically reduced and the culture is pale green in colour. The photosystem 2 activity of these cells was measured as a function of the irradiance using 2,6-dichlorophenol-indophenol as the electron acceptor after partial fixation of the cells in formaldehyde using the procedure devised by HALLIER & PARK (1969). The green cells yielded results similar to those given by bundle sheath chloroplasts or cells (*cf.* Figs. 3, 4), whereas the dark-red control cells, obtained from cultures grown at low illuminances, gave results similar to those shown by maize mesophyll chloroplasts (*cf.* Fig. 3). For activity measurements, red light which is absorbed only by chlorophyll and not by phycoerythrin was used.

Photosystem 2 Activity and Illuminance in Agranal Chloroplasts of C_3 Plants

As indicated in Table 1 grana formation is light dependent. The use of flashing light in particular provides an experimental means of producing in C_3

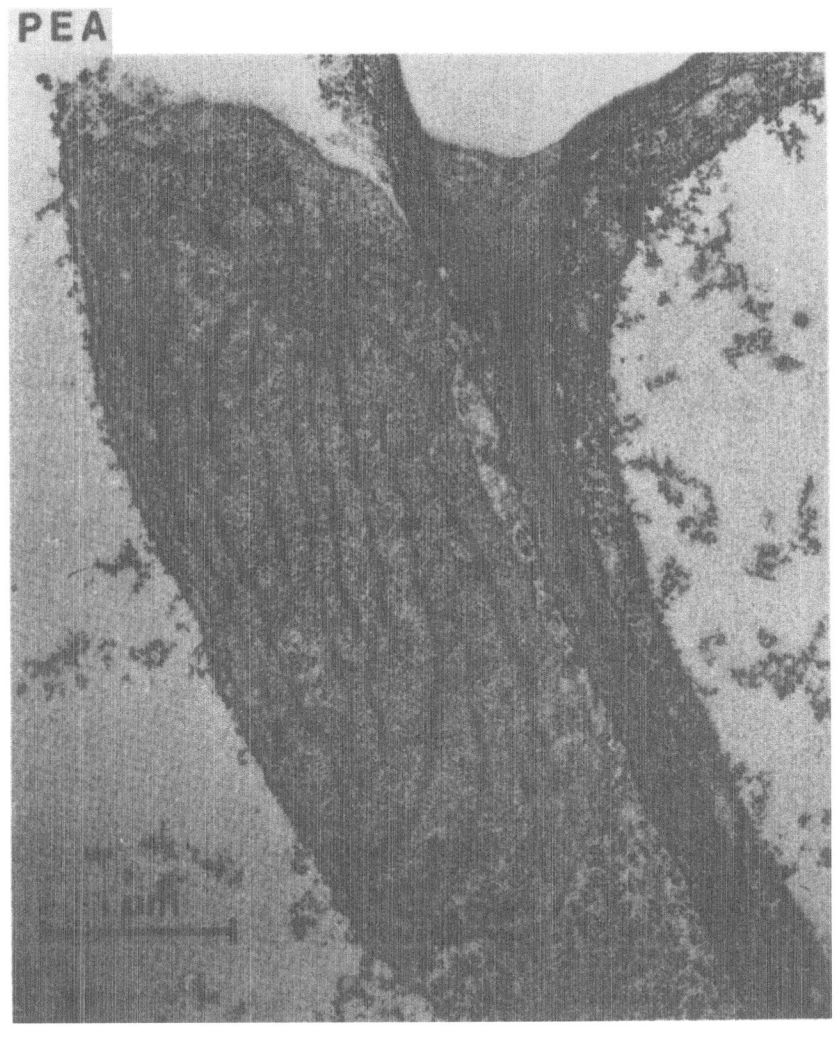

Fig. 5. An electron micrograph of a pea leaf chloroplast. Pea plants were grown in the dark for 6 days and then exposed to a cycle of 2 min of white light (200 W tungsten lamp at a distance of 1 m) followed by 118 min of darkness for a total of 48 h.

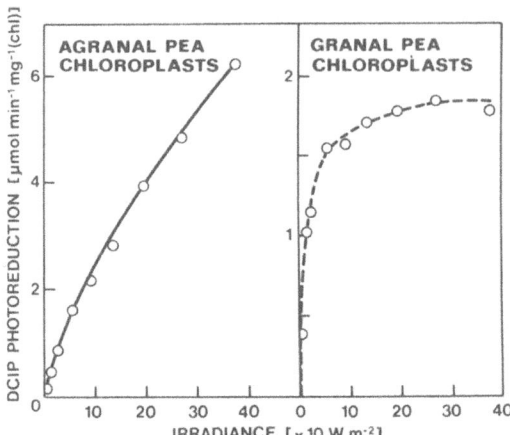

Fig. 6. The photoreduction of 2.6-dichlorophenol-indophenol (DCIP) by pea chloroplasts. Isolated agranal chloroplasts were prepared from plants treated as described in Fig. 5. Granal chloroplasts were isolated from plants grown in the dark for 7 days and then exposed to 24 h continuous white light. The reaction mixtures (0.75 ml) contained pea chloroplasts (0.63 μg chlorophyll), 50 mM tricine buffer pH 7.5, 2.5 mM KH_2PO_4, 100 mM NaCl, 5mM $MgCl_2$, 0.5% (w/v) bovine serum albumin and 8.3 μM 2,6-dichlorophenol-indophenol.

plants chloroplasts which are almost devoid of grana (SIRONVAL et al. 1969, ARGYROUDI-AKOYUNOGLOU et al. 1971). In collaborative experiments with Dr. R. HILLER of Macquarie University, Sydney, a cycle of 2 min of white light followed by 118 min of darkness was employed to allow the development of chlorophyll and individual lamellae in a pea chloroplast but not of grana. Some of the lamellae overlap but appression of lamellae is rare (Fig. 5) and the ratio of chlorophyll *a* to chlorophyll *b* is high, around twelve. This procedure also works with C_4 plants and in green maize leaves grown under the flashing light, even the mesophyll chloroplasts have very few appressed lamellae.

Fig. 6 compares photosystem 2 activities as a function of the irradiance in agranal pea chloroplasts and normal granal pea chloroplasts. Photosystem 2 activity, measured by DCIP reduction, is present in the agranal chloroplasts and high activities are obtained provided the irradiance is sufficiently high. More significantly however, the photosystem 2 activity of these granal pea chloroplasts responds in an almost identical way to increasing irradiance as do maize bundle sheath chloroplasts. Similar results were obtained using agranal chloroplasts from bean leaves. The photosystem 1 activity of both types of pea chloroplast saturated at relatively low irradiances. Presumably, in chloroplasts containing grana, those lamellae which do not become appressed (that is, the stroma lamellae) retain the properties of the non-appressed lamellae of the agranal chloroplast (*cf.* HALL et al. 1971). Thus the efficiency of radiant energy capture and utilisation at different irradiances will vary depending on the relative degree of the structural elaboration of the lamellar membranes. It might be envisaged that under certain environmental conditions, for example in the case of plants exposed to full tropical sunlight, the ability to limit the extent of grana formation in some cells and therefore the efficiency of radiant energy usage for photosystem 2 may be advantageous to the plant by circumventing otherwise excessive high rates of photosynthesis.

Summary

Two distinct types of lamellae are found in chloroplasts, appressed lamellae (grana lamellae) and non-appressed lamellae (stroma lamellae). In some chloroplasts, such as those found in the bundle sheath cells of certain C_4 plants and in some algae, appressed lamellae are rare or absent. The type of chloroplast structure found in a plant is genetically controlled, but is responsive to physiological and environmental influences. The significance of these structural variations to the optimization of photosynthetic activity has been examined by comparing the photochemical activities of several appressed and non-appressed lamellar systems and their relative dependence upon the illuminance. The following chloroplast systems which contain essentially only non-appressed lamellae were examined: the bundle sheath chloroplast of maize, red algal chloroplasts depleted of the photosynthetic accessory pigment phycoerythrin, and agranal pea chloroplasts obtained by growing pea plants in flashing light. In all cases photosystem 2 activity, measured by the reduction of ferricyanide or 2,6-dichlorophenol-indophenol, was present but saturation was not reached even at very high irradiances (200-400 W m^{-2} of red light). In contrast, the photosystem 2 activity of chloroplasts containing appressed lamellae approached saturation at much lower irradiances. These results have led us to the concept that genetic control of the formation of appressed lamellae in chloroplasts provides the basis of an adaptive mechanism involved in the optimization of photosynthetic activity under different environmental conditions. Although there are two light reactions in photosynthesis, photosystem 2 appears to be the main one involved in this adaptive mechanism.

Acknowledgement

We should like to thank Dr. J. BAIN for the electron micrograph.

REFERENCES

ANDERSEN, K. S., BAIN, J. M., BISHOP, D. G. & SMILLIE, R. M., *Plant Physiol.* 49: 461, 1972.

ANDERSON, J. M., BOARDMAN, N. K. & SPENCER, D., *Biochim. biophys. Acta* 245: 253, 1971.
ARGYROUDI-AKOYUNOGLOU, J. H., FELEKI, Z. & AKOYUNOGLOU, G., *Biochem. biophys. Res. Commun.* 45: 606, 1971.
BALLANTINE, J. E. M. & FORDE, B. J., *Amer. J. Bot.* 57: 1150, 1970.
GOODENOUGH, U. W., ARMSTRONG, J. J. & LEVINE, R. P., *Plant Physiol.* 44: 1001, 1969.
GRAHAM, D., GRIEVE, A. M. & SMILLIE, R. M., *Phytochemistry* 10: 2905, 1971.
DE GREEF, J., BUTLER, W. L. & ROTH, T. F., *Plant Physiol.* 47: 457, 1971.
HALL, D. O., EDGE, H. & KALINA, M., *J. Cell Sci.* 9: 289, 1971.
HALLIER, U. W. & PARK, R. B., *Plant Physiol.* 44: 535, 1969.
HATCH, M. D. & SLACK, C. R., *Biochem. J.* 112: 549, 1969.
HOMANN, P. H. & SCHMID, G. H., *Plant Physiol.* 42: 1619, 1967.
LYTTLETON, J. W., BALLANTINE, J. E. M. & FORDE, B. J., in: BOARDMAN, N. K., LINNANE, A. W. & SMILLIE, R. M. (ed.): Autonomy and Biogenesis of Mitochondria and Chloroplasts, p. 447, North-Holland Publ. Comp., Amsterdam-London 1971.
NASYROV, Yu. S., ALIEV, K. A., ABDULLAEV, Kh. A. & MUZAFFAROVA, S., in: *Geneticheskie Aspekty Fotosinteza*, p. 5, Donish, Dushanbe 1971.
PUNNETT, T., *Science* 171: 284, 1971.
SCOTT, N. S., NAIR, H. & SMILLIE, R. M., *Plant Physiol.* 47: 385, 1971.
SIRONVAL, C., MICHEL, J.-M., BRONCHART, R. & ENGLERT-DUJARDIN, E., in: METZNER, H. (ed.): Progress in Photosynthesis Research, Vol. I, p. 47, Tübingen 1969.
SMAÏLI, R. M., SKOTT, N. S., GREKHEM, D. & PATTERSON, V. D., in: Teoreticheskie Osnovy Fotosinteticheskoï Produktivnosti, p. 133, Nauka, Moskva 1972.
SMILLIE, R. M., ANDERSEN, K. S., TOBIN, N. F., ENTSCH, B. & BISHOP, D. G., *Plant Physiol.* 49: 471, 1972a.
SMILLIE, R. M., BISHOP, D. G. & ANDERSEN, K. S., in: FORTI, G., AVRON, M. & MELANDRI, A. (ed.): Photosynthesis, Two Centuries after Its Discovery by Joseph Priestley, Vol. 1, p. 779, Junk, The Hague 1972b.
SMILLIE, R. M., SCOTT, N. S. & BISHOP, D. G., in: LEE, J. W. & POLLAK, J. K. (ed.): The Biochemistry of Gene Expression in Higher Organisms, p. 479, Australia and New Zealand Book Co. Pty Ltd., Sydney 1973.
TAGEEVA, S. V., GENEROSOVA, I. P. & SEMEYENOVA, G. A., in: METZNER, H. (ed.): Progress in Photosynthesis Research, Vol. I, p. 21, Tübingen 1969.
TAGEEVA, S. V., GENEROZOVA, I. P., DEREVYANKO, V. G., LADYGIN, V. G., SEMENOVA, G. A., in: ZALENSKIÏ, O. V. (ed.): Fotosintez i Ispol'zovanie Solnechnoï Energii, p. 126, Nauka, Leningrad 1971.
TAYLOR, A. O. & CRAIG, A. S., *Plant Physiol.* 47: 719, 1971.
TAYLOR, A. O. & ROWLEY, J. A., *Plant Physiol.* 47: 713, 1971.
WALLES, B., in: GIBBS, M. (ed.): Structure and Function of Chloroplasts, p. 51, Springer-Verlag, Berlin-Heidelberg-New York 1971.
WEISSNER, W. & AMELUNXEN, F., Arch. Mikrobiol. 67: 357, 1969.
VON WETTSTEIN, D., in: SAN PIETRO, A., GREER, F. A. & ARMY, T. J. (ed.): Harvesting the Sun, p. 153, Academic Press, New York-London 1967.
WOO, K. C., ANDERSON, J. M., BOARDMAN, N. K., DOWNTON, W. J. S., OSMOND, C. B. & THORNE, S. W., *Proc. nat. Acad. Sci. U.S.A.* 67: 18, 1970.
ZIEGLER, H., ZIEGLER, I., SCHMIDT-CLAUSEN, H. J., MÜLLER, B. & DÖRR, I., in: METZNER, H. (ed.): Progress in Photosynthesis Research, Vol. III, p. 1636, Tübingen 1969.

THE GENETICS OF PHOTOSYNTHESIS AND RATIONAL MEANS OF BREEDING HIGHLY PRODUCTIVE PLANTS

A. A. NICHIPOROVICH

K. A. Timiryazev Institute of Plant Physiology, Academy of Sciences of the U.S.S.R., Moscow, U.S.S.R.

Introduction

The genetics of photosynthesis is a complex and important problem of general biological significance. The ultimate aim of work in this field is to learn how to control the photosynthetic process and to improve the photosynthetic apparatus on a genetical basis and in this way obtain new plant cultivars with high photosynthetic activity.

However, the relationship between plant productivity and photosynthesis is a complex and multifarious one. It depends on the size of the photosynthetic apparatus, on the general level of the photosynthetic rate, on the specificity of the transformations of carbon, nitrogen, phosphorus and sulphur and on the qualitative composition of photosynthates, the conditions and rate of their flow and utilization in metabolic and growth processes, etc. Besides, there are many other important traits determing the level of plant productivity which should be altered in breeding new plant cultivars: immunity, resistance to high and low temperatures, resistance to soil salinization, characteristics defining the quality of crops, etc.

Any new plant cultivar must possess a number of obligatory traits of high productivity which can be combined into a single logical system forming one part of the general theory of plant productivity, the 'quantitative, complex theory of the photosynthetic productivity of plants and principles of its optimization' (see, e.g. WATSON 1952; MONSI & SAEKI 1953; NICHIPOROVICH 1956, 1963, 1966, 1972; NICHIPOROVICH et al. 1961; ROSS 1967, 1969, 1972; SAN PIETRO et al. 1967; PP-Photosynthesis and Utilization of Solar Energy 1968; EASTIN et al. 1969; ZALENSKIĬ & RODIN 1969; Prediction and Measurement of Photosynthetic Productivity 1970; ŠESTÁK et al. 1971; TOGARI & YOKENDO 1971; ZELITCH 1971). In the following exposition of the theory generalized data are presented which are based on the papers mentioned above as well as on many other studies.

It is assumed here that the basis of high crop yields is photosynthesis, the decisive process of plant nutrition and activity. In the formation of 1×10^3 kg dry biological mass, plants consume during photosynthesis approximately 2×10^3 kg CO_2, evolve at least 1.5×10^3 kg free oxygen, absorb about $170 - 340 \times 10^6$ kJ solar energy and store in the biological mass of the crops nearly 17×10^6 kJ. Part of the energy absorbed is expended in evaporation of approximately $200 - 400 \times 10^3$ kg water (NICHIPOROVICH 1956, 1963; NICHIPOROVICH et al. 1961). At periods of most intense photosynthetic activity and plant growth 500 to 1 200 kg CO_2 is absorbed from the air as a result of turbulent transfer and diffusion; this may be compared with the amount of CO_2 in a crop volume of 1 m high, which is only 5 kg CO_2 per 10^4 m^2. In order to carry out this vast photosynthetic work a number of important conditions must be met. These include: (a) specific features of the structure and activity of the ultimate productive photosynthesizing systems, the phytocoenoses, crop stands and green plantations; (b) certain physiological and morphological traits of plants forming the phytocoenoses; (c) optimal supply of the plants with radiant energy, warmth, carbon dioxide, water and mineral elements.

These conditions and processes are related to each other qualitatively and quantitatively in a very definite manner. To obtain a high crop yield requires that the best combination of conditions should be ensured. This may differ for different plants and depends on the aim and conditions of their cultivation.

In order to be maximally productive, crop stands should form photosynthesizing organs (leaves) of appropriate size and structure and at an optimal rate. This optimal leaf area (L_{opt}) should ensure that most of the solar energy is absorbed and that the efficiency of photosynthesis is as high as possible, the result being a maximal rate of photosynthate formation. Along with this optimal optical density of the crop its optimal ventilation should ensure rapid turbulent transfer of large amounts of CO_2-containing air within the stand.

The photosynthates should be utilized in a most favorable way for growth of nutritive organs (roots, leaves) and then for the ripening of the reproductive and storage organs.

An important characteristic of the productive level of plants is the efficiency (ϵ) of utilization by the plants of the photosynthetically active solar radiation (PhAR) incident on the plants during the whole vegetative period. Values of ϵ of 0.5-1% pertain to a low productivity, those of 1-2% indicate an average productivity, 2-3% a good, and 4-5% a high productivity. Theoretically, values as high as 6-8% or even 10% should be attainable as a result of further work on optimization of the photosynthetic activity of plants and on the productive processes within them.

The aim of the following exposition is to characterize and assess the main factors which determine and limit the levels of photosynthetic productivity and to analyze the possibility of overcoming the existing limitations by altering the photosynthetic characteristics.

Yields as a Function of Daily Increases of Dry Biomass, Types of Growth, Leaf Area, and Leaf Photosynthesis

Economic crop yields of, say, fruits, grain, roots and tubers comprise an appreciable fraction of the total biological yield, Y_{biol}. The economic value of a crop can therefore be characterized by the coefficient $K_{econ} = Y_{econ}/Y_{biol}$. High economic values can therefore be obtained by increasing the magnitude of Y_{biol} or K_{econ} (NICHIPOROVICH et al. 1961; NICHIPOROVICH 1963, 1966) (see Fig. 1).

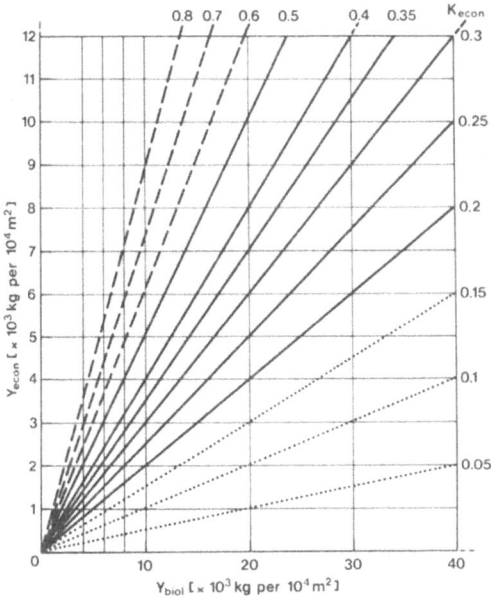

Fig. 1. Possible relations between economic (Y_{econ}) and biological yield (Y_{biol}) and the economic coefficient (K_{econ}) for cereals (———), root, tuber and fodder crops (– – – –), and technical plants (wood, fibers, sugar, etc.) (··········).

Biological crops are formed as a result of photosynthetic activity, and the yield (Y_{biol}) is the sum of the daily increases of organic biological mass (C) over the vegetation period of t days: $Y_{biol} = \sum_{0}^{t} C$. It can also be defined as the product of the mean daily increase (C_m) by the number of vegetation days (t), $Y_{biol} = C_m \, t$.

For biennial or perennial plants the values of C increase after germination and the appearance of the first shoots (phase I), attain a maximum (C_{max}), remain at this level for a certain period of time (phase II) and then begin to decrease during the period of intense formation and maturation of reproductive organs (phase III); at the end of the vegetative period values may vanish or even become negative.

The magnitudes of the biological and (on the condition that the values of K_{econ} are optimal) economic crop yields closely correlate with the mean (C_m) and maximal (C_{max}) daily increases of dry biological mass (Fig. 2). High yields are obtained if the total photosynthetic effect during the optimal state is large in the crop stand and the daily dry mass increase is of the order of 300-600 kg per $10^4 \, m^2$.

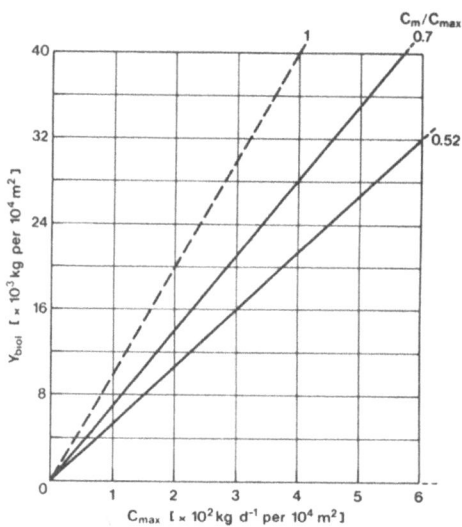

Fig. 2. Possible relations between Y_{biol}, C_{max} and C_m in crop stands with a 100-d period of vegetation.

Daily Increases of Dry Biomass as a Result of CO_2 Transfer and Assimilation from Air

The magnitudes of C_{max} and C_m, on the other hand, are closely related to the area of the leaves, L_{opt}, and activity of the photosynthetic apparatus of the crop plants during the period of maximal growth.

In phytocoenoses of most cultivated plants the optimal leaf area (LAI_{opt}) usually varies between 4 to 6 m^2 per m^2. High values of C_{max} and C_m in this case are possible only if the photosynthetic apparatus is highly efficient as assessed by the net photosynthetic productivity, NA (g m^{-2} d^{-1}) or total daily (P_d) or mean daily ($P_{d, \, m}$) or maximal daily ($P_{d, \, max}$) photosynthetic rates of CO_2 consumption per unit leaf area.

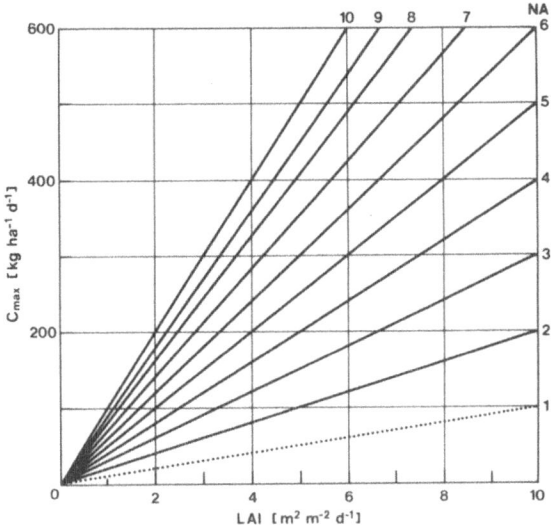

Fig. 3. The dependence of possible C_{max} levels on LAI and NA levels.

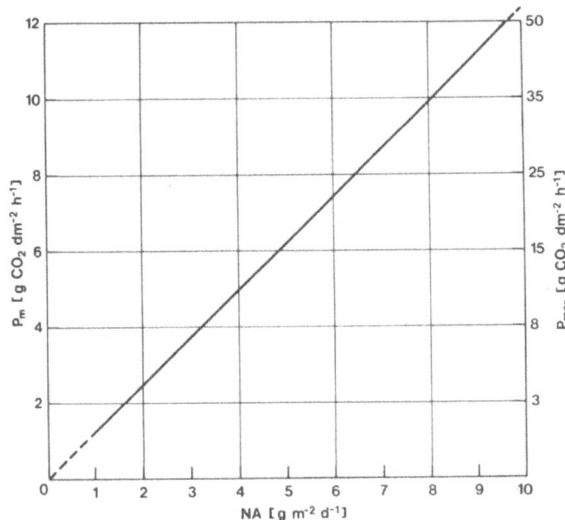

Fig. 4. The dependence of possible NA levels on P_m (mean daily photosynthesis) and P_{max} (maximal photosynthetic rate during the day). Day length 15 h, LAI = 5 m^2 m^{-2}; Pd/NA = 1.8; P_{max}/Pd = 4 : 1 when levels of NA are high, and 1.5 : 1 when levels of NA are low. P_m/P_{max} [mg CO_2 dm^{-2} h^{-1}].

Possible dependences of C on LAI and NA and their various combinations are shown in Fig. 3. As a rule NA decreases with the growth of LAI due to mutual shading of the leaves and impairment of airing of the crops. The best results are therefore obtained for certain mean values of LAI (which depend on the plant species and illuminance) but only if the values of L_{max} (or L_{opt}), NA and their product are high. High values of NA can be attained only if the optical density and structure of the phytocoenosis are optimal and air movement within it is not impeded; high activity of the photosynthetic apparatus is also required.

Possible relations between the NA values and mean (P_m) and maximal (P_{max}) daily rates of photosynthesis of a leaf (Fig. 4) show that high values of NA can be obtained even when the values of LAI are average and rates of photosynthesis are those usually observed.

In summary, improvement of the structure of phytocoenoses and enhancement of the efficiency of the photosynthetic apparatus are means of increasing photosynthetic productivity; for the solution of this problem not only the techniques of plant physiology, agronomy and agricultural chemistry should be applied but also those of genetics and plant breeding.

Theoretical Energetic Levels of Productivity

According to the classical equation of photosynthesis ($CO_2 + H_2O \rightarrow [CH_2O] + O_2$) the formation of 1 kg of carbohydrates requires the consumption of 1.46 kg CO_2. The daily increase in dry mass of plants mentioned above is the algebraic sum of the increase due to gross photosynthesis and mineral nutrition and the decrease due to respiration and exosmosis. As a consequence, the ratio of CO_2 consumed to dry mass increase (C) is not 1.46 but rather 1.75 to 2.0 (NICHIPOROVICH 1956, 1963, 1966; NICHIPOROVICH et al. 1961).

High crop yields can therefore be obtained if, during the period of most rapid growth (phase II), from 300 to 600 kg of dry biological mass are formed per $10^4 m^3$ of the crop stand, and correspondingly from 600 to 1 200 kg of CO_2 are consumed in gross photosynthesis per day.

In a crop stand 1 or 2 m high the amount of CO_2 per $10^4 m^3$ is 5-10 kg. This should not be reduced by more than 10 or 20% (or be lower than 270-240 vpm) since photosynthesis will otherwise be greatly reduced (Fig. 13). Therefore, within the crop stand the volume of air will be renewed by at least 600 to 1 500 times per 'working' day as a result of convention and turbulent exchange. This corresponds to a mean air velocity of 2-3 cm s^{-1} within the coenosis, and when the photosynthetic rate is maximal and equals 150-200 kg CO_2 h^{-1} per $10^4 m^3$ in the middle of the day the air must be exchanged at least 300-400 times per hour and move with a velocity of 8-11 cm s^{-1} (Fig. 5). At 2-3 m above the crops the air velocity is usually ten to one hundred times greater than that mentioned above. However, near the surface of the crop, and especially within it the velocity markedly decreases depending on the density and structure of the coenosis (Fig. 6).

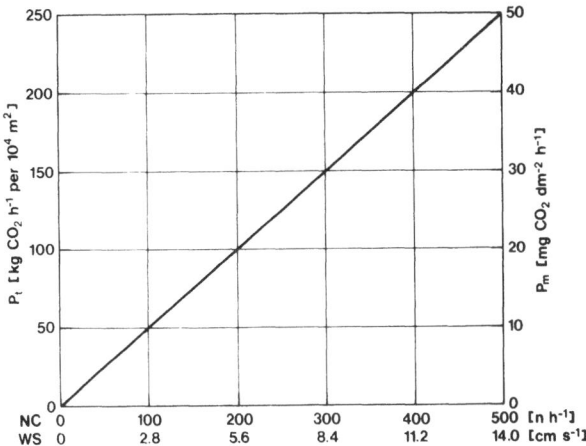

Fig. 5. Number of air exchanges (NC), wind speed (WS) and average photosynthetic rate of leaves (P_m) within a crop stand necessary for supplying the different levels of total daily photosynthesis (P_t). Height of crop stand 1 m; LAI = 5 m² m⁻²; CO_2 concentrations not lower than 270 vpm.

This factor frequently restricts the total photosynthetic activity of phytocoenosis (Fig. 7) (SAN PIETRO et al. 1967; ZALENSKIĬ & RODIN 1969; Prediction and Measurement of Photosynthetic Productivity 1970; ŠESTÁK et al. 1971; TOGARI & YOKENDO 1971; NICHIPOROVICH 1972; MONTEITH 1973).

This is further confirmation that a highly productive phytocoenosis must possess: (1) an optimal optical density for optimal absorption and utilization of PhAR, (2) good ventilation (i.e. a small diffusion resistance), and (3) high photosynthetic activity of leaves and hence maximal absorption of CO_2 from the air (SAN PIETRO et al. 1967; ROSS 1969; ZALENSKIĬ & RODIN 1969; TOGARI & YOKENDO 1971; NICHIPOROVICH 1972).

Photosynthesis and Efficiency of Radiant Energy

About 8-10 x 10⁶ kJ of energy are stored when 300-600 kg of dry biological mass are formed per day. Plants growing rapidly in the middle of summer are irradiated by 60-125 x 10⁶ kJ of PhAR energy per 10⁴ m² each day. The actual efficiencies of energy storage (ϵ_{st}) by crops under most favorable conditions are (for maize and sorghum) 10-12% (Fig. 8, levels I and II).

However, only a part of the incident energy utilized in gross photosynthesis during the day is stored. Thus in C_4-plants (HATCH et al. 1971; BLACK 1972) approximately 20% of the energy consumed in photosynthesis is lost during a

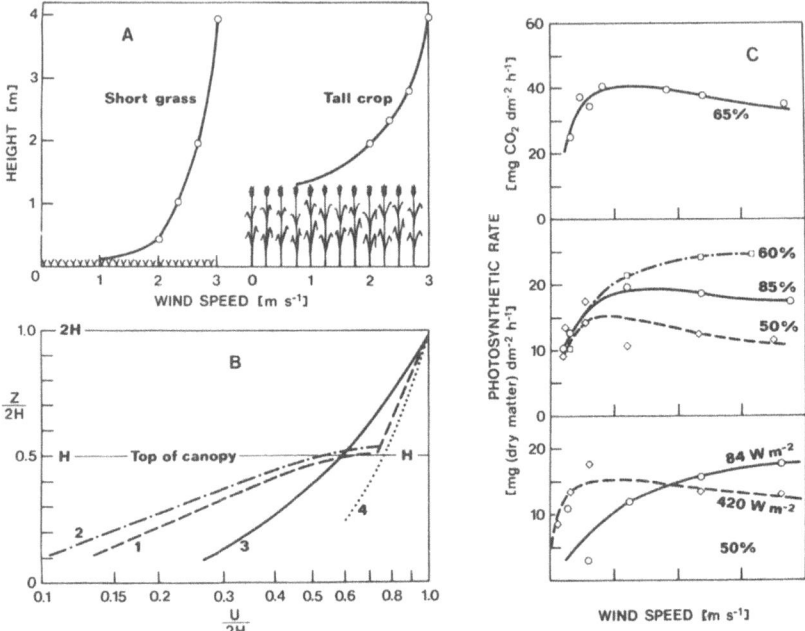

Fig. 6. A. Wind speed dimishing above and within phytocoenoses of different height.
B. Dimishing of wind speed within phytocoenoses of different plants (LEMON 1969). Wind speed at the double height (2H) of phytocoenosis is taken as a basic unit. 1−maize, 2−wheat, 3−Christmas trees, 4−bushel baskets.
C. Dependence of photosynthetic rate determined as CO_2 influx (above) or dry matter increment (middle and bottom) on wind speed, relative air humidity (% at the curves) and irradiance (from TOGARI & YOKENDO 1971).

24 h-day as a result of dark respiration; in C_3-plants another 20% are lost as a result of light respiration.

Thus the actual efficiencies for utilization of incident PhAR are, for C_{max}, only 80-60% of gross photosynthesis during the period of maximal photosynthesis and growth. Therefore under these favorable conditions the true daily efficiency of photosynthesis in phytocenoses may be 12 or even 15% with respect to the incident radiation (Fig. 8, levels VI and VII).

These values are so high that it seems desirable to compare them with theoretical values. According to present concepts, 8 einsteins (or approximately 1 600 kJ of solar radiation) must be absorbed in order to permit one mole (44 g) of CO_2 to be consumed under the most favorable conditions. About 470-480 kJ are stored in the direct products of photosynthesis which represent about 28% of the absorbed radiation (Fig. 8, level III) or $\approx$ 22% of the incident radiation since $\approx$ 15% is reflected and 3-5% is transmitted (level IV).

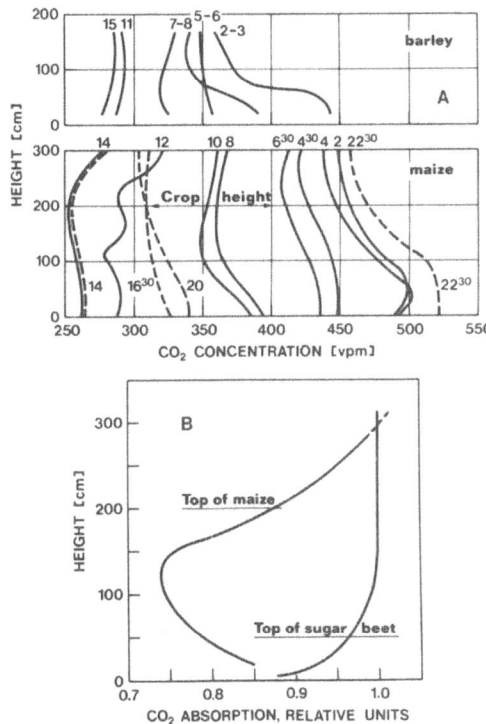

Fig. 7. Characteristics of CO_2 regime in crop stands: A — according to CO_2 profiles measured at different hours of the day (figures at the curves) (from MONTEITH 1962 and KARPUSHKIN 1966); B — according to the rate of CO_2 absorption by 2 N sodium hydroxide in shallow vessels placed at different heights in maize and sugar beet stands. (From NICHIPO-ROVICH et al. 1973.)

These values are the maximal energy efficiencies of gross photosynthesis of leaves subjected to short time exposures and not possessing photorespiration. Since 20% of the energy stored is expended in dark respiration during a 24 h day it may be assumed that the maximal efficiency of utilization of incident light energy in dry mass production per day is $\approx$ 18% and for plants of the C_3 group, which possess light respiration, about 14-15% (Fig. 8, levels V and VI).

These are the highest possible efficiencies of utilization of incident PhAR by plants assuming that throughout the day the quantum requirement of photosynthesis is 8 for all leaves or that the energy efficiency for absorbed radiation is 28%.

Energy efficiencies of gross photosynthesis close to those compatible with 8 quantum requirements ($\epsilon = 28\%$) are observed only under the most favorable conditions and at relatively low irradiances. With increase of irradiance the rate of

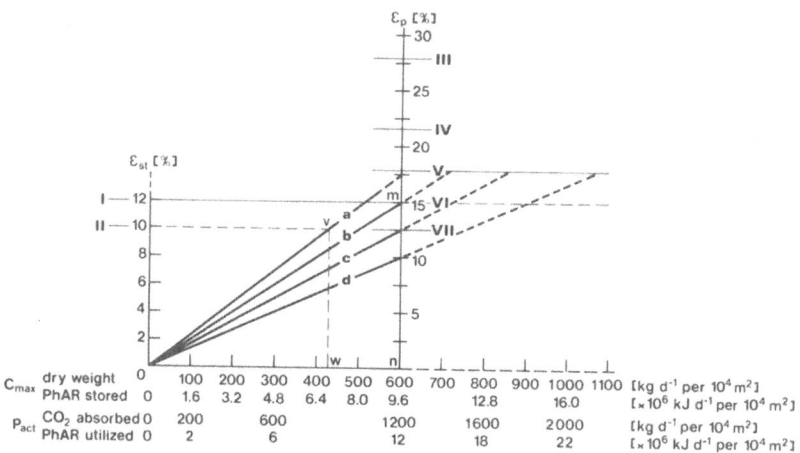

Fig. 8. Relative amounts (full lines) of radiant energy (PHAR) stored (ϵ_{st}, %) in daily increments of dry biomass (C_{max}) and used for actual photosynthesis (ϵ_p, %) at different C_{max}, actual photosynthesis (P_{act}), and incoming radiant energy (PHAR), i.e. a – 63, b – 84, c – 105, d – 126 x 10^6 kJ d^{-1} per 10^4 m^2. Zone 10-v-w-0 contains the real values for the majority of C_3-plants. Zone 12-m-n-0 represents C_4-plants and the most productive C_3-plants Broken parts of the straight lines a-d show the zone of possible increases of C_{max} and ϵ % with their approach to level VI and then to the theoretically possible maximum level V.

photosynthesis increases but the energy efficiency decreases, in some cases down to 7-5% and usually to 4-2% (Fig. 9).

It is therefore interesting to ask how phytocenoses manage to store radiant energy with such high efficiencies approaching the theoretical values permitted by an 8 quantum mechanism. The basis of this remarkable effect is that foliage of an appropriate geometry in phytocenoses not only completely absorbs the entering energy, PhAR$_{inc}$ (Fig. 10), but is able to more or less uniformly distribute the energy over the leaf surface; the area of the leaf blades (LAI) is usually several times larger than that of the soil covered by the crop stand. In phytocenoses of most plants LAI$_{opt}$ is about 4-6 m^2 m^{-2} but may reach 10 m^2 m^{-2}, for example, in fodder grass communities with small, narrow leaves (NICHIPOROVICH et al. 1961; NICHIPOROVICH 1966, 1972; ROSS 1967, 1972; SAN PIETRO et al. 1967; EASTIN et al. 1969; ŠESTÁK et al. 1971; TOGARI & YOKENDO 1971; ZELITCH 1971). In undergrowth with vertical leaves (*Typha*) the ratio may even be as high as 16-18 m^2 m^{-2} (DYKYJOVÁ 1971).

With such high LAI values and optimal crop structure the plant leaves are irradiated by weak radiation during most of the day and especially in the morning and evening. The photosynthetic rate in the leaves is low but this is compensated by a large leaf area and by higher values of ϵ. The result is that the integral rate of photosynthesis, the ϵ values of the crop stand as a whole and the daily increases of dry biological mass (C_{max}) are all high.

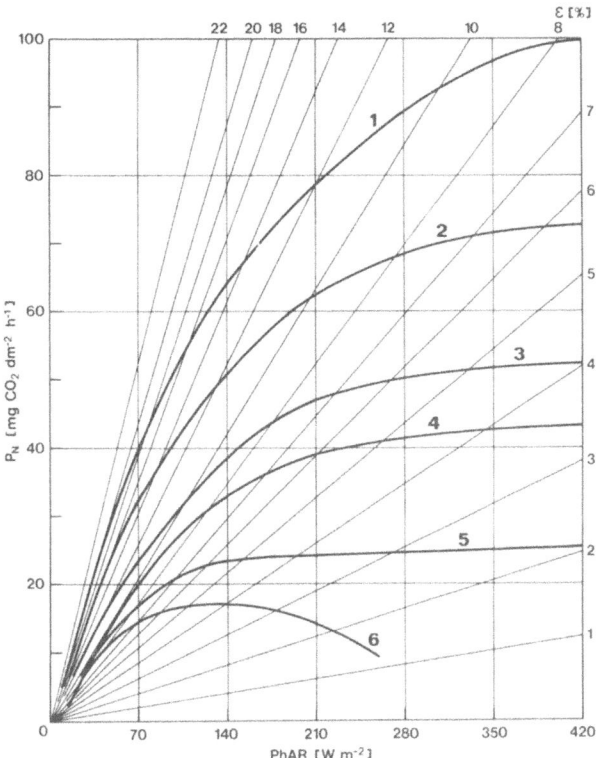

Fig. 9. Dependence of net photosynthesis (P_N) on irradiance (PhAR) in maize, sorghum and sugar cane (1,2), sunflower and sugar beet (3,4) many cereals and dicotyledonous C_3 plants (5), and trees and shade plants (6). Straight lines correspond to photosynthesis at different ϵ % of incoming radiation.

High photosynthetic productivities approaching theoretically maximal values are rather an exception and occur in crops of the most productive plants such as maize, sugar cane and sugar beet (see below). In crops of most cultivated plants the daily increase of dry mass C_{max} is ordinarily 100 to 250 per 10^4 m² and most often 50 kg per 10^4 m² instead of the 300-600 kg per 10^4 m² mentioned above.

Density and Structure of Crop Stands as a Factor of Photosynthetic Productivity

One of the causes for the above discrepancy is that crops are frequently deficient in water or mineral nutrition (NICHIPOROVICH 1972). In such cases the leaf

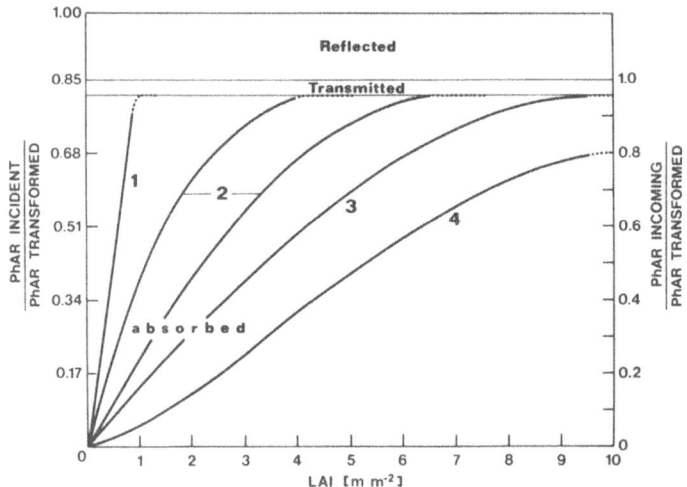

Fig. 10. Coefficients of transformation (reflection, absorption, transmission) of energy of incident PhAR in phytocoenoses as affected by their density and structure:
1. Phytocoenoses with monolayer of horizontal leaves (*Lemna, Nuphar, Victoria*, etc.).
2. Phytocoenoses of majority of cultivated plants with 'spherical' type of leaf position and orientation.
3,4. Phytocoenoses with small or long vertical leaves (*Typha*, etc.) (DYKYJOVÁ 1971).

area of the crop is usually insufficient, not all PhAR is absorbed or favorably distributed over the assimilatory surface.

In this case the main task of plant breeding should be to select plants with an enhanced growth rate and larger leaf area. Crops of such plants should be more efficient even under conditions of insufficient water and mineral element supply (KUMAKOV 1972). However, due to the mutual shading of leaves, the average rate and net yield of photosynthesis decreases, airing of crops selected for large leaf area is impeded, and the supply of carbon dioxide is reduced. Not infrequently, the ratio between the economically valuable and the vegetative organs changes. Thus, although the crop yield may be higher, the efficiency of fertilizers and watering decreases. Such negative characteristics must be suppressed by further breeding.

Be as it may, phytocoenoses are ultimately formed with an optimal leaf surface (L_{opt}) if the crops are provided sufficiently with water and mineral nutrition, and growth processes are stimulated. These positive changes (increase in leaf surface and amount of absorbed PhAR energy) balance the negative changes. In this case the plant is capable of carrying out net photosynthesis at a maximal rate and forms maximal amounts of dry mass (C_{max}) with a high efficiency of utilization of solar energy.

Further attempts to increase crop yields by the usual means of improving water conditions or supplying fertilizers do not lead to positive results. The reason for this is that an increase of mineral nutrient level results in an increase of growth processes and, in particular, growth of leaf area, but the amount of absorbed energy does not increase, and hence the conditions for photosynthesis are greatly impaired. The lower leaves which do not receive sufficient amounts of radiant energy and CO_2 wither and die. In their struggle for light plants raise their active leaf level to greater heights. The stems are greatly weakened and this may be the cause of lodging, impairment of crop structure, sharp drop of K_{econ} and the reduction and impairment of economical crop yield.

Hence it is important to breed varieties with moderate growth, genetically stunted and compact; the plants should utilize the photosynthates for optimal, moderate growth of vegetative organs and then subsequently for rapid growth and supply of the capacious reproductive and storage organs. These properties of the new varieties must be stable with respect to supply of water and fertilizers (especially nitrogen) which should not stimulate excessive growth of vegetative organs but rather stimulate the photosynthetic apparatus and the growth of genetically large and capacious reproductive and storage organs of economic value. In some cases it may be useful to breed plants with vertical leaves (e.g. in cereals).

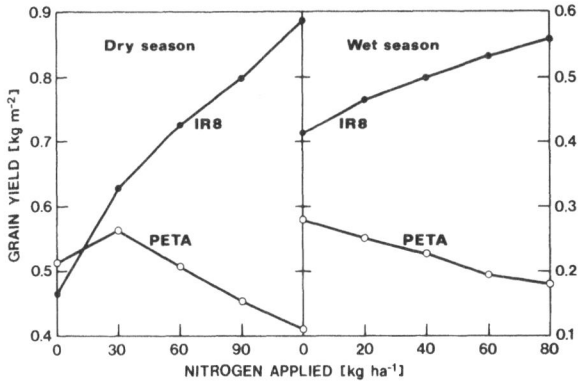

Fig. 11. Changes in grain yields of rice cultivars IR 8 and PETA induced by increasing doses of nitrogen fertilisers. (From CHANDLER 1969.)

Good examples of solutions to this problem have been the breeding of stunted cultivars of wheat, maize, and especially rice of the IR group. Thus the stunted rice cultivar IR 8 with compact bushes, vertical leaves and large panicles (CHANDLER 1969), in contrast to its predecessor PETA, responds favourably to strong doses of nitrogen fertilizers (Fig. 11).

Improvements of Plant Morphology and Photosynthetic Activity as Main Aims of Increasing Yields at a Genetic Level

The example just described is not a universal case. The height, shape and morphology of plants differ from plant to plant. Moreover, the size, shape and activity of the leaves may also be different, as are their position on the carrier organs and their spatial orientation.

To a large extent the result of this is that plants form phytocoenoses of various heights and requiring various sowing and planting densities.

If the dependence of all plants on irradiance were the same and linear the problem regarding the optical and spatial structure of phytocoenoses (providing the aerodynamic conditions are suitable) would be unimportant. For efficient utilization of radiant energy it would be sufficient that light absorption be complete; any sufficient leaf area would be suitable, providing it were not excessive, and any crop structure would also be acceptable. In this case the energy efficiency (ϵ %) of gross photosynthesis would be the same for all leaves irrespective of their position and equal to that of the phytocenosis as a whole (Fig. 9). The productivity would then depend on the initial slope of the curves and on dark respiration losses.

In reality the dependence of photosynthesis on irradiance and CO_2 concentration is more complicated (Fig. 9). As a consequence there is a strict dependence between the light and carbon dioxide curves of photosynthesis, on the one hand, and the optimal structure of phytocoenoses, on the other (KARPUSHKIN 1966; NICHIPOROVICH et al. 1973). The higher the activity of the photosynthetic apparatus and the closer the 'light curves' of photosynthesis of the leaves are to the straight lines, the more efficient and successful will be the solution of the problem of optimization of phytocoenosis structure.

This defines three equivalent but closely related problems in breeding directed towards the increase in plant productivity: (1) Improving structures of phytocoenoses with respect to the morphology, biology and activity of the photosynthetic apparatus of the given plant. (2) Improving growth functions of plants and obtaining plants which expend the optimal amount of photosynthates on growth of vegetative organs and a maximal amount for the growth, formation and maturation of economically valuable reproductive, storage and other organs. (3) Improving photosynthetic apparatus itself: 'straightening out' 'light curves' of photosynthesis, increasing their slope, and making them less dependent on the amount of available carbon dioxide (high photosynthesis at low CO_2 concentrations, higher saturation plateau at high concentrations).

Genetic Diversity of Photochemical Functions as Basis for Improving the Photosynthetic Apparatus

In ordinary breeding work, in which highly productive forms are selected, the growth function and morphological features of the plant are the properties which are primarily altered and which determine the high photosynthetic productivity of the phytocoenoses. Parallel variations in the activity of the photosynthetic apparatus are as a rule not observed. This has been noted for a number of plants such as beet (WATSON & WITTS 1959), wheat (EVANS & DUNSTONE 1970; KUMAKOV 1972) and rice (CHANDLER 1969).

In investigations in which the time course of the crop yield was studied it was found that the number or size of the storage and reproductive organs or of other economically valuable organs increased and also that K_{econ} strongly increased (NICHIPOROVICH 1963, 1966; KUMAKOV 1972). In some cases the increase in crop yield was related to a higher degree of compactness of the plant and to less branching; however the plants were more leafy and possessed a more profitable (vertical) leaf orientation. In cereals, tillering was found to increase simultaneously with the growth, fructification and maturing of the tillers. An important factor is dwarfness of cereals, a good example being that of rice (ISHIZUKA 1969).

On the other hand there are no reliable facts which prove that the productivity increase involved in ordinary breeding is related to an increase of the activity of the photosynthetic apparatus. Thus in the high-yielding rice cultivar IR8 mentioned above the photosynthetic rates are similar to those of other plants, varying between 35 and 67 mg CO_2 dm^{-2} h^{-1}. Moreover, in cultivars of certain plants (e.g. wheat) the photosynthetic rate characteristics are very much the same (BELIKOV et al. 1961) and no reliable differences in this property can be correlated with the differences in yield (EVANS & DUNSTONE 1970; KUMAKOV 1972).

This should not be regarded as meaning that in genetic work on crop yield improvement no increase in activity of the photosynthetic activity of plants should be possible. However, for a solution of this problem and that regarding the further direction of work and techniques for breeding highly productive plants, it seems desirable to have an explanation of these facts and to draw correct conclusions from them.

During the long evolutionary process plant taxons were formed differing with respect to details of organization of the photosynthetic apparatus and photosynthetic productivity. There is no need to discuss the principle differences in the photosynthetic apparatus of such organisms as bacteria, blue-green algae and other algae but it should be mentioned that there are large differences (EL-SHARKAWY & HESKETH 1965) in the higher plants which also contain various types.

The most striking example is the existence of C_3- and C_4-plants (HATCH et al. 1971; BLACK 1972) with strongly differing photosynthetic properties. A peculiar group consists of plants of the *Crassulaceae* family and also many woody plants (BLACK 1972). Moreover, pronounced differences in photosynthetic properties and activity are observed in ecologically different species of plants of a single genus (BJÖRKMAN & HOLMGREN 1963; BJÖRKMAN 1968) or even in different cultivars of cultivated plants such as rice (CHANDLER 1969), alfalfa (PEARCE et al. 1969), bean (IZHAR & WALLACE 1967), oat (CRISWELL & SHIBLES 1971) and soybean (DORNHOFF & SHIBLES 1970). In many cases there is a pronounced relation between the productivity and activity of the photosynthetic apparatus of natural forms. Examples are C_3 and C_4 cultivated plants or species of plants grown under high and low illuminance (BJÖRKMAN & HOLMGREN 1963; BJÖRKMAN 1968), etc.

The most reliable explanation is that differences in plant productivity depend not only on the activity of the photosynthetic apparatus, but also on the absolute and relative rates of utilization of photosynthates in the growth of photosynthesizing and other organs. Productivity differences also depend on the magnitude of respiratory losses. An important factor is the plant morphology which largely determines the structure and hence the photosynthetic productivity of plant communities. Variation of the relative importance of various photosynthetic organs (e.g. flag leaf and other leaves, stems and parts of cereal ears) during ontogenesis is also important in determining the total and economic crop yields. Depending on conditions, different characters may be in the first and major minimum.

The predominant direction of evolution may depend on environmental conditions as well as on the genetic predisposition of various forms with respect to change in a certain direction. Hence the appearance and development of C_4 plants may be a result of adaptation of certain plant forms to atmospheres (created by the plants themselves) with extremely low CO_2 concentrations and high O_2 concentrations.

Other plants adapted themselves to other conditions such as increasing drought or cooling, or seasonal periodicity of the climate. In such plants the biology of growth and development mainly changed (e.g. increase in number of juvenile or slowly-growing and, in particular, woody plant types), the photosynthetic process remaining fundamentally the same. However it became delicately adapted to the specific conditions of the environment in the ecological sense: the response of photosynthesis and growth to irradiance, temperature, humidity, etc. changed.

From this viewpoint it is interesting that C_4-plants predominantly inhabit the tropical regions. It is possible that the set of favorable conditions for both photosynthesis and growth resulted in the improvement of these two processes.

It is also possible that the improved supply of water and mineral elements to cultivated plants resulted in greater stimulation of the growth processes than of photosynthesis.

Genetically, the very organization of the photosynthetic apparatus (NASYROV 1971, and the present volume) and the regulatory systems which determine its activity are extra-ordinarily complicated. The activity of the photosynthetic apparatus depends on the anatomical structure of the leaves, on their optical properties, on the number and size of chloroplasts, their distribution in various tissues, and on the activity of components related to the photooxidation of water, electron transfer, primary CO_2 fixation and subsequent transformation of carbon in the chain of product formation. Also important are the systems of interactions between the chloroplasts, cell and transport systems, responsible for the flow of photosynthates, and their numerous enzyme systems regulating the processes mentioned above and controlled by the genetic apparatus of both the nucleus and the chloroplast, consisting of nucleic acids and protein synthesizing systems, etc. (SAN PIETRO et al. 1967; ZELITCH 1971; NICHIPOROVICH 1972).

Genetically, the photosynthetic apparatus of various plant genera, species or cultivars is such a complex, polygenic and balanced system that simple selection of plants with higher yields does not automatically lead to new cultivars with enhanced photosynthetic activities.

On the other hand growth, morphogenesis and morphology are genetically simpler. As an example we might mention that dwarfness of wheat plants depends on only three independent genes. It is precisely because of this that in breeding work, aimed at raising productivity, plant forms are produced with altered morphology and growth rate characters involving higher productivity.

If differences between the photosynthetic activity of different cultivars are observed, possibly due to causes uncontrollable by man (different geographic or phylogenetic origin, different breeding conditions) it is not infrequent that the differences do not correlate with the productivity.

From the foregoing one can draw the following important conclusion: it cannot be expected that an automatic improvement of photosynthesis or of the photosynthetic apparatus will occur in ordinary breeding work if they are not subjected to special control and study and if special methods are not applied.

The possibilities of raising the photosynthetic productivity of plants by improving growth and morphology are not unlimited; there are foreseeable limits to plant dwarfness, increase of K_{econ}, variation of the morphology, etc. In the future, and in some cases quite soon (e.g. for rice and wheat) these possibilities will be exhausted. Then the only remaining possibility for further increase of crop yield will be the enhancement of activity of the photosynthetic apparatus in which we have a reserve of untapped opportunities (see Fig. 9). Other conditions being equal (e.g. under conditions of optimal plant community structure) those plants will be more productive which possess a higher photosynthetic rate. Such plants not only utilize radiant energy and consume carbon dioxide at a higher rate but also assimilate mineral nutrition elements, primarily nitrogen, more efficiently (NICHIPOROVICH et al. 1972).

Breeding work with photosynthetically efficient plants should be directed towards obtaining larger economically valuable reproductive and storage organs since photosynthates will not be the limiting factor. As noted above, optimization of the plant community structure can also be attained with greater ease.

Work on improvement of the photosynthetic apparatus is important in those agricultural regions which do not receive enough rainfall and in which artificial watering is difficult. The absence of an optimal-size photosynthetic apparatus in such cases should be compensated by a higher photosynthetic activity.

However, of no less importance is such work on improvement of photosynthesis in conditions in which the soil is fertile and the leaf area is sufficient or even excessive. Vegetative growth may be too rapid in such cases and the conditions for photosynthesis may be impaired, the values of K_{econ} will be lower and further measures for raising soil fertility will be of no consequence.

Some Promising Lines of Breeding Photosynthetically Efficient Plants

At present no artificially induced mutants are known which possess an enhanced photosynthetic activity. However, artificially induced mutagenesis may be feasible for obtaining new plant forms with highly active photosynthetic machinery or parts thereof (BLACK 1972). Some chlorophyll mutants, for example (BLACK 1972), possess a lower rate of photosynthesis but in some cases contain larger amounts of 'small' photosynthetic units with higher assimilation numbers. Such favorable alterations of certain parts of the photosynthetic apparatus could conceivably be used in further synthetic breeding.

Artificially induced polyploidy also apparently does not favorably affect the photosynthetic apparatus (NICHIPOROVICH et al. 1941; FRYDRYCH 1971), but alters the anatomical structure of the leaves and other organs and also growth.

Some data indicate that the heterosis effect of linear hybrids may involve an increase in the activity of the photosynthetic apparatus. However the effect is suppressed in later generations and hence is a phenomenon of another kind.

From these scarce data the following preliminary conclusion can be drawn: methods which may be efficacious but are not directed toward the alteration of definite genes or characters will probably not rapidly yield positive results in such a genetically complex function as the polygenic photosynthetic function.

No conclusions can be drawn, on the basis of available data, regarding the feasibility of applying direct selection of plants for favorable integral photosynthetic characteristics, such as high gas exchange rates (CO_2 consumption or O_2 evolution). Gas exchange, as an integral characteristic of photosynthesis and the result of interaction of various structures and regulatory systems, is extremely labile with respect to the phenotype and physiological state of the plant. Therefore it is not very useful for assessment of the original material employed in breeding nurseries.

Thus if selection is based on high photosynthetic activity it is necessary that comparative assessment of the material be carried out under controlled stationary conditions. The response of the photosynthetic apparatus to various factors, and particularly to irradiance, CO_2 and O_2 concentrations, temperature, etc., should continually be recorded.

There are no simple characters or processes which define the integral ultimate activity of the photosynthetic apparatus. It is controlled by many genes (probably tens if not hundreds) and it would be difficult to carry out breeding work if they were combined in a single stable system.

In reality, however, the situation is much more favorable. First of all, in the integral system of the photosynthetic apparatus where the overall activity consists of genetically correlated characters, many traits are to a great extent variable and independent and this may be profitable for the system. It is also important that not all genetically conditioned active systems and traits are equivalent from the viewpoint of integral activity of the photosynthetic apparatus. For example, in C_4-plants (in contrast to C_3-plants) the high activity of the photosynthetic apparatus is primarily determined by the cooperative interaction between two carboxylating enzymes located in cells of different types of tissues and also by the peculiar anatomical structure of the plants.

These differences between C_4- and C_3-plants are so fundamental that when the plants are crossed (e.g. C_4- and C_3-species of *Atriplex*; BJÖRKMAN et al. 1971; HATCH et al. 1972) the hybrids retaining the C_4-type anatomy do not acquire a higher photosynthetic activity, whereas hybrids without a C_4-type structure possess a lower activity than the original C_3-plants.

A different result is obtained when species of a single genus of C_3-plants differing ecologically are crossed, such as inhabitants of strongly illuminated and shady locations (BJÖRKMAN & HOLMGREN 1963). The first generation hybrids are more active photosynthetically than the shady plants. This signifies that, if the initial form were valuable in some certain respect and was required to raise the photosynthetic activity and productivity, then crossing of this sort could play a positive role.

It is significant that improvement of an overall characteristic, the photosynthetic rate as determined by gas exchange measurements, was correlated with the improvement of activity of the carboxylating enzyme, ribulose di-phosphate carboxylase (BJÖRKMAN 1968). This is particularly important since the decisive role of carboxylating enzymes has been noted in many other investigations in which the plants differed genetically or phenotypically. In some cases the photosynthetic apparatus was activated by improving conditions of illumination, or water supply, nitrogen nutrition, etc. (MOSS et al. 1969; MOSS 1970).

The genetic differences in activity of the photosynthetic apparatus (as assessed by CO_2 exchange) have also been found to correlate with such characteristics as xeromorphic structure of leaves (small cells, better venation, larger number of stomata), low total gas diffusion resistance in leaf and specific surface weight of leaf blades (EL-SHARKAWY & HESKETH 1965; IZHAR & WALLACE 1967; PEARCE et al. 1969).

It is less clear what significance characteristics such as the compensation point of C_3-plants have with respect to the activity of the photosynthetic apparatus (MOSS et al. 1969; MOSS 1970; DVOŘÁK & NÁTR 1971). Since the compensation point is practically zero in C_4-plants it would seem logical to assume that it probably reflects some important peculiarities of the organization and activity of the photosynthetic apparatus and deserves serious attention in future work.

A more efficient method for differentiating material in breeding work is probably the screening technique employed by American investigators (MENZ et al. 1969) as an index of the relative rate of light respiration.

At present the data are too scarce to attempt an assessment of the problem as a whole. However some conclusions can nevertheless be drawn:

First of all extensive studies of the fundamental processes and traits defining the level of the integral photosynthetic activity should be carried out with the aim of systematically improving them and ultimately of raising the integral activity of the photosynthetic apparatus. The main method of work in this direction should not consist in comparing large amounts of initial breeding material in field nurseries, but rather in a detailed quantitative investigation of photosynthesis in a restricted number of rationally chosen initial forms cultivated in stationary, controllable conditions. Such material should give an idea as to the main causes limiting the level of photosynthetic activity of a given type of plant and of the 'bottlenecks' which should primarily be changed.

Another important aspect is a detailed study of photosynthesis in the most characteristic representatives (cultivars) and particularly in inbred strains of a given species of genus of plants grown under various geographic, ecological conditions or differing with respect to their systematic or genetic origin. Such forms should be studied in stationary but varying conditions of cultivation and of formation of the photosynthetic apparatus. The purpose of such investigations should consist in the elucidation of the localization of the component characters of the photosynthetic activity in various branches and groups of the general phylogenetic system of a given genus, species or set of cultivars of plants which are being improved by breeding. In other words, as a result of such work the genetic funds of those photosynthetic traits should be determined which, by means of further hybridization and selection, can be introduced into the genomes of future cultivars.

The next step should determine the role of various traits in the integral activity of the photosynthetic apparatus and in the productivity of plants. The genetic nature of the most important and significant characters should be determined by hybridological analysis. The purpose of this work should be the assessment of these traits in breeding work and their subsequent introduction into the genotypes of future plants by hybridization.

The fourth stage should be the devising of models of an 'ideal plant of the future productive cultivar' and of a general plan of its synthetic breeding based on detailed knowledge of the localization of the favorable traits in the genetic funds of the given systematic group.

At present much has been accomplished in the study of the traits and laws of highly productive plant communities and of the relation between the characteristics of the communities and the biological, morphological and growth peculiarities of the component plants and also the integral photosynthetic activity.

A set of such characters is sufficient to set up a model of 'ideal cultivars' and to carry out work for creating them. In some cases this type of work has yielded outstanding results (e.g., the dwarf wheat plants bred in Mexico, or the dwarf rice and maize cultivars).

However we still do not have sufficient material which would permit us to devise detailed 'ideal models' of a highly active photosynthetic apparatus of a given plant and propose how the model could be realized in breeding work.

Increase of Plant Productivity on the Genetic Level

In work designed to find the limits of the integral activity of the photosynthetic apparatus, allowance should be made for the fact that, even in a given plant under different conditions, the decisive traits may be different.

As an example, Fig. 12 shows the chlorophyll contents (top) and photosynthetic rates (bottom) in leaves of maize and broad bean plants grown in Knop solution with various nitrogen doses. A threefold increase of the nitrogen dose results in an increase of the chlorophyll content in each plant species. However, in maize this is accompanied by an increase in the photosynthetic rate, whereas in the broad bean plants photosynthesis increases only so long as the nitrogen dose does not exceed 2 norms.

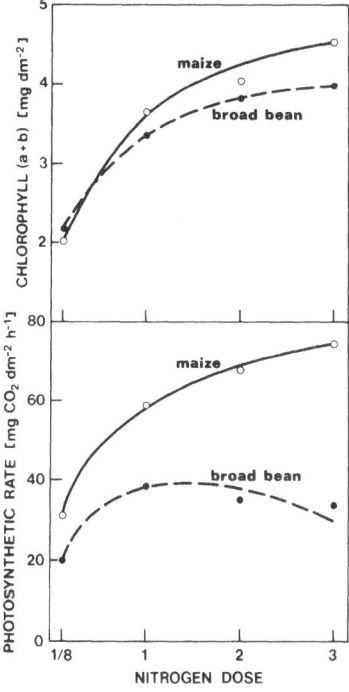

Fig. 12. Effect of increasing the dose of nitrogen fertilizers on chlorophyll contents and photosynthetic rates in leaves of maize and broad bean.

It may therefore be concluded that the chlorophyll content in maize is one of the important characters determining the photosynthetic rate at all nitrogen levels tested. In broad bean plants, on the other hand, other traits become the decisive ones when nitrogen doses exceed the normal by more than two times. The decisive trait in this case is the limited activity of the carboxylating enzymes, primarily of RuDP-carboxylase (AVDEEVA & ANDREEVA 1973). Maize, as a C_4-plant, is in this respect in more favorable conditions, as will be seen below.

From the foregoing it follows that the search for decisive traits which determine the organization of the photosynthetic apparatus and limit its activity, and which should be improved at a genetic level, should be carried out under conditions corresponding to the conditions of actual cultivation. Another important point is that the plants should be amply fed during the formation of the photosynthetic apparatus as well as during its investigation to find out the factors determining the maximally attainable level of the integral activity of the photosynthetic apparatus.

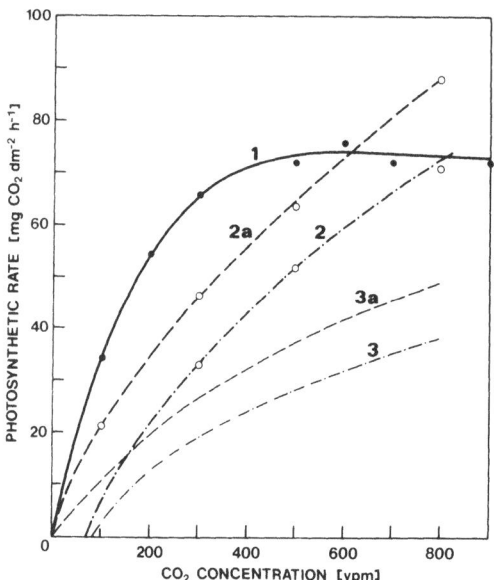

Fig. 13. CO_2-curves of photosynthesis for maize (1), sugar beet (2), and broad bean (3) measured at 21% (1, 2, 3) and 2% O_2 in air (1, 2a, 3a) in 'white light' of 400 W m^{-2}.

Let us consider another example of a 'bottleneck' in the photosynthetic activity which should subsequently be improved at a genetic level. The rates of photosynthesis in maize, sugar beet and broad bean (Fig. 13) were measured at various CO_2 concentrations in the air and at an irradiance of 400 W m^{-2}. The C_4 plant maize, which fixes CO_2 with the aid of PEP carboxylase, is capable of rapidly assimilating CO_2 at comparatively low concentrations of this gas and at both low and normal concentrations of oxygen in the air.

The beet plant (a C_3 plant in which primary fixation of CO_2 involves RuDP carboxylase) assimilates low concentrations of CO_2 at a much lower rate, particularly when the oxygen content is high. However, at high CO_2 concentrations the rate of photosynthesis in beet plants may exceed that of maize, especially when the O_2 concentration is low. At 300 vpm of CO_2 the photosynthetic rate was saturated by radiant energy in the beet and broad bean plants. In maize the rate of photosynthesis was only close to the saturation level. At the irradiances employed an increase in CO_2 concentration strongly increased the photosynthetic rate in all three plants.

From these data it may be concluded that under normal conditions of growth the photosynthetic apparatus formed possesses large potentialities with respect to electron transport, pigment and enzyme systems. A rapid improvement of the conditions for photosynthesis, for example an increase of the CO_2 concentration, results in a rapid and strong increase of the photosynthetic rate without any significant alterations of the systems occurring.

This leads us to another conclusion. Under the given conditions the main factor limiting the photosynthetic activity is the diffusion resistance and particularly the 'chemical resistance' which depends on the activity of the carboxylating enzymes (see Table 1).

Table 1. Activity of carboxylating enzymes in leaves of various plants (data of T.A. AVDEEVA).

Plant	Enzyme activity [μmol CO_2 dm^{-2}]	
	RuDP–C	PEP–C
Sugar beet	80 – 127	6.45
Bean	49 – 60	4.20
Maize	20 – 30	60 – 70

Hence the high photosynthetic rate of maize at low CO_2 concentrations is primarily determined by the fact that RuDP carboxylase, through which practically all assimilated CO_2 passes, is actively assisted by PEP carboxylase; the height of the plateau of the carbon dioxide curve of photosynthesis, however, is determined by the restricted activity of RuDP carboxylase. Consequently in breeding work with maize carried out with the aim of raising the photosynthetic rate, it is particularly important to increase the activity of RuDP carboxylase.

In the beet plant it is most important to lower the light respiration rate and the inhibiting effect of oxygen. If this could be done the beet plant, despite the fact that it is a C_3 plant, could be at least as photosynthetically active as maize.

This is an important conclusion and it signifies that possession of C_3 metabolism is not an obstacle to attaining photosynthetic rates as high as those observed in C_4 plants.

What has been said above should not be taken as meaning that the whole problem of improving the photosynthetic apparatus of plants can be reduced to that of raising the activity of carboxylating enzymes, especially since their role is complex and they are located in various tissues and cells and consist of isoenzymes and allosteric forms (TING & OSMOND 1973a, b). Finally, much remains to be explained about the genetic regulation mechanisms of their activity. Another complicating factor is that the activity of the photosynthetic apparatus is controlled by many other active systems of no less importance.

In conclusion, the close relation between the properties of the photosynthetic apparatus and those of the plant community which they form should also be mentioned. Maize, for example, thrives in tall communities which are poorly aired and hence may be deficient in CO_2 (NICHIPOROVICH et al. 1973). However, this is compensated by the capacity of this plant to consume CO_2 from low concentrations in the atmosphere.

Beet plant communities are low in height and the airing conditions are such that even CO_2 from the soil can be utilized. As a result, this plant is able to carry out intense photosynthesis, even though the latter depends on the CO_2 concentration to a greater extent than it does in maize; moreover, the plant may be as productive as maize. Photosynthesis of beet plants is also stimulated by the active attracting role of its roots which possess a highly active growth and strong function.

Perspectives

The material presented above illustrates the point that various internal factors affecting the activity of the photosynthetic apparatus may be decisive depending on the plant itself and on the intensity of the vital processes. It is quite possible that, by improving one property, a second will be converted into the limiting factor by the breeder. New approaches and efforts will then be required in order to make further progress. The problem is to be able to assess and then improve the main limiting factor either by special selection or by systematic hybridization.

Best results can be derived from the afore mentioned considerations only if plant physiologists, geneticists and plant breeders keep in mind that the activity of photosynthesis is one of the most important but not the sole factor of plant productivity and it would be hopeless to try to improve the photosynthetic function in plants possessing a poor biology or morphology and incapable of forming highly satisfactory phytocoenoses.

Work on improvement of the photosynthetic function cannot be effective if growth is poor and measures are not taken to improve it and also if capacious, economically valuable organs which utilize large amounts of the assimilates formed by the highly active photosynthetic apparatus are not created by genetic means.

Improvement of the photosynthetic activity for purposes of increasing plant productivity should be carried out both analytically and synthetically; the whole problem should be assessed from the standpoint of those characters which an 'ideal plant' of a given type and an ideal phytocoenosis, as the ultimate photosynthesizing productive system, should possess. The result of such work may be an increase of photosynthetic productivity of future plant cultivars such that, during periods of intense growth, when $C = C_{max}$, the plants will be able to utilize PhAR energy with an efficiency ϵ of 14-15% and for the complete vegetation period $\epsilon = 5\text{-}6\%$. In the future the vegetation period efficiency may be raised to 8% and in controllable conditions even to 10% which may be contrasted to the present day 2% and, in exclusive cases, 4-5%.

REFERENCES

AVDEEVA, T. A. & ANDREEVA, T. F., *Photosynthetica* 7: 140, 1973.
BELIKOV, P. S., MOTORINA, M. V. & KURKOVA, E. B., *Izv. timiryaz. sel'.-khoz. Akad.* 5: 44, 1961.
BJÖRKMAN, O., *Physiol. Plant.* 21: 1, 1968.
BJÖRKMAN, O. & HOLMGREN, P., *Physiol. Plant.* 16: 889, 1963.
BJÖRKMAN, O., PEARCY, R. W. & NOBS, M. A., Carnegie Inst. Year Book 69: 640, 1971.
BLACK, C. C. Jr. (ed.): Net Carbon Dioxide Assimilation in Higher Plants, South. Sect. amer. Soc. Plant Physiologists, Cotton Inc., Raleigh, Univ. South Alabama, Mobile 1972.
CHANDLER, R. F., in: EASTIN, J. D., HASKINS, F. A., SULLIVAN, C. Y. & van BAVEL, C. H. M. (ed.): Physiological Aspects of Crop Yield, p. 265, Amer. Soc. Agron. & Crop Sci. Soc. Amer., Madison, Wisc. 1969 a.
CHANDLER, R. F., New Horizons for an Ancient Crop, XII int. bot. Congr., All Congr. Symp. World Food Supply 1969 b.
CRISWELL, J. G. & SHIBLES, R. M., *Crop Sci.* 11: 550, 1971.
DORNHOFF, G. M. & SHIBLES, R. M., *Crop Sci.* 10: 42, 1970.
DVOŘÁK, J. & NÁTR, L., *Photosynthetica* 5: 1, 1971.
DYKYJOVÁ, D., *Photosynthetica* 5: 329, 1971.
EASTIN, J. D., HASKINS, F. A., SULLIVAN, C. Y. & van BAVEL, C. H. M. (ed.): Physiological Aspects of Crop Yield. Amer. Soc. Agron. & Crop Sci. Soc. Amer., Madison, Wisc. 1969.
EL-SHARKAWY, M. & HESKETH, J., *Crop Sci.* 5: 517, 1965.
EVANS, L. T. & DUNSTONE, R. L., *Aust. J. biol. Sci.* 2: 725, 1970.
FRYDRYCH, J., *Photosynthetica* 5: 38, 1971.
HATCH, M. D., OSMOND, C. B. & SLATYER, R. O. (ed.): Photosynthesis and Photorespiration, Wiley-Interscience, New York-London-Sydney-Toronto 1971.
HATCH, M. D., OSMOND, C. B., TROUGHTON, J. H. & BJÖRKMAN, O., Carnegie Inst. Year Book 71: 135, 1972.
ISHIZUKA, Y., in: EASTIN, J. D., HASKINS, F. A., SULLIVAN, C. Y. & van BAVEL, C. H. M. (ed.): Physiological Aspects of Crop Yield, p. 15, Amer. Soc. Agron. & Crop Sci. Soc. Amer., Madison, Wisc. 1969.
IZHAR, S. & WALLACE, D. H., *Crop Sci.* 7: 457, 1967.
KARPUSHKIN, L. T., in: NICHIPOROVICH, A. A. (ed.): Fotosinteziruyushchie Sistemy Vysokoï Produktivnosti, p. 149, Nauka, Moskva 1966.
KUMAKOV, V. A., in: NICHIPOROVICH, A. A. (ed.): Teoreticheskie Osnovy Fotosinteticheskoï Produktivnosti, p. 500, Nauka, Moskva 1972.

LEMON, E., in: EASTIN, J. D., HASKINS, F. A., SULLIVAN, C. Y. & van BAVEL, C. H. M. (ed.): Physiological Aspects of Crop Yield, p. 117, Amer. Soc. Agron. & Crop Sci. Soc. Amer., Madison, Wisc. 1969.
MENZ, K. M., MOSS, D. N., CANNELL, R. O. & BRUN, W. A., *Crop Sci.* 9: 692, 1969.
MONSI, M. & SAEKI, T., *Jap. J. Bot.* 14: 22, 1953.
MONTEITH, J. L.: *Neth. J. agr. Sci.* 10: 333, 1962.
MONTEITH, J. L., Principles of Environmental Physics, Edward Arnold (Publishers), London 1973.
MOSS, D. N., in: Prediction and Measurement of Photosynthetic Productivity, p. 323, PUDOC, Wageningen 1970.
MOSS, D. N., KRENZER, E. G. Jr. & BRUN, W. A., *Science* 164: 187, 1969.
NASYROV, Yu. S.: Geneticheskie Aspekty Fotosinteza [Genetic Aspects of Photosynthesis], Donish, Dushanbe 1971.
NICHIPOROVICH, A. A.: Fotosintez i Teoriya Polucheniya Vysokikh Urozhaev [Photosynthesis and Theory of Obtaining of High Yields], Izd. Akad. Nauk SSSR, Moskva 1956.
NICHIPOROVICH, A. A. (ed.): Fotosintez i Voprosy Produktivnosti Rastenii [Photosynthesis and Problems of Plant Productivity], Izd. Akad. Nauk SSSR, Moskva 1963.
NICHIPOROVICH, A. A. (ed.): Fotosinteziruyushchie Sistemy Vysokoi Produktivnosti [Photosynthesis of Productive Systems], Nauka, Moskva 1966.
NICHIPOROVICH, A. A. (ed.): Teoreticheskie Osnovy Fotosinteticheskoi Produktivnosti [Theoretical Bases of Photosynthetic Productivity], Nauka, Moskva 1972.
NICHIPOROVICH, A. A., CHMORA, S. N., SLOBODSKAYA, T. A. & AVDEEVA, T. A., *Fiziol. Rast.* 20: 300, 1973.
NICHIPOROVICH, A. A., NGUEN-TKHYOK & ANDREEVA, T. F., *Fiziol. Rast.* 19: 1066, 1972.
NICHIPOROVICH, A. A., OSTAPENKO, L. A. & VASIL´EVA, N. G., *Izvest. Akad. Nauk SSSR, Ser. biol.* 2: 300, 1941.
NICHIPOROVICH, A. A., STROGONOVA, L. E., CHMORA, S. N. & VLASOVA, M. P.: Fotosinteticheskaya Deyatel´nost´Rastenii v Posevakh [Photosynthetic Activity of Plants in Canopies], Izd. Akad. Nauk SSSR, Moskva 1961.
PEARCE, R. B., CARLSON, G. E., BARNES, D. K., HART, R. H. & HANSON, C. H., *Crop Sci.* 9: 423, 1969.
PP-Photosynthesis and Utilization of Solar Energy. Level III Experiments. – JIBP/PP-Photosynthesis Level III Group, Jap. nat. Subcomm. PP, Tokyo 1968.
Prediction and Measurement of Photosynthetic Productivity, Proc. IBP/PP Tech. Meet. Třeboň, PUDOC, Wageningen 1970.
ROSS, Yu. K. (ed.): Fitoaktinometricheskie Issledovaniya Rastitel´nogo Pokrova [Phytoactinometric Studies of a Plant Community], Valgus, Tallin 1967.
ROSS, Yu. K. (ed.): Voprosy Effektivnosti Fotosinteza [Problems of Photosynthesis Efficiency], Izd. Akad. Nauk Est. SSR, Tartu 1969.
ROSS, Yu. K. (ed.): Solnechnaya Radiatsiya i Produktivnost´Rastitel´nogo Pokrova [Solar Radiation and Productivity of Plant Canopy], Izdat. Akad. Nauk Est. SSR, Tartu 1972.
SAN PIETRO, A., GREER, F. A. & ARMY, T. J. (ed.): Harvesting the Sun. Photosynthesis in Plant Life, Academic Press, New York-London 1967.
ŠESTÁK, Z., ČATSKÝ, J. & JARVIS, P. G. (ed.): Plant Photosynthetic Production. Manual of Methods, Junk, The Hague 1971.
TING, I. P. & OSMOND, C. B., *Plant Physiol.* 51: 439, 1973 a.
TING, I. P. & OSMOND, C.B., *Plant Physiol.* 51: 448, 1973b.
TOGARI, I. & YOKENDO (ed.), [Photosynthesis and Matter Production of Crop Plants], Tokyo 1971.
WATSON, D. J., *Advanc. Agron.* 4: 101, 1952.
WATSON, D. J. & WITTS, K. J., *Ann. Bot.* 23: 431, 1959.
ZALENSKII, O. V. & RODIN, L. E. (ed.): Obshchie Teoreticheskie Problemy Biologicheskoi

Produktivnosti [General Theoretical Problems of Biological Productivity], Nauka, Leningrad 1969.
ZELITCH, J.: Photosynthesis, Photorespiration, and Plant Productivity, Academic Press, New York-London 1971.

GENETIC VARIATION OF PHOTOSYNTHETIC RATE IN LEAF DISCS OF ZEA MAYS L.

NADĚŽDA AVRATOVŠČUKOVÁ* & STANISLAVA FOUSOVÁ**

Department of Genetics, Faculty of Natural Sciences, Charles University, Viničná 5, 120 00 Praha 2, Czechoslovakia, and Institute of Experimental Botany, Czechoslovak Academy of Sciences, Flemingovo n. 2, 160 00 Praha 6, Czechoslovakia

The intraspecific genetic variation of the photosynthetic rate of leaf discs (PRLD) is controlled by polygenic systems. A hybridologic analysis of this character has been performed with a large number of hybrid combinations in the species *Nicotiana tabacum* L., *Zea mays* L. and *Cucumis sativus* L. (FOUSOVÁ 1964, 1972; AVRATOVŠČUKOVÁ 1967; FOUSOVÁ & AVRATOVŠČUKOVÁ 1967; SLAVÍKOVÁ 1969; AVRATOVŠČUKOVÁ in BJÖRKMAN 1970).

In most of the analysed cases the genetic variation in PRLD was found to bear an additive character. Quite frequently various degrees of positive or negative dominance appeared. And in several hybrids, characterised by heterosis in growth and yield parameters, the heterotic effect was found in PRLD as well. The non-allelic gene interactions of epistatic character may well participate in this phenomenon.

In this paper we summarize the results of the genetic analysis of several selected maize hybrids exhibiting heterosis in PRLD to various extents. To specify the genetic bases of this phenomenon we inserted the results of measurements into HAYMAN'S six parameter model (HAYMAN 1958, 1960). This procedure allows estimation of the additive component of genetic variation (parameter d), dominance component (parameter h) and, particularly important, the three kinds of non-allelic gene interactions. The latter include interactions between two genes with additive effects (i), between one gene with an additive and another with a dominance effect (j), and interaction of two genes with dominance effects (l). HAYMAN's method permits assessment of the significance of these parameters and their participation in the total genetic variation of a character. At the same time their relative frequency in the various hybrid combinations is evaluated.

For the genetic analysis we chose five maize hybrids obtained by crossing the inbred lines: $A_{111} \times LD_{115}$, $A_{111} \times WH$, $LR_3 \times LR_4$, $A_{188} \times 153\,R$, $A_{171} \times A_{188}$. Thirty plants were analysed in each P_1, P_2, F_1, F_2, B_1 and B_2 generation.

The photosynthetic rate of leaf discs was measured as their increase in dry weight when irradiated under standard conditions (BARTOŠ et al. 1960; ŠETLÍK et al. 1960). The 5 h long irradiation of the discs by 280 W m^{-2} of photosynthetically active radiation (400 to 700 nm) took place in an annular irradiation chamber

Table 1. Average values and standard errors of PRLD [mg dm^{-2} h^{-1}] of the parental lines and their hybrid progenies and % heterosis in F_1. The heterosis is expressed by recalculation to the midparent values.

	$A_{111} \times LD_{115}$	$A_{171} \times A_{188}$	$A_{188} \times 153R$	$LR_3 \times LR_4$	$A_{111} \times WH$
P_1	11.63±0.7	4.35±0.2	9.75±0.5	8.34±0.3	12.47±0.2
P_2	5.64±0.4	9.01±0.3	10.03±0.5	5.20±0.5	13.50±0.3
F_1	18.33±0.1	11.63±0.3	16.64±0.4	10.26±0.4	17.73±0.1
F_2	11.67±0.7	8.69±0.5	13.43±0.5	8.31±0.5	15.26±0.7
B_1	14.90±0.7	7.01±0.4	12.84±0.4	8.10±0.7	14.89±0.5
B_2	12.21±0.1	8.46±0.4	13.70±0.4	7.53±0.6	15.36±0.1
[%]	112.4	74.1	68.3	51.6	36.5

with a turntable (AVRATOVŠČUKOVÁ 1967; ŠETLÍK et al. 1967) at 30°C and in air containing 2% CO_2.

Although the differences in PRLD between the parent inbred lines were of various magnitudes, the heterotic effects in the character were observed in all F_1 hybrids. The heterotic effects were significant both when referred to the midparent values or to the value of the parent with the higher PRLD (Table 1). The magnitude of the heterotic effect seems to be related to the difference in the values characterizing the parental lines, but the relation is not apparent in all cases as shown for the hybrid $A_{188} \times 153R$. Reciprocal crosses were in all combinations identical. In the F_2 generation the values of PRLD were lower in comparison to the F_1. This lowering was much more apparent in hybrids in which the F_1 heterotic effect was high. The average values for the F_2 generation lie, as a rule, between the average value of the F_1 and the midparent value. The difference between F_1 and F_2 average values was always highly significant. The F_2 generation also had a larger range of the phenotypic variation than P_1, P_2 and F_1 generations. Back-crosses revealed situations similar to the F_2 generations and the differences between the F_2 and B values were mostly insignificant (Fig. 1).

The analysis of the mean values of six parent and hybrid populations compared by HAYMAN's six parameter model has proved that the genetic causes of heterosis in PRLD may differ in the various hybrid combinations (Table 2).

In the hybrid $A_{111} \times LD_{115}$, in which the highest heterosis effect was observed, the sources of heterosis most probably stem from a combination of dominance and epistasis. The additive effects are not important and their influence is significant only in the period of grain development. The presence of epistasis distorts to a certain extent the estimates of the remaining components of the genetic variation. It can therefore only be concluded that epistatic effects are present and at least as important as those of dominance. Among the two-gene interactions the most important in this hybrid are those between the additive and between the dominance effects. An interaction between additive and dominance effects has not been demonstrated.

The hybrid $A_{171} \times A_{188}$ presents some problems. Although the χ^2-test dis-

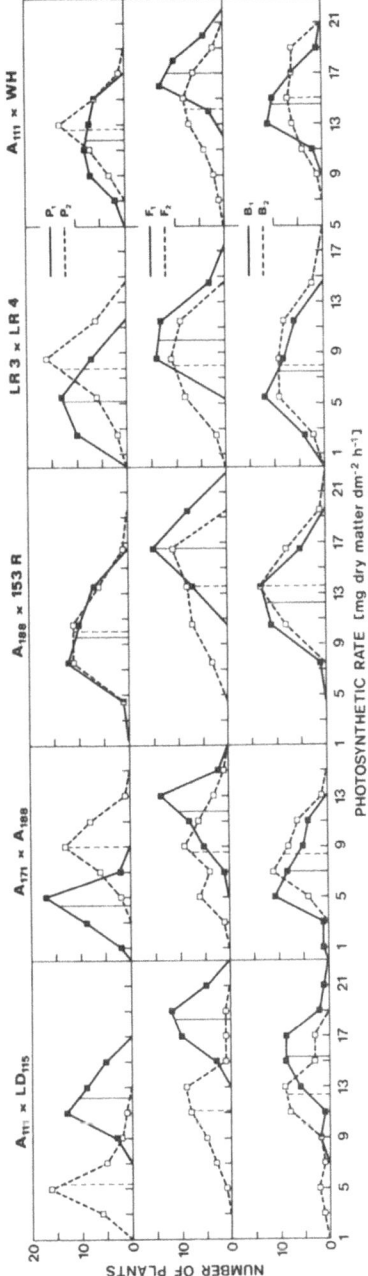

Fig. 1. Hybrid combinations $A_{111} \times LD_{115}$, $A_{171} \times A_{188}$, $A_{188} \times 153R$, $LR_3 \times LR_4$ and $A_{111} \times WH$. Frequency distribution curves of photosynthetic rate of leaf discs from parent lines and their progenies in F_1, F_2, B_1 and B_2 generations.

Table 2. Genetic effects on PRLD in five hybrid combinations of maize.

	$A_{111} \times LD_{115}$	$A_{171} \times A_{188}$	$A_{188} \times 153R$	$LR_3 \times LR_4$	$A_{111} \times WH$
m	11.67±0.7++	8.69±0.3++	13.43±0.3++	8.31±0.4++	15.26±0.7++
d	2.69±0.6++	−1.45±0.2++	−0.86±0.3++	0.57±0.6++	−0.47±0.6
h	17.23±2.4++	1.14±1.3	6.12±1.4++	1.51±2.1++	4.21±2.4
i	7.54±2.4++	−3.82±1.1++	−0.64±1.4	−1.98±2.0	−0.54±0.2
j	−0.31±0.7	0.88±0.3+	−0.72±0.3	−1.00±0.7	0.05±0.6
l	−7.83±3.2+	9.50±1.8++	0.62±1.7	4.78±3.0	1.47±3.2
$x^2 (3)$	10.210	0.832	0.003	0.979	0.625
p	0.02	0.80	0.99	0.80	0.80

+ significant at 0.05 level
++ significant at 0.01 level

proved any presence of epistatic effects, the parameters i, j and l were, nevertheless, significant. The parameter l for interaction of the dominance effects has the highest value, but the parameter i for interaction of the additive and dominance effects is also fairly high. Moreover, the additive and dominance components of the genetic variation (parameters d and h) are also highly significant. The participation of epistasis in heterosis of this hybrid combination requires further verification, even if the strong positive dominance plays an important role in this heterosis.

The other three remaining hybrid combinations $A_{188} \times 153R$, $LR_3 \times LR_4$ and $A_{111} \times WH$, in which the percentage of heterosis in PRLD is lower than in the preceding hybrid combinations, have yielded mutually similar results in genetic analysis. The presence of epistasis was not proved in any combination by the χ^2-test, but the parameter h is highly significant as demonstrated by the relatively strong dominance in all three hybrids. Heterosis in PRLD evidently stems from dominance in the polygenic systems. All three combinations also display a highly significant additive component of the genetic variation.

Summary

The genetic analysis of five hybrid combinations of maize by HAYMAN's six parameter model has proved that heterosis in the photosynthetic rate of leaf discs may result from different genetic causes in the various hybrid combinations. In some hybrids it stems from the dominance potency of polygenic systems of variable strength, in others ($A_{111} \times LD_{115}$) from non-allelic gene interactions of the epistatic type.

References

AVRATOVŠČUKOVÁ, N., CSc. Thesis, Fac. nat. Sci., Charles Univ., Praha 1967.
BARTOŠ, J., KUBÍN, Š. & SETLIK, I., *Biol. Plant.* 2: 201, 1960.
BJÖRKMAN, O., in: Prediction and Measurement of Photosynthetic Productivity, p. 369. PUDOC, Wageningen 1970.
FOUSOVÁ, S., Thesis, Fac. nat. Sci., Charles Univ., Praha 1964.
FOUSOVÁ, S., CSc. Thesis, Inst. exp. Bot., Czech. Acad. Sci. Praha 1972.
FOUSOVÁ, S., AVRATOVŠČUKOVÁ, N., *Photosynthetica* 1: 3, 1967.
HAYMAN, B.I., *Heredity* 12: 371, 1958.
HAYMAN, B.I., *Genetica* 31: 133, 1960
ŠETLÍK, I., AVRATOVŠČUKOVÁ, N. & KŘITEK, J., *Photosynthetica* 1: 89, 1967.
ŠETLÍK, I., BARTOS, J. & KUBÍN, Š., *Biol. Plant.* 2: 292, 1960.
SLAVÍKOVÁ, D., Thesis, Fac. nat. Sci., Charles Univ., Praha 1969.

BIOCHEMICAL AND GENETICAL BASIS FOR THE TEMPERATURE SENSITIVITY OF PHOTOSYNTHESIS AND GROWTH IN CHILLING-SENSITIVE PLANTS

A. SHNEYOUR, R. M. SMILLIE & J. K. RAISON

Plant Physiology Unit, C.S.I.R.O. Division of Food Research and School of Biological Sciences, Macquarie University, North Ryde 2113, Sydney, Australia

Introduction

Many plants show a dramatic decrease in photosynthetic rate and a cessation of growth once the temperature falls below 10 °C to 12 °C. A wide variety of plants exhibit this sensitivity to chilling temperatures, that is temperatures in the range of 0 to about 12 °C (see for example Table 1). A number of important food crops and most tropical plants are sensitive to chilling temperatures. This sensitivity to chilling temperatures is important not only because it limits the growth of plants to climates where chilling temperatures are infrequent or do not occur, but also because the storage life of fresh food commodities obtained from these plants cannot be increased by storage at low temperatures.

Table 1. Some chilling-sensitive and cold-tolerant plants.

Chilling-sensitive (No growth and deterioration below 12 °C)	Cold-tolerant (Growth below 12 °C)
Bean	Pea
Tomato	Lettuce
Cucumber	Beetroot
Sweet potato	Cauliflower
Maize	Potato
Sorghum	Wheat
Banana	Barley
Avocado	Pear

LYONS & RAISON (1970) have shown that there is a large increase in the activation energy of respiratory enzymes at chilling temperatures in mitochondria isolated from chilling-sensitive plants. Mitochondria from cold-tolerant plants do not show this temperature-induced effect. This increase in the activation energy of enzymes of mitochondria from chilling-sensitive plants is related to and is considered a direct consequence of a temperature-induced phase change in the lipids of the mitochondrial membranes (RAISON *et al.* 1971a). As a further approach to the problem of chilling sensitivity of certain plants, physical properties and photochemical activities of chloroplast membranes at different temperatures have been studied using chloroplasts isolated from chilling-sensitive plants and cold-tolerant plants.

Results and Discussion

The Photoreduction of NADP from Water at Different Temperatures

Fig. 1. shows the effect of temperatures on the photoreduction of NADP from water by chloroplasts isolated from four different plants. Two of these plants, bean and tomato, are chilling-sensitive, while the other two, lettuce and pea, are cold-tolerant. The data are presented by plotting the logarithm of the activity of the photoreduction of NADP as a function of the reciprocal of the absolute temperature. In this Arrhenius plot a straight line relationship might be expected and this is obtained in the case of the two cold-tolerant plants, lettuce and pea. In the case of bean and tomato there is a distinct change in slope at around the critical temperature, that is around $12°C$. The actual slope indicates the activation energy of the reaction and the calculated values are given on the graphs. Thus

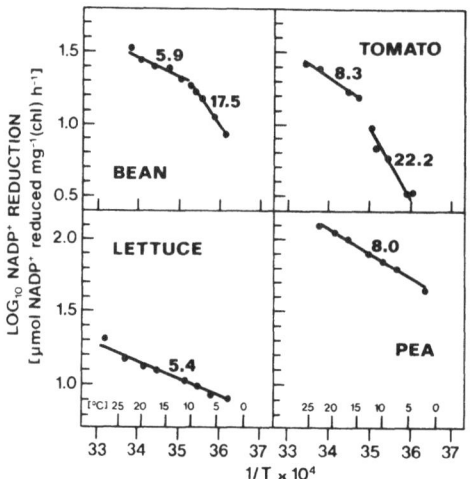

Fig. 1. Arrhenius plot of photoreduction of NADP from water by isolated chloroplasts of bean and tomato (chilling-sensitive) and pea and lettuce (chilling-resistant). The values adjacent to the plotted lines in this and the following figures are the calculated activation energies [kcal mol^{-1}]. The reaction mixture contained in a total volume of 1 ml [µmol]: phosphate at pH 8.0, 9; $MgCl_2$, 0.9; sorbitol, 27; NADP, 0.25; a saturating amount of Swiss chard ferredoxin; chloroplasts containing 5 µg chlorophyll. (Data from SHNEYOUR et al. 1973.)

below the chilling temperature there is an increase in activation energy of this enzyme system and a progressively greater decrease in activity for each degree drop in temperature. This decreased capacity of the chloroplasts to reduce NADP in the light below 12 °C might be expected to adversely affect the ability of the plants to carry out photosynthesis. It should be emphasised that the chilling-sensitive plants received no exposure to cold before the chloroplasts were isolated. Healthy plants which had not been exposed to cold temperatures were used in all of these experiments.

A Temperature-Induced Phase Change in Membrane Lipids of Chloroplasts from Chilling-Sensitive Plants

Earlier studies (see Introduction) had indicated that the increase in activation energy of membrane-bound enzymes at chilling temperatures is related to a temperature-induced phase change in the membrane lipids. Thus the change in activation energy of the enzymes is viewed not as an intrinsic property of the enzyme protein but rather as a conformational change in the protein of the membrane-bound enzyme resulting from a temperature-induced phase change in the lipids of the membrane (RAISON 1973, RAISON *et al.* 1971a, b). Direct evidence for this phase change in membrane lipids comes from electron spin resonance (ESR) measurements of the mobility of nitroxide spin-labelled compounds incorporated into membranes. A phase change in the membrane lipids then can be inferred from a change in the activation energy of motion of the incorporated spin-label (compare the experiment shown in Fig. 2). A nitroxide spin-labelled compound, in this case the methyl ester of 12-nitroxide stearic acid, is added to chloroplasts after they have been isolated. Again two chilling-sensitive plants are used, tomato and maize, and two cold-tolerant plants, beet and pea. ESR measurements at different temperatures are then made on these chloroplasts. The data is plotted in the same way as was done for measurements of enzymic activity. That is, the logarithm of the correlation time ($\tau \times 10^{10}$ s) obtained from the ESR measurements is plotted as a function of the reciprocal of the absolute temperature. A straight line relationship is obtained for the cold-tolerant plants. In contrast, in the case of tomato and maize chloroplasts, there is a distinct change in slope at around 12 °C. This indicates that there is a phase change in the lipids of the membranes of these chloroplasts at chilling temperatures that does not occur in the chloroplast from cold-tolerant plants. The phase change which occurs at chilling temperatures is reversible when the temperature is again raised above the chilling temperature. Ways in which the phase change of the lipids of the membranes and the associated increase in activation energy of membrane-bound enzymes may lead to secondary irreversible changes and symptoms of chilling injury in chilling-sensitive plants during prolonged exposures to cold temperatures are discussed elsewhere (RAISON 1973).

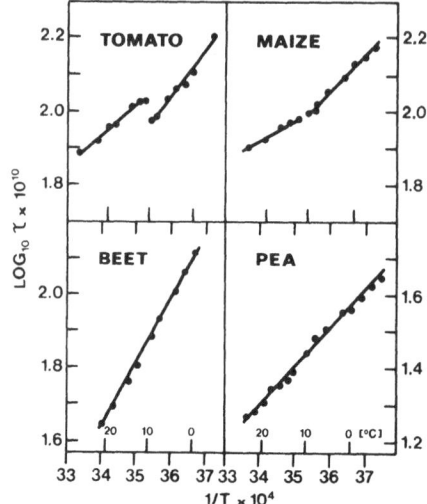

Fig. 2. Arrhenius plot of the effect of temperature on the motion of 12-nitroxide methyl stearate with chloroplast membranes. The spin-label was added to suspensions of chloroplasts (1-2 mg of chlorophyll per ml) to give a bulk concentration of 2×10^{-4} M. Spectra were recorded using a *Varian E4* spectrometer fitted with a temperature control unit which maintained the sample at ± 0.1 °C of the desired temperature. τ_0 was calculated as described by RAISON *et al.* (1971a).

The Site of the Temperature-Induced Effect on NADP Photoreduction

Results of experiments designed to localize the site of the temperature-induced change of NADP photoreduction in chloroplasts from chilling-sensitive plants indicate that the temperature-induced effect is located in the last step of the electron transfer chain that is catalyzed by the enzyme ferredoxin NADP reductase (see

Table 2. Temperature-induced changes in photosynthetic electron transfer reactions in chloroplasts isolated from chilling-sensitive plants.

Reaction	Plant		
	Tomato	Bean	Maize
1. $H_2O \rightarrow$ NADP	+	+	+
2. $H_2O \rightarrow$ DCIP		–	–
3. $DCIPH_2 \rightarrow$ NADP	+	+	
4. $H_2O \rightarrow$ diquat		–	–
5. NADPH $\rightarrow$ diaphorase	+	+	+

Note: +: change in activation energy of reaction below 12 °C; –: no change in activation energy over entire temperature range.

the first four reactions in Table 2). This was confirmed by measurements of the NADPH diaphorase activity of this enzyme at different temperatures (Table 2 and Fig. 3). The enzyme can be extracted from tomato chloroplast (see below) and the extracted enzyme still shows an increase in activation energy at chilling temperatures. The same enzyme in chloroplasts from pea, a cold-tolerant plant, does not show a temperature-induced change at around 12 °C (Fig. 3).

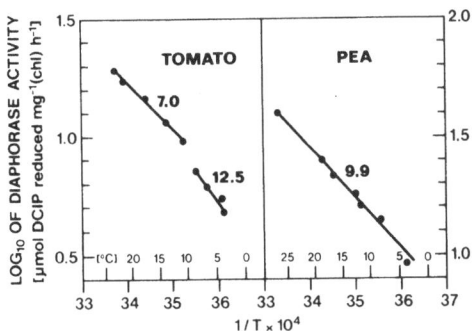

Fig. 3. Arrhenius plot of ferredoxin NADP reductase (measured as NADPH-diaphorase activity) of isolated chloroplasts of tomato and pea. The reaction mixture contained in a total volume of 1 ml [μmol]: phosphate at pH 8.0, 9; $MgCl_2$, 0.9; sorbitol, 27; DCIP, 0.02; chloroplasts containing 5 μg chlorophyll. To initiate the reaction 0.25 μmol NADPH from a freshly prepared solution were added. (Data from SHNEYOUR et al. 1973.)

Rendering Chloroplasts from a Cold-Tolerant Plant Chilling-Sensitive in NADP Photoreduction

If the temperature-induced change affecting the photoreduction of NADP in chloroplasts from chilling-sensitive plants is restricted to the enzyme ferredoxin NADP reductase, then it may be possible experimentally to make isolated chloroplasts from a cold-tolerant plant behave like chloroplasts from a chilling-sensitive plant in NADP photoreduction by replacing the ferredoxin NADP reductase of the cold-tolerant chloroplasts with the same enzyme extracted from chloroplasts of a chilling-sensitive plant (Fig. 4). Chloroplasts are isolated from the leaves of pea, a cold-tolerant plant. The ferredoxin NADP reductase is then extracted from these chloroplasts. The chloroplasts are now deficient in this enzyme and cannot photoreduce NADP. This activity can be regained by adding back the ferredoxin NADP reductase which has just been extracted. The reconstituted chloroplast system will now photoreduce NADP and since all components come from a cold-tolerant plant there is no change in activation energy for NADP photoreduction at chilling temperatures in the reconstituted chloroplast system (Fig. 5). However, if instead of adding enzyme extracted from pea chloroplasts to get the reconstituted system, ferredoxin NADP reductase extracted from chloroplasts of a chilling-sensitive

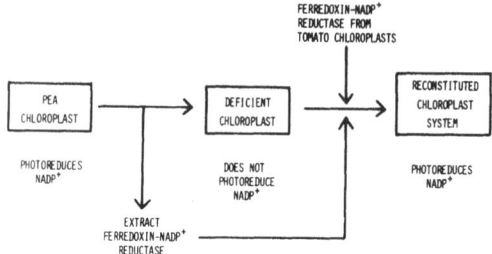

Fig. 4. Outline of an experiment designed to demonstrate the localization of temperature-induced effect on NADP photoreduction to the step catalyzed by ferredoxin NADP reductase. This enzyme was extracted from isolated pea or tomato chloroplasts by 50 mM Tris-HCl buffer pH 8.0 and concentrated by precipitation between 42% and 60% saturated ammonium sulphate solution containing 50 mM sodium pyrophosphate.

plant such as tomato is added, then the effect of temperature on this reconstituted system, that is, extracted pea chloroplasts and enzyme extracted from tomato chloroplasts, can be examined. The reconstituted pea chloroplast system now behaves like chloroplasts obtained from a chilling-sensitive plant (Fig. 5).

Genetic Implications

Now that the basic mechanism responsible for chilling injury is known what are the possibilities of rendering economically important chilling-sensitive crops more tolerant to cold? Examples of adaptation of chilling-sensitive plants to cold con-

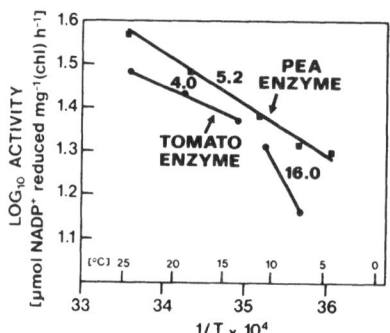

Fig. 5. Arrhenius plots of photoreduction of NADP by reconstituted pea chloroplasts systems. See Fig. 4 and the text for details of the experimental procedure. The reaction mixture is given in Fig. 1 and ferredoxin NADP reductases extracted from pea and tomato chloroplasts were used at concentrations sufficient to saturate the reaction. (Data from unpublished experiments of SHNEYOUR, SMILLIE, RAISON.)

ditions are known, e.g. some species of *Sorghum* have adapted to cold climates. In association with Prof. J. R. McWILLIAM of the University of New England at Armidale, New South Wales, we are investigating a common native grass, known locally as kangaroo grass. This grass is a C_4-plant and is widespread in tropical Australia and New Guinea where it is a chilling-sensitive plant. However, it has also adapted to the much colder environment of Tasmania and the New South Wales tablelands. The same species growing in these locations behaves as a cold-tolerant plant. Investigations of the chilling sensitivities and lipid compositions of these and similar grasses which have adapted to cold conditions and the genetic control of membrane composition may point to ways of enlarging the geographic range of growth of cold-sensitive food crops.

Summary

The effect of temperature has been studied on the activity of chloroplasts isolated from plants whose rates of photosynthesis and growth are adversely affected by chilling temperatures (from $0°C$ to about $12°C$). Chloroplasts isolated from the chilling-sensitive plants tomato and bean showed a large increase in activation energy for the photoreduction of NADP at temperatures below $12\,°C$. This temperature-induced effect was localized to the last step of the chloroplast electron transfer pathway catalyzed by the enzyme ferredoxin NADP reductase. No temperature-induced effects below $12\,°C$ were found with chloroplasts isolated from the cold-tolerant plants lettuce and pea. However, if pea chloroplasts were extracted with buffer to remove ferredoxin NADP reductase and the extracted chloroplasts then mixed with ferredoxin NADP reductase from tomato, the pea chloroplasts then showed a temperature-induced increase in activation energy below $12\,°C$ for the photoreduction of NADP. The motion of nitroxide spin-labels in chloroplasts indicated that for the chilling-sensitive plants a temperature-induced phase change of lipids of the chloroplast membranes occurs at precisely the same temperature as the change in activation energy of the chloroplast enzyme. These results indicate that the change in enzyme activation energy is a consequence of the thermal change in lipid components of the chloroplast membrane.

REFERENCES

LYONS, J. M. & RAISON, J. K., *Plant Physiol.* 45: 386, 1970.
RAISON, J. K., *Soc. exp. Biol. Symp.* 27: 485, 1973.
RAISON, J. K., LYONS, J. M., MEIILIIORN, R. J. & KEITH, A. D., *J. biol. Chem.* 246: 4036, 1971a.
RAISON, J. K., LYONS, J. M. & THOMSON, W. W., *Arch. Biochem. Biophys.* 142: 83, 1971b.
SHNEYOUR, A., RAISON, J. K. & SMILLIE, R. M., *Biochim. biophys. Acta* 292: 152, 1973.

PHOTOSYNTHETIC ADAPTATION IN THE AUTUMNIZATION PROCESS

M. DÉVAY & S. RAJKI

Agricultural Research Institute of the Hungarian Academy of Sciences, Martonvásár, Hungary

Introduction

In order to study the relationship between metabolism and heredity the character of the photosynthetic process proceeding at low temperatures and the formation of the ability to photosynthesize at similar temperatures were investigated in wheat.

As was shown earlier, in the winter wheat cv. Bánkuti 1201 the photosynthetic process taking place at low (0 °C) temperatures differs from that taking place at high (20 °C) temperatures with respect to its metabolic pathway and/or isoenzyme pattern. In winter wheat the chlorophyll complex P_{685} enables photosynthesis to proceed at a temperature of ca. 0 °C. At high temperatures the P_{695} complex is involved in the photosynthetic process. Decreasing temperature increases the level of P_{685} and below the freezing point decomposition of the P_{695} complex occurs. The P_{685} seems to be more thermostable than P_{695}. The irradiance-induced increase in the level of chlorophyll complexes and thus its role in photosynthesis itself is regulated by the night temperature (DÉVAY 1972a).

The temperature-induced decomposition of pigment complexes could be inhibited by low-energy red irradiance. Far-red irradiance nullifies the positive effect of red radiation (DÉVAY 1972a).

The spring wheat cultivars Lutescens 62* and Penjamo 62 do not photosynthesize at low temperature (DÉVAY 1965, 1972b). Their initial stocks were shown by both conventional and aneuploid genetical analysis to be pure spring wheat. However, under conditions produced by appropriate autumn sowing combinations, a gradual transition from spring to winter type took place in accordance with the quality and quantity of autumn cultivation (RAJKI 1967).

The changes in metabolism observable in the course of autumnization are gradual. Their first step is the adaptation of the photosynthetic system to the decreasing temperature (RAJKI et al. 1970).

Thus in recent years one branch of our research has mainly centered on the

---

Abbreviations: L 62 (Lutescens 62), P 62 (Penjamo 62), SD (short day), LD (long day), S (September), O (October), N (November), M (March), MM (pure spring varieties), MS (spring wheat sown in September), MO (spring wheat sown in October), MN (spring wheat sown in November).

mechanism of photosynthetic adaptation, e.g. on the adaptive significance of the major forms of chlorophyll in the autumnization process.

Material and Methods

The experiments were carried out on 2-leaf plants of spring wheat cv. Lutescens 62, Penjamo 62 and on the autumnizing and autumnized plants of both cultivars developed in the Martonvásár autumnization experiments (RAJKI 1967; RAJKI & RAJKI 1972). The plants were grown in conditioned chambers for 14 d at 15 °C and 85 W m^{-2} during the 8 h-short day (SD) and 16 h-long day (LD). The plants were first analysed, and then kept at a temperature of 2 °C until the second sampling, i.e. for 7 d, in order to determine their temperature reactions.

The photosynthetic activity of plants was determined as dry matter accumulation [mg dry matter accumulated in 100 plants per 7 d at 2 °C]

The chlorophyll in vivo state was examined according to SHIBATA (1958). The absorption spectra were measured with a UNICAM SP 800 spectrophotometer at room temperature and the difference spectra were calculated using the method of CHAPMAN et al. (1971). The chlorophyll *a* forms were identified according to FRENCH & BERRY (1971) and BROWN et al. (1972).

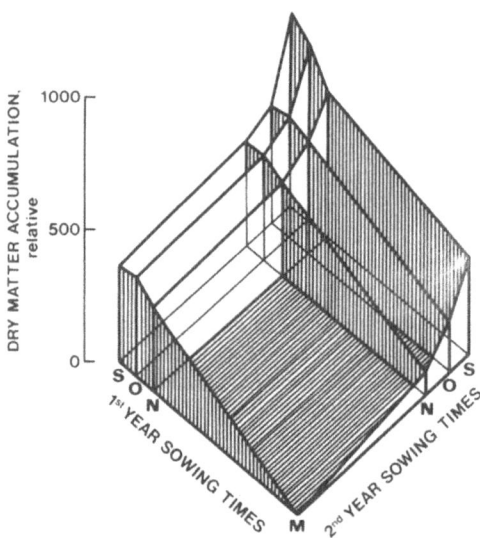

Fig. 1. The influence of sowing times of wheat cv. P-62 on dry matter accumulation at 2 °C.

Results

A decrease in temperature resulted in a decreased rate of photosynthesis in spring wheat L 62 and P 62, so that at 2 °C only decomposition of dry matter could be detected. The ability for photosynthesis to proceed actively at low temperature was produced by a proper combination of autumn sowing times, i.e. by autumnization (Fig. 1). The optimum sowing time was September.

The spectral differences in the leaves of spring and autumnized plants induced by a decrease in the temperature from 15 to 2 °C indicate the changes in chlorophyll complexes produced by autumnization (Fig. 2). A strong decomposition of the chlorophyll complexes appeared in the spring wheat leaves, but in the variants sown in earlier autumn synthesis had taken place. In plants sown in September P_{685} and in plants sown in October P_{675} were synthesized preferentially.

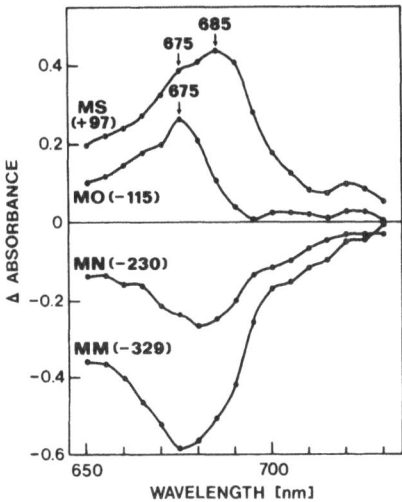

Fig. 2. Spectral differences induced in leaves of pure spring wheat P-62 (MM) and plants affected by various sowing combinations (MS, MO, MN) by decrease of the growing temperature from 15 to 2 °C. The numbers in parentheses indicate the relative rates of dry matter accumulation.

The photosynthetic activity in any variant studied was proportional to the changes in the chlorophyll complex levels (figures in Fig. 2).

The difference spectra between the non-treated spring wheat and the autumn sown variants grown at temperatures of 15 and 2 °C indicate that the increase in the level of P_{685} following September sowing (Fig. 3) or autumnization (Fig. 4) could only be realized at a low temperature. On the other hand, the increased

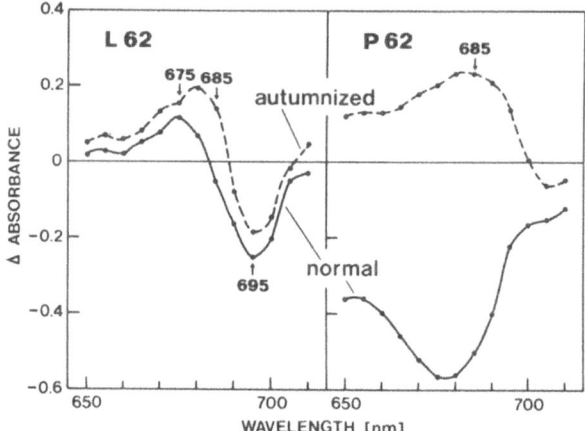

Fig. 3. Spectral differences in wheat leaves of spring and autumnized cv. L-62 and P-62 induced by lowering the growing temperature from 15 to 2 °C.

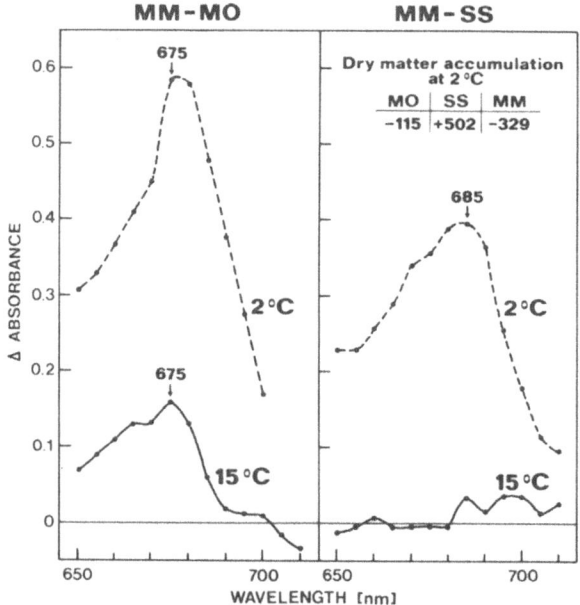

Fig. 4. Synthesis of P_{675} and P_{685} in the autumnization process of wheat cv. P-62 influenced by different growth temperatures (15 and 2 °C). Difference spectra MM-MO (left) and MM-SS (right).

P_{675} level following October sowing seems to be effective at both low and high temperatures (Fig. 4).

A cold treatment (2 °C) at the beginning resulted in decomposition of the chlorophyll complexes in the leaves of spring wheat cultivars under both SD and LD conditions. Low temperature for 2-3 weeks induced an increase in the P_{685} level under LD conditions and in the P_{675} complex under SD conditions. Hence the changes in chlorophyll complexes characteristic of September or October sowing could be associated primarily with daylength. The beginning of the synthesis of both complexes at low temperatures resulted in an increase in the photosynthetic efficiency (Fig. 5).

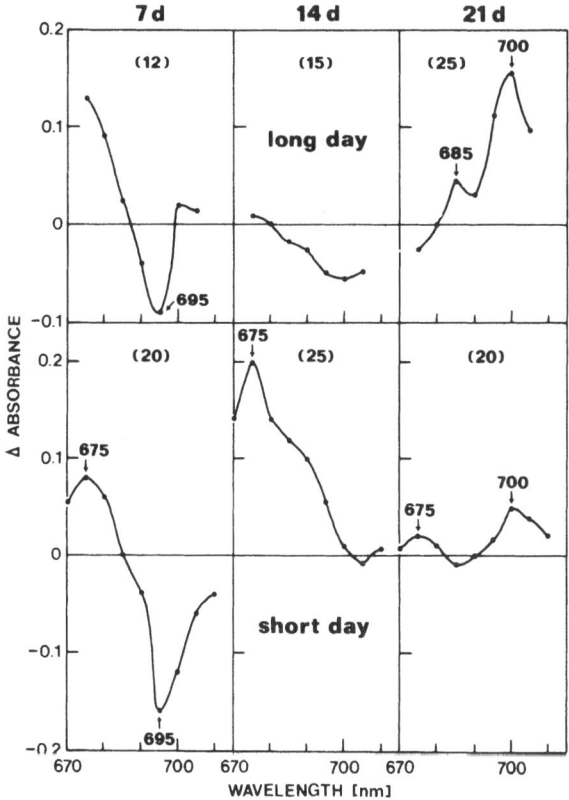

Fig. 5. Spectral differences in spring wheat cv. L-62 induced by 7 to 21 d long treatment at 2 °C under LD and SD. Numbers in parentheses give the rate of dry matter accumulation at 2 °C.

Conclusions

(1) The ability for active photosynthesis at low temperature (2 °C), lacking in spring wheat cv. Lutescens 62 and Penjamo 62, could be induced by a proper combination of autumn sowing times (autumnization), mainly by September sowing.
(2) Synthesis of the chlorophyll a complexes P_{685} and P_{675} is associated with an increase in the photosynthetic efficiency at a low temperature.
(3) The changes in the pattern of chlorophyll complexes characteristic of the sowing times could be associated with the effect of daylength.

REFERENCES

BROWN, J. S., CHAPMAN, G. S. & FRENCH, C. S., in: FORTI, G., AVRON, M. & MELANDRI, A. (ed.): Photosynthesis, Two Centuries after Its Discovery by Joseph Priestley, Vol. 1, p. 731, The Hague 1972.
CHAPMAN, G. S., BROWN, J. S., LAWRENCE, M. C. & FRENCH, C. S., Carnegie Inst.Year Book 70: 498, 1971.
DÉVAY, M., Bot. Közlem. 52: 215, 1965.
DÉVAY, M., in FORTI, G., AVRON, M. & MELANDRI, A. (ed.): Photosynthesis, Two Centuries after Its Discovery by Joseph Priestley, Vol. 3, p. 2137, Junk, The Hague 1972a.
DÉVAY, M., in: Proceedings of Colloquium on the Winter Hardiness of Cereals, p. 105, Publ. agr.Res.Inst. hung.Acad.Sci., Martonvásár 1972b.
FRENCH, C. S. & BERRY, J. A., Carnegie Inst. Year Book 70: 495, 1971.
RAJKI, S., Autumnization and Its Genetic Interpretation, Publ. Akadémiai Kiadó, Budapest 1967.
RAJKI, S., DÉVAY, M. & RAJKI, E., in: *Eucarpia*, Proc. Meeting Sect. Cereals Physiol., p. 103, Dijon, France 1970.
RAJKI, S., RAJKI, E., in: Proceedings of Colloquium on the Winter Hardiness of Cereals, Publ. agr. Res. Inst. hung. Acad. Sci., Martonvásár 1972.
SHIBATA, K., J. Biochem. 45: 599, 1958.

PHOTOSYNTHESIS, ASSIMILATION NUMBERS, AND PHOTOSYNTHATE TRANSLOCATION IN WHEATS OF DIFFERENT STALK LENGTH

I.A. TARCHEVSKIĬ, V.I. CHIKOV, ANISAT YU. SULEĬMANOVA, ANNA P. IVANOVA & YULIYA E. ANDRIYANOVA

Kazan State University and Tatar Research Agricultural Institute, Kazan, U.S.S.R.

During recent years much attention has been paid to the role of photosynthesis of organs other than leaves in the general productivity of cereals (THORNE 1963; KRIEDEMANN 1966; VOLDENG & SIMPSON 1967; MITROFANOV et al. 1969; EVANS & RAWSON 1970; etc.): these organs have been supposed to contribute to the grain yield by as much as 10-70% (depending on conditions, object and methods of studying). Methods used for studying this problem in cereals included the removal of leaves of different insertion positions, leaf or ear shading, studying the structure of photosynthetic potentials, etc. The surface of all photosynthesizing organs (including the stalk and the ear) was often measured for this purpose (KUMAKOV 1968). All these methods have a number of drawbacks.

In our experiments (TARCHEVSKIĬ 1971; TARCHEVSKIĬ & CHIKOV 1972) the production processes and the contribution of individual photosynthetic organs are characterized with several other indices. The estimation of the capacity of the photosynthetic apparatus requires determination of the chlorophyll content in different organs and calculation of the chlorophyll indices and the chlorophyll photosynthetic potentials. The contribution of individual organs to grain yield is determined by their photosynthetic rates and the rate of photosynthate translocation into grain. Chlorophyll $(a + b)$ content is estimated colorimetrically in extracts in 96% ethanol. The photosynthetic rate is measured with $^{14}CO_2$ using 0.03% CO_2 concentration in the assimilation chamber, and natural illuminance (usually higher than 50 klx). The temperature in the chamber is recorded during the day. After a 5 min plant exposure to $^{14}CO_2$ the plant is dismembered and fixed with boiling 80% ethanol. Radioactivity of homogenized samples is measured according to TARCHEVSKIĬ (1958) and CHIKOV et al. (1972). The rate of translocation of photosynthates from the leaves is studied according to IVANOVA (1971).

Our investigation of indices of photosynthetic activity carried out with more than 10 cultivars of wheat showed a specific correlation between the organs' contribution to the photosynthesis of a whole plant and the length of the stalk. Differences between the cultivars were observed mainly after ear formation. Hence we present here results obtained in the milk-ripeness phase of the grain.

As a rule, the ears of plants with shorter stalks contributed more to the formation of economically valuable yield than the ears of plants with longer stalks in which the contribution of the stalk was more important. The plant leaves' contribution to photosynthesis of the whole plant varied sometimes among different wheats, but its value depended on the individual features of each cultivar. This index is likely to be a genotype characteristic of a cultivar.

The two cultivars of spring wheat described in this paper differ in the lengths of their stalks — Kubanka 100-110 cm and Italiya 178 35-40 cm; in the chlorophyll ($a + b$) content in individual organs (Table 1) — Italiya has more chlorophyll in the ear (24%), and Kubanka in the stem (55%); and in the photosynthetic rate in individual organs (Table 1) — the ear contribution is higher in the short-stalk cultivar Italiya. The efficiency of chlorophyll was considerably lower in the cv. Italiya (4.4 mg CO_2 mg^{-1} (chl) h^{-1}) than in Kubanka (6.6 mg CO_2). The stalk had the lowest assimilation number (4.1) in cv. Kubanka; that for the ear was 7.7 while the highest was for leaves (11.0). In cv. Italiya 178 the assimilation numbers were practically equal in the ear and leaves (6.1 and 6.0). Plants with more pigmented tissues have, as a rule, lower assimilation numbers.

Table 1. Chlorophyll ($a + b$) content [mg per plant] and photosynthetic rate P_N [mg CO_2 h^{-1} per plant] in spring wheat cultivars differing in stalk length.

Characteristics	Organ	cv. Kubanka (stem length 100 cm)		cv. Italiya 178 (stem length 35 cm)	
		[mg or mg h^{-1}]	[%]	[mg or mg h^{-1}]	[%]
Chlorophyll	Ear	0.3	17	0.7	24
	Stalk	1.0	55	1.4	48
	Leaves	0.5	28	0.8	28
	Plant average	1.8	100	2.9	100
Photosynthetic rate	Ear	2.3	19.5	4.3	33.7
	Stalk	4.1	34.5	3.8	29.7
	Leaves	5.5	46.0	4.7	36.6
	Plant average	11.9	100	12.8	100

The photosynthetic activity of whole plants and of individual organs varies greatly during the day (Table 2). At midday photosynthesis is to some extent inhibited, mainly in the leaves, but to some extent also in the stalk. Ear photosynthesis is almost unchanged. Hence observations made during plant ontogenesis must always be completed by measurements done during the day.

Studies on assimilate flow in different plants showed that, when the rates of translocation of assimilates were equal, the ears of dwarf plants received much more photosynthates from the flag leaf than the ears of long stalked wheats because of a shorter path between the photosynthesizing and consuming organs (Table 3). Moreover, there is some information that dwarf wheats have larger cross

Table 2. Daytime changes in the photosynthetic rate P_N [mg CO_2 h^{-1} per plant], in individual organs of spring wheat Italiya 178.

Organ	Time of day					
	9.00		13.00		18.30	
	P_N	[%]	P_N	[%]	P_N	[%]
Ear	4.30	100	4.25	99	1.90	44
Stalk	3.80	100	2.45	65	0.80	21
Leaves	4.70	100	2.90	62	1.80	38
Plant average	12.8	100	9.6	75	4.5	35
Illuminance [klx]	50		50		22-23	
Temperature in the assimilation chamber [°C]	26		34		25	

Table 3. Photosynthate flow rate from photosynthesizing leaves in different spring wheats [^{14}C content as % of general plant radioactivity].

Organ	^{14}C exposure of the 1st leaf		^{14}C exposure of the 2nd leaf	
	Short-stem (35-40 cm) wheats	Long-stem (80-100 cm) wheats	Short-stem (35-40 cm) wheats	Long-stem (80-100 cm) wheats
The leaf exposed to $^{14}CO_2$	38.5	38.5	56.6	62.1
Its sheath	3.7	2.2	1.4	1.8
The stalk above the flag-leaf node or the 2nd leaf	12.3	33.9	11.2	13.7
The rest of the leaves	1.0	1.0	4.8	3.6
Ear	27.5	11.3	14.5	3.3
The stalk below ear	17.0	13.1	11.5	15.5

sectional area of vessels of the conductive system; their stalks are usually thicker and have the same mass as the stalks of long-stalk wheats (RAWSON & EVANS 1971).

Summary

The present data on the chlorophyll (*a* + *b*) content in individual organs, photosynthetic rate and photosynthates translocation from the organs permit us to conclude that the short-stalk spring wheats (e.g. cv. Italiya 178), owing to their structure and characteristics of photosynthetic apparatus, have a greater potential possibility for yield formation in comparison with long-stem wheats (cf. cv. Kubanka). In addition to this, the dwarfs are less susceptible to falling when cultivated on a high agrophone.

REFERENCES

CHIKOV, V.I., BULKA, M.E. & LOZOVAYA, V.V., *Fiziol. Rast.* 19: 190, 1972.
EVANS, L.T. & RAWSON, H.M., *Austr.J.biol.Sci.* 23: 245, 1970.
IVANOVA, A.P., in: *Materialy Konferentsii Tatarskogo Nauchno-Issledovatel'skogo Instituta Sel'skogo Khozyaïstva*, p. 132, Kazan 1971.
KRIEDEMANN, P., *Ann.Bot.* 30: 349, 1966.
KUMAKOV, V.A., *Sel'skokhoz.Biol.* 3: 362, 1968.
MITROFANOV, B.A., GULYAEV, B.I., MAKHOVSKAYA, M.A., LAVRENTOVICH, D.I., POCHINOK, Kh.N. & OKANENKO, A.S., in: *Puti Povysheniya Intensivnosti i Produktivnosti Fotosinteza*, Vol. 3, p. 69, Naukova Dumka, Kiev 1969.
RAWSON, H.M. & EVANS, L.T., Austr.J.agr.Res. 22: 851, 1971.
TARCHEVSKII, I.A., *Uch.Zap.kazan'.gos.Univ. Lenina* 118: 111, 1958.
TARCHEVSKII, I.A., in: *Materialy Konferentsii Tatarskogo Nauchno-Issledovatel'skogo Instituta Sel'skogo Khozyaïstva*, p. 130, Kazan 1971.
TARCHEVSKII, I.A. & CHIKOV, V.I., in: *Tezisy Respublikanskoï Nauchnoï Konferentsii po Fiziologicheskim Osnovam Povysheniya Produktivnosti i Ustoïchivosti Zernovykh Kul'tur*, p. 11, Alma-Ata 1972.
THORNE, G.N., *Ann.Bot.* 27: 155, 1963.
VOLDENG, H.D., & SIMPSON, G.M., *Can.J.Plant Sci.* 47: 359, 1967.

CONCLUSIONS

ACTUAL PROBLEMS OF GENETICS OF PHOTOSYNTHESIS

YU.S. NASYROV *(Dushanbe)*, K.V. KVITKO *(Leningrad)* & Z. ŠESTÁK *(Praha)*

The necessity of studying genetic bases of the most important process in plants, i.e. photosynthesis, became evident many years ago. Nevertheless, up to now the practical selection of agricultural, horticultural and forest plants, such that mankind obtains maximum yields under proper agrotechnics and favourable vegetation seasons, is as a rule not based on the results of genetic research of photosynthesis and productivity. This is why the Dushanbe symposium attempted to present and discuss the majority of important problems and results in this field and why this publication aspires to consider and disseminate the accumulated knowledge.

Chloroplast Deoxyribonucleic Acid

One of the more discussed aspects was chloroplast genophore structure in relation to the plastid autonomy. The structure of chloroplast DNA, its reproduction and molecular size were analyzed in the reports of M.S. TURISHCHEVA (U.S.S.R.) and K. TEWARI (U.S.A.). The chloroplast DNAs have ordinarily a ringlike configuration. K. TEWARI's finding of groups consisting of three rings attached to the membrane of the chloroplast at one point led to the assumption that reproduction of the chloroplast DNA might be analogous to the mechanism postulated by N. SUEOKA for bacteria. This is in good agreement with the evidence from N. SUEOKA and K.S. CHIANG about the semi-conservative type of chloroplast DNA reproduction in *Chlamydomonas*.

In fact, we do not know exactly how DNA complexes with protein. A survey presented from the laboratory of V.V. PINEVICH (U.S.S.R.) has established the presence in chloroplasts of a protein strongly associated with DNA, possibly explaining the formation of DNA-protein complexes in chloroplasts. The investigations on the nature of these membrane proteins which determine constant reduplication of DNA during cell reproduction are of great importance.

Origin of Chloroplasts

Active discussions of current concepts on the origin of chloroplasts in higher plants lend support to the theory of their symbiotic origin. It was noted that the Russian evolutionists K.S. MEREZHKOVSKIĬ from Kazan' and A.S. FAMINTSIN from Petersburg initiated this hypothesis in the last century. Later in 1909 K.S. MEREZHKOVSKIĬ in his monograph 'Theory of Two Plasms as the Basis for Symbiogenesis, the New Doctrine in the Origin of Organisms' and in 1924 B.M. KOZOPOLYANSKIĬ in 'New Principle in Biology' developed these ideas to the level of a general theory of symbiotic evolution. Unfortunately, the process of symbiotic origin of organelles can not be demonstrated experimentally, though there is sufficient information about symbiosis between photosynthesizing and heterotrophic organisms.

Genetic Function of Chloroplasts

The efforts of future researchers in this field will probably be turned to studying the size of the chloroplast genome, the degree of its homology estimated by the method of nucleic acid hybridization, and the mechanism of regulation of the reduplication and transcription of the plastid genetic information.

The genetic function of chloroplasts is a very complex aspect of the problem due to the present limited possibility of studying only RNAs transcribed from the plastid genome. L. BOGORAD (U.S.A.) demonstrated the existence of nuclear and organelle genes which control the structure of peptides and plastid ribosomes. Generally it is still not clear whether the extranuclear organelle genes are localized in the chloroplast or mitochondrial genophores. It is beyond any doubt that information for RNA-type components of the plastid ribosomes resides in the genome of chloroplasts. K.A. ALIEV's finding in pea etioplasts of a specific form of mRNA which triggers the protein-synthesizing system, provides an experimental source for studying the genetic function of chloroplasts.

In the Symposium numerous facts were discussed supporting the concept of nuclear-plastid co-operative control of chloroplast development and functional activity (YU.S. NASYROV, U.S.S.R.).

Importance of Chloroplast Structure

The efficiency of biochemical methods for selecting mutation systems has been well demonstrated on *Arabidopsis thaliana* and other model objects. Studies of the ultrastructure of chloroplasts in pigment mutants of higher plants (from measurements of the morphometry of membranes) provides a quantitative estimation of the variability of plastid structure components in relation to their development and functional activity.

Plastid mutants of higher plants are fine objects for studying the genetic information of chloroplasts. According to R. HAGEMANN (G.D.R.) the reproduction of plastids may proceed independently of the synthesis of plastid proteins, but protein synthesis in the plastid is required for the normal differentiation of chloroplasts. The development of chloroplast structure and cellular metabolism are two closely related processes as was shown by J. SCHIFF (U.S.A.) in experiments with *Euglena*. A.A. SHLYK (U.S.S.R.) demonstrated that disturbances of the protein-synthesizing system of chloroplasts evoke changes in the reactions of chlorophyll biosynthesis.

Carbon Fixation Pathways

Lively debates on the existence of the C_4 carboxylation pathway took place during consideration of the genetic control of photosynthetic assimilation of carbon. M. GIBBS (U.S.A.) presented experimental data that chloroplasts of maize assimilate carbon mainly through the reductive pentose phosphate cycle. R.M. SMILLIE (Australia) supported the hypothesis of HATCH & SLACK on the existence of the C_4 pathway of carbon metabolism and claimed that all disagreements on this problem are due to the methods used for experiments. O.V. ZALENSKIĬ (U.S.S.R.) presented data on the carbon metabolism of desert Kara-Kum plants which have the pentose phosphate cycle of CO_2 fixation. According to his view there is only one pathway of carbon reduction during photosynthesis. Failure to prove the intercellular transport of malate and aspartate is one of the main arguments against the HATCH-SLACK theory.

Consideration of both evolutionary and ecological aspects of the problem is a promising way for elucidating the photosynthetic carbon assimilation pathways. The potential of adaptive properties of the photosynthetic apparatus was created in the course of strict selection during periods when ecological conditions changed in the range of continents. That may have caused the shift of adaptive types of carbon metabolism in individual plant groups within different taxons.

Pigment Mutants and Photosynthesis

Information from D. VON WETTSTEIN (Denmark) was devoted to the genetic nature of photosynthetic mutations and their phenogenesis. His laboratory is investigating barley mutants with variations in 86 nuclear loci controlling chlorophyll synthesis.

The predominant role of the nuclear genome was also demonstrated by A.T. MOKRONOSOV (U.S.S.R.) who pointed out the change of normal reactions of genotype in the course of ontogenesis of potato plants. According to P.D. USMANOV (U.S.S.R.) the plastom controls chloroplast size while plastid number per cell is under nuclear control. His information is based on work with the collection of *Arabidopsis thaliana* species and mutants in the Institute of Plant

Physiology and Biophysics of the Tajik Academy of Sciences in Dushanbe.

Presentation of carotene-deficient mutants of maize by ÁGNES FALUDI-DÁNIEL (Hungary), and chlorophyll mutants of barley by HERTA SAGROMSKY (G.D.R.) stimulated discussions on the role of pigments in photosynthesis.

Mutation Analysis of Green Algae

As shown by the classical experiments of R.P. LEVINE and co-workers with acetate-dependent mutants, unicellular green algae appear to be efficient models for studying the genetic nature of photosynthetically significant structures. Convincing evidence has been obtained by V.G. LADYGIN, G.A. SEMENOVA and S.V. TAGEEVA (U.S.S.R.) on retention of membrane structures of the two chloroplasts carried by gametes in the zygote of *Chlamydomonas*. This led to discussions on the differences between the experimental data from the laboratories of K.S. CHIANG and RUTH SAGER (U.S.A.). Data on the retention of chloroplast structures in gametes of mating type (−) are opposite to SAGER's conclusion concerning DNA destruction in this chloroplast. These discrepancies may well be reconciled, as suggested by L. BOGORAD, since the retention of radioactive label in all types of zygote DNA, as previously shown by K.S. CHIANG, might be related to reutilization of the products of DNA destruction (which was demonstrated by R. SAGER) when further synthesis of DNA occurs during zygote development. K.V. KVITKO proposed that these discrepancies cannot explain the essence of the debate, because if one takes into consideration the existence of 26 DNA copies in a chloroplast, then any proof of the destruction of nearly all of them in the zygote does not exclude the retention of at least one genetically valuable copy from each gamete, thus leading to the biparental inheritance of chloroplast genes according to the Mendelian type.

Mutation analysis of green algae genotype structure, exercised in the Leningrad University by A.V. STOLBOVA and K.V. KVITKO, is considered to be the only genetic method of investigation for the agamic species. Parallel studies of such sexually reproduced forms as *Chlamydomonas*, and agamic forms of *Chlorella* and *Scenedesmus* revealed the great similarity of their mutation variability. Mutation systems 'gene-character' have already been created for such characteristics as light-sensitivity and the response to antibiotics.

Attempts are made to map the genome of chloroplasts in *Chlamydomonas*. The factors have been determined which control the light-sensitivity and change of colour under visible radiation.

Synthetic Pigment-Lipoprotein Complexes

Studies of primary mechanisms of photosynthesis made on pigment mutants in the laboratories of HERTA SAGROMSKY (G.D.R.), A.A. KRASNOVSKIĬ and

YU.E. GILLER (U.S.S.R.) were supported by parallel investigations on the synthetic pigment-lipoproteid complexes. YU.E. GILLER's conclusion about the predominant role of protein components in the molecular organization of the chloroplast pigment structures is in agreement with the data of B.T. MUKHAMA-DIEV and K.V. KVITKO (U.S.S.R.) obtained with *Chlorella* mutants resistant to DCMU.

Genetics and Optimization of Photosynthetic Activity

Principles and methods of optimization of the photosynthetic activity of plants were considered in the final sessions of the Symposium. A.A. NICHIPOROVICH (U.S.S.R.) pointed out the role of synthetic breeding as a method of creating forms combining the best properties of the world gene fund.

However, O.V. ZALENSKIĬ and K.V. KVITKO noted that there is no evidence that the characters responsible for maximal potential photosynthetic rate have a simple, monogeneous type of inheritance.

As reported by NADĚŽDA AVRATOVŠČUKOVÁ and STANISLAVA FOUSOVÁ (Č.S.S.R.), the heritability of the photosynthetic activity of leaves is controlled mainly by polygenic systems with additive effects, by systems of respectively different strengths of dominance, mainly in the positive direction. Hybrid combinations characterised by heterosis in growth and yield are also heterotic in the photosynthetic activity. Hybrid combinations differ in the relative importance of components of genetic variability, i.e. additivity, dominance and epistasis. When additivity is the most important component, the selection of genotypes with highest yields is preferable. In the case of dominance and epistasis, combinatory hybridization should be applied taking into account the necessary growth parameters, yield coefficients, etc.

Many participants of the discussion (K. TEWARI, D. VON WETTSTEIN, R. HAGEMANN) expressed the idea that experimental mutagenesis, as a method of breeding, is rather perspective as it may produce forms never met in nature before. Great potentialities of methods of phenotypical optimization of photosynthetic activity were also stressed by R.M. SMILLIE. For example, a change in the content of Mg ions might convert oppressed thylakoids into single ones, thus making light curves of photosynthesis more linear.

In order to program agricultural yields and to find highly productive plant forms, we should more often test the properties of the photosynthetic apparatus in our breeding work. Joint efforts of geneticists, biochemists, biophysicists and plant physiologists might help in the realization of these tasks.

Conclusion

The specialized Dushanbe symposium reflected rather well the present knowledge of photosynthesis and the recent lines of research in this field. Problems similar to those presented at the symposium are studied in other world laboratories and — as may be seen in the bibliographies of photosynthesis — the number of papers and solved questions concerning the genetics of photosynthesis is still far from satisfactory.

At present, the problem is studied at two levels. Research at the molecular level concentrates on plastid deoxyribonucleic acid, its composition, dimension, and importance. The results show that, in the course of plant phylogenesis, the importance of plastid DNA continuously decreases, its informational functions being passed on to the nuclear DNA.

The level of genetic analysis deals mostly with mutations, polyploidy, and quantitative analysis of the complex 'photosynthetically active leaf'. Mutants of algae and higher plants enable detailed studies on the formation of the pigment- and chloroplast (chromatophore) apparatus, photosynthetic electron transfer chain and cycles of carbon dioxide reduction, their individual steps and enzymes. These findings are supported by studies on individual plant species and their hybrids. The most desirable (but unfortunately rare) quantitative genetic analyses of the complex 'photosynthetically active leaf' and its components together with the mutant and polyploid studies underline the primary importance of polygenic systems, localized in the nucleus. Nevertheless, the extranuclear genes probably co-operate with nuclear genes in controlling some processes important for photosynthesis, e.g. the development of chloroplast structure and function. The importance of plastid DNA for the genetics of such a complex phenomenon as photosynthesis still remains to be solved. Another important problem is localization of the individual genes or gene-complexes which control the radiant energy absorbing and transforming structures, the electron transport chains, the carbon fixation paths, and the optimum internal structures of leaves (deciding the appropriate transport of gases) and plants (predetermining the appropriate structure of the crop canopy).

The symposium reflected certain progress towards an understanding of the genetic systems controlling photosynthesis. Co-operation of workers from different scientific schools leads us to believe that in the near future we will understand all the fascinating aspects of the organization of 'the miniature laboratories of photosynthesis', and it may be possible to regulate the process of this most effective solar energy utilization for the benefit of all mankind.

AUTHORS' INDEX

Numbers in italics refer to pages where the references appear.

A

Abdullaev, Kh.A. (= Abdullaev, H.A.) *103*, 135, *143*, 189, 196, *200-1*, *314*
Abdullaeva, S.K. *144*, *208*, 271, *283-4*
Abdurakhmanova, Z.N. *145*, *268*
Adler, K. *284*
Aghion, J. 137, *143*, 278, *283-4*
Akhmedov, A. Ya. *201*
Akoyunoglou, G. *314*
Aksentsev, S.L. *174*
Alamuti, N. *284*
Aliev, K.A. 93-4, *103*, *314*, 370
Alina, B.A. 105, *113*
Aliyev, K.A. (= Aliev, K.A.) 94, *103*
Allen, M.B. 203, *208*, 230, *236*
Allfrey, V.G. *36*
Amelunxen, F. 307, *314*
Amesz, J. *261*
Andersen, K.S. 307-8, *313-4*
Anderson, I.C. 217-8, *223*
Anderson, J.M. 239, *244*, 308, *314*
Anderson, L. 79, *89*
Andreeva, T.F. 336, *339-40*
Andrews, T.J. 173, *174*
Andriyanova, Yu.E. 363
Androshchuk, A.F. 209, *215*
Ansevin, A.T. 32-3, *35*
Apel, P. 251, *253*
Argyroudi-Akoyunoglou, J.H. 307, 312, *314*
Arkhipov, V.N. *184*
Armstrong, J.J. *48*, *314*
Army, T.J. *145*, *314*, *340*
Arnold, W.G. 263, *268*
Arnon, D.I. 32, *35*, 80, *89*, 162, 207, *208*, 264, *268*
Arthur, W.E. *268*
Ashida, J. *261*
Asoeva, L.M. *208*, *268*, 271
Attardi, B. *103*
Attardi, G. 101, *103*
Avadhani, N.G. 67-8, *89*
Avdeeva, T.A. 177, 336-7, *339-40*
Averina, N.G. 119-20, 125, *131-2*, *145*
Avetisov, V.A. 209
Avratovšcuková, N. 343-4, *347*, 373
Avron, M. *90*, *145*, *244*, *261*, *314*, *362*
Azi, T. *208*

B

Bain, J.M. *313*
Bainbridge, R. *166*
Baldry, C.W. 177, *185*
Baleva, E.F. 119, *132*, *145*
Ballantine, J.E.M. 307, *314*
Baranov, A.A. *236*
Barnes, D.K. *340*
Barnett, W.E. 69, *89-90*
Bartels, P.G. 108, *112*
Barthe, F. *175*
Bartoš, J. 343, *347*
Bastia, D. 43, 47, *48*
Batalov, R.B. 135, *143*, *145*, *201*
Bavel, C.H.M. van *339-40*
Beletskiĭ, Yu.D. 209, *215*
Belikov, P.S. 329, *339*
Belyaeva, O.B. 137, 142, *144*, 271, *284*
Belyatskaya, O.Ya. *201*
Belozerskiĭ, A.N. *36*
Bendix, S. 203, *208*, 230, *236*
Benitez, A. 274-5, *284*
Ben-Shaul, Y. 69, 87, *89*, *91*
Bergfeld, R. 169, *174*
Berry, J.A. 358, *362*
Bershteĭn, B.I. 287, 289, *293*
Bertsch, W. *144*, *283-4*
Bezsmertnaya, I.N. *103*, 105, *113*
Bianchi, A. 239
Bikasiyan, G.R. 135, 138, *143*, 264, 267, *269*
Bil', K.Ya. 178, *184-5*
Bisalputra, A.A. 47, *48*
Bisalputra, T. 6, 47-8, *48*, 111, *112*
Bishop, D.G. *145*, 305, *313-4*
Bishop, N.I. 74, *89*, 203, *208*
Björkman, O. 178, *185*, 330, 333, *339*, 343, *347*
Black, C.C. Jr. *156*, 321, 329, 332, *339*
Boardman, N.K. 79, 89, *89*, *91*, *103*, 125, *132*, *145*, *201*, *236*, *285*, 272-3, 281, *283*, *314*
Bobodzhanov, V.A. *201*
Bogorad, L. 2, *8*, 39, *41*, 51-3, 55-8, *62*, 74, *89*, 119, 124, 126, *131*, 295, *300*, 370, 372
Bokány, A. *244*
Bold, H.C. 63, *89*

375

Bolyakina, Yu.P. *132*
Bonner, J. 31-2, 34, *35-6, 284*
Borisova, O.F. *284*
Börner, T. 115-7, *118*
Borshchevskaya, T.N. 203, *208*
Borst, P. 28, *30*
Böttger, M. *118*
Bottomley, W. 34, *36*, 51-2, *62*, 111, *113*
Boublík, M. *36*
Bovarnick, J.G. 76-9, 81, 84, *89*
Bowes, G. 172, *175*
Boyadzhiev, P.H. 230, *236*
Boynton, J.E. 134-5, *144-5, 201, 285*
Boyton, J.E. van *236*
Brändle, E. 93, *103*
Brandt, A.B. 274-5, *283*
Braten, T. 43, *48*
Brawerman, G. 74-5, 84, *89*
Britten, R.J. 23, *30*
Brodmann, W. *284*
Brody, M. 276-7, *283*
Brody, S.S. 276-7, *283*
Bronchart, R. *314*
Brown, B.W. 32-3, *35*
Brown, J.S. 136, *143*, 358, *362*
Brun, W.A. *340*
Buetow, D.E. 67-8, *89, 91*
Bukovac, M.J. *301*
Bulka, M.E. *366*
Burton, H. 2, 6, *8*, 111, *112*
Burton, K. 32, *36, 113*
Butenko, R.G. 209-10, *215*
Butler, J.A.V. 32, *36*
Butler, W.L. 258, *261, 314*
Butterfass, T. 195, *200*
Bystrova, M.I. 137, *143*, 271, 276, *283*

C
Calvin, M. 149, 152, *157*
Cannell, R.O. *340*
Carell, E.F. 69, 79, *89*
Carlson, G.E. *340*
Carr, N.G. 41, *41*
Čatský, J. 159-61, *166, 340*
Cavalier-Smith, T. 43, 47, *48*
Cerff, R. *175*
Chalkley, G.R. *35*
Chandler, R.F. 327, 329-30, *339*
Chang, S.W. *89*
Chapman, D. 276, *283*
Chapman, G.S. 358, *362*
Chargaff, E. *89*
Chartier, M. *166*

Chartier, P. 161, 165, *166*
Chen, J.L. 111, *113*
Chen, T.M. *156*
Chernitskiï, E.A. *174*
Chiang, K.-S. *48*, 232-3, *236*, 369, 372
Chikov, V.I. 363, *366*
Chmora, S.N. *340*
Chulanovskaya, M.V. 204, *208*
Chunaev, A.S. 225, 230-1, *236*
Claes, H. 217-8, *223*
Cleland, R.E. 67, *89*
Cocucci, S. *145*
Cohen 67
Cole, C.V. 162, *166*
Colowick, S.P. *36, 268*
Conroy, J. 305
Coombs, J. 177, *185*
Correns 64
Cosbey, E. *48*
Cowan, C.A. *89*
Cowan, I.R. 161, *166*
Craig, A.S. 307, *314*
Cramer, M. 63, *89*
Criswell, J.G. 330, *339*

D
Dahmus, M. *35*
Daletskaya, I.A. 204, *208*
Das, M. *245*
Davidson, J. 54
Davies, J. 76, *91*
Davis, B.D. 76, *90*
Demchenko, S.I. 209
Demeter, S. 241, *244*
Derevyanko, V.G. *314*
Dévay, M. 357, *362*
Dhont, E. *103*
Diamond 84
Dickinson, H.G. 47, *48*
Dinant, M. 137, *143*
Djavanchir, A. 161, *166*
Döbel, P. 135, *143*
Doebel 252
Doman, N.G. *185, 244*
Dontsova, S.V. 142, *144*
Dornhoff, G.M. 330, *339*
Dörr, I. *314*
Doty, P. 31, *36*
Downton, W.J.S. *314*
Draber, W. *268*
Drozdova, I.S. 167, *175, 268*
Dubinin, N.P. 190, *200*
Duffus, J.H. 31, *36*

Dunstone, R.L. 329, *339*
Duysens, L.N.M. 247, *253*, 255, *261*, 280, *283*
Dvořák, J. 333, *339*
Dwyer, M.R. *103*
Dykyjová, D. 324, 326, *339*

E
Eastin, J.D. 315, 324, *339-40*
Eckardt, F.E. *166*
Edelman, M. 74-5, 88, *89*, *91*
Edge, H. *314*
Edwards, G.E. *166*
Egan, J.M.Jr. 77, 79, 84, *89*
Egle, K. 251, *253*
Ehrenberg, L. *201*, 209, *215*
Eisenstadt, J.M. 68, 74-5, *89-90*
Ellis, R.J. 93, *103*
El-Sharkawy, M. 329, 333, *339*
Emmons, C.W. 190, *200*
Englert-Dujardin, E. *314*
Entsch, B. *314*
Epler, J.L. *90*
Epstein, H.T. 66, 69, *89-91*, 119, 124, *132*, 295, *301*
Erokhin, Yu.E. 137, *143-5*, 271, 277-80, *283-4*
Evans, G.C. *166*
Evans, H.J. *132*
Evans, L.T. 329, *339*, 363, 365, *366*
Evans, W.R. 71, 83, *89*
Eveleigh, D. 210, *215*

F
Fager, E.W. 149, *156*
Fairfield, S.A. *89-90*
Falk, H. 108, 111, *113*
Faludi, B. 239, *244*
Faludi-Dániel, Á. 239-40, 242, *244-5*, *261*, 372
Fambrough, D. *35*
Famintsin, A.S. 370
Fam Tkhan Kho (= Pham Than Ho) 233, *236*
Farr, A.L. *36*
Fast, P.G. 276, *283*
Feleki, Z. *314*
Fernández-Morán, H. 2, *8*, 111, *113*
Ferrari, A.G. 267, *268*
Filippovich, I.I. 93, *103*, 105-6, 111, *113*
Fjeld, A. 43, *48*
Flight, W.F.G. 277, 279, *284*
Forde, B.J. 307, *314*

Forti, G. 89, *90*, *145*, *244*, *261*, *314*, *362*
Fousová, S. 343, *347*, 373
Fradkin, L.I. 125, *131-2*, *244-5*, *261*
Franck, J. 272, *283*
Frank, G.M. *48*
Freedman, Z. *89*
French, C.S. 136, *145*, 241, *244*, 358, *362*
Frenkel, A.W. 250, *253*
Frenster, J.H. 34, *36*
Frey-Wyssling, A. 196, *200*
Fridvalszky, L. *244*
Friedman, R.M. 93, *103*
Fry, K.E. 264, *268*
Frydrych, J. 332, *339*
Fujimura, F. *35*
Fuller, R.C. 80, *90*, 170, *174*, *301*

G
Gaffron, H. *156*, *268*
Gamborg, O. 210-5, *215*
Gaponenko, V.I. 119, *132*, *145*
Garay, A. 240, *244*
Gardisky, R.L. *91*
Gassman, M. 126, *131*, 295, *300*
Gauhl, E. 178, *185*
Generosova, I.P. *314*
Generozova, I.P. *314*
Gesteland, R.F. 55, *62*
Gibbons, G.G. *145*
Gibbs, M. 79, *91*, *145*, 149, 153, *156-7*, 177, *185*, *314*
Gibbs, S.P. 67, 76, *90*, 371
Gibor, A. 74, *90*
Gichner, T. *145*, 209, *215*
Giller, Yu.E. 133, 137, 140, *144*, 207, *208*, 230, *236*, 263, *268*, 271-2, 277-8, 280, *283-5*, 373
Gillham, N.W. 43, 47, *48*, 221, *223*, *236*
Glagoleva, T.A. 204, *208*
Gnanam, A. 105, *113*
Godnev, T.N. 133, *144*, 295-6, *300*
Goedheer, J.C. 258, *261*
Gofshteĭn, L.V. 31, *36*
Gojdics, M. 63, *90*
Goodchild, D.J. *185*
Goodenough, U.W. *48*, 93, *103*, 233, *236*, 307, *314*
Goodwin, T.W. *62*, *91*, 133, *144-5*, *223*, 243, *244*, *301*
Gorozhanin, P.P. *36*
Gostimskiĭ, S.A. *268-9*
Gottschalk, W. 135, *144*
Govindjee 255, *261*, 280, *284*

Graham, D. *103, 145,* 168, *174, 301,* 306, *314*
Granick, S. 43, *48,* 74, *90,* 119, 124, 126, 129, *131-2,* 133-5, *144-5,* 295, *301*
Greef, J. de 307, *314*
Green, B.R. 2, *8,* 111, *113*
Greer, F.A. *145, 314, 340*
Gregory, R.P.F. 137, *144,* 271, *283*
Grekhem, D. (= Graham, D.) *314*
Grieve, A.M. *103, 145, 174, 314*
Grozovskaya, M.S. *132*
Gulyaev, B.A. 271, 276-7, *283-4*
Gulyaev, B.I. *366*
Gurinovich, G.P. 272, 276, *283*
Gustaffson, A. *201*
Gyurján, I. 239, 242, *244,* 255, 260, *261*

H
Haberer, G. *200*
Hagemann, R. 115, 117, *118,* 135, 139, *143-4,* 189, *200,* 371, 373
Hall, D.O. 312, *314*
Hallier, U.W. 310, *314*
Hanawalt, P.C. 74-5, *90*
Hanson, C.H. *340*
Hart, R.H. *340*
Harth, F. *268*
Hartley, M.R. 93, *103*
Hase, E. 74, *90*
Haskins, F.A. *339-40*
Hastings, P.J. 43, *48*
Hatch, M.D. 149, 152, *157,* 170, 172, *174, 185,* 306, *314,* 321, 329, 333, *339,* 371
Hayman, B.I. 343-4, 346, *347*
Heber, M. *185*
Heber, U. 178, *185*
Heizmann, P. 67, *90*
Henningsen, K.W. 134-5, *144-5, 201,* 272, 275-6, *283*
Herrmann, F. 115, *118, 144,* 263, *268*
Hesketh, J. 329, 333, *339*
Heslop-Harrison, J. 47, *48*
Hess, J.L. 173, *174*
Hew, C.-S. 157, *185*
Highkin, H.R. 135, *144,* 250, *253*
Hiller, R. 312
Hines, G. 149
Hirono, G. 135, *144*
Hirsch, A. 255, *261*
Hoare, T.A. 32, *36*
Hogetsu, D. 149, *157*
Holdsworth 87
Hollaender, A. 190, *200, 301*

Holmgren, P. 330, 333, *339*
Holowinsky, A.W. 82-4, *90-1*
Holt, A.S. *283*
Homann, P.H. 307, *314*
Hoober, J.K. 93, *103*
Horváth, G. 240, *244*
Horváth, G.I. 240, *244*
Huang, R.-C.C. 34, *35-6*
Huberman, J. *35*
Hudock, G.A. 80, *90*
Hudock, M.O. *48*
Hufnagel, D.A. *89*
Hug, O. 190, *200*
Hunter, J.A. *90*
Hutner, S.H. 63, *90*
Huzisige, H. 208, *208*
Hwang, M.-I.H. *103*

I
Inge-Vechtomov, S.G. 225-6, *236*
Ingle, J. 29, *30,* 76, *90*
Inhaber, E. *91*
Ishizuka, Y. 329, *339*
Israfilova, U. *201*
Ivanishcheva, S.Yu. *293*
Ivanov, V.I. 190, *200-1*
Ivanova, A.P. 363, *366*
Ivanova, S.B. 31
Iwai, K. 31, *36*
Izhar, S. 330, 333, *339*

J
Jacobs, E.E. 276, *283*
Jagendorf, A.T. 289, *293, 301*
Jarvis, P.G. 160-1, *166, 340*
Jensen, R. *35*
Jensen, R.G. 274-5, *284*
Johns, E.W. 32, *36*
Jones, R.F. 232, *236*
Jourdan, F. *90*

K
Kahn, J.S. 105, *113*
Kaler, V.L. 119, 124, 129, *131,* 295-7, 299-300, *300*
Kalina, M. *314*
Kamyshenko, L.K. 119, *132, 145*
Kanivets, N.P. 287
Kannangara, G.C. *145, 201, 285*
Kaplan, N.O. *36, 268*
Kapler, R. 272-3, *284*
Karako, P.S. *132*
Karapetyan, N.V. 255-6, 258, *261*

Karimov, Kh.Kh. 142, *144*
Karnaukhov, V.N. *185*
Karneeva, N.V. *284*
Karpilov, Yu.S. 177-8, 184, *185*
Karpilova, I.F. *184*
Karpushkin, L.T. 323, 328, *339*
Kas'yanenko, A.G. 135, *144-5, 201, 268*
Kautsky, H. 255, *261*
Kawashima, N. 88, *90*
Keith, A.D. *355*
Kelemen, G. *244*
Kellerer, A.M. 190, *200*
Kellner, G. 111, *113*
Khaitova, L.T. 271-2, *283*
Khodzhiev, A.Kh. 167, 173, *175*
Kholmatova, M. 93
Khramova, G.A. 263, *269*
Khropova, V.I. 137-8, *144*, 227
Kikuti, T. *208*
Kirichenko, E.B. *143, 200, 208, 236, 283-4*
Kirk, D.T.O. (= Kirk, J.T.O.) 189, *200*
Kirk, J.T.O. 93, *103*, 135, *144*, 295, *301*
Kissimon, J. *244*
Klein, S. 63, 67-9, 71, 74, 81-4, 87, *90*
Klein, W.H. 119, 124, *131-2*, 295, *301*
Klimov, V.V. 255-6, 258, *261*
Klyachko, N.L. *132*
Knoth, R. 117, *118*
Kobozev, N.I. *284*
Kohne, D.E. 23, *30*
Kolodner, R. 9-10, 13-4, 27, *30*
Kolyago, V.M. *131*
Komissarov, G.G. *284*
Konev, S.V. 167, *174*
Konigsberg, N. 84, *89*
Konishi, K. 239, 243, *244*
Korogodin, V.J. *201*
Kosobutskaya, L.M. 279, *283*
Kostyuk, N.N. 119, 129-30, *132, 145*
Kovács, K. *244*
Kowallik, W. 167-8, *175*
Kozo-Polyanskiĭ, B.M. 370
Krakhmaleva, I.N. *261*
Kranz, A.R. 134-5, *144*
Krashenninikov, I.A. 31, *36*
Krasichkova, G.V. *144, 208*, 271-2, *283*
Krasnovskiĭ, A.A. 137, *144*, 255, *261*, 264, 267, *268*, 271, 277, 279, *283-5*, 372
Krasnovsky, A.A. (= Krasnovskiĭ, A.A.) 239, *244*
Krawiec, S. 67, *90*
Krenzer, E.G.Jr. *340*

Kriedemann, P. 363, *366*
Krinsky, N.J. 217, *223*
Křítek, J. *347*
Kromhout, R. *283*
Krotkov, G. 167, *175*
Kubín, Š. *347*
Kudinova, S.V. 142, *144*
Kulaeva, O.N. 120, *132*
Kul'pina, A.I. *215*
Kumakov, V.A. 326, 329, *339*, 363, *366*
Kung, S.D. 1, *8*, 31, *36*
Kurkova, E.B. *339*
Kuroedov, V.A. *132*
Kursanov, A.L. 120, *132*
Kvitko, K.V. 137-8, *143-4*, 189, *200*, 203, *208*, 225, 227, 230-1, *236*, 369, 372-3

L
Laber, L.J. 149
Ladygin, V.G. 43-4, 47, *48*, 195, *200, 314*, 372
Laïsk, A. 165, *166*
Lane, D. 43, 47, *48*
Láng, F. *244*, 255, 260, *261*
Lascelles, J. 119, 124, *132*, 295, *301*
Latzko, E. 149, *157, 185*
Laudenbach, B. 173, *175*
Läuger, P. *284*
Lavrentovich, D.I. *366*
Lawrence, M.C. *362*
Ledoigt 67
Lee, J.W. *314*
Lee, R.W. 232, *236*
Lee, S.S. *156*
Leedale, G.F. 64, *90*
Leff, J. 74-5, *90*
Lefort-Tran, M. *90*
Lemon, E. 322, *340*
Levine, E.E. *48*
Levine, R.P. 43, *48*, 74, *90*, 93, *103*, 203, *208*, 233-4, *236, 314*, 372
Lewin, S. 33, *36*
Lichtenthaler, H.K. 208, *208*
Lightfoot, D. *30*
Linnane, A.W. 89, *91, 103, 145, 201, 236, 285, 314*
Lipkind, B.I. *144, 208*, 271
Lippincott, J.A. 278, *284*
Lipskaya, A.A. 31
Litvin, F.F. 137, 142, *144, 245*, 263-4, 267, *268*, 271-2, 275-7, *283-4*
Locke, M. *90, 132*
Loeb, J.E. 32, *36*

Loginov, M.A. *201*
Loppes, R. *48*
Lorimer, G.H. *174*
Los, S.I. *284*
Losev, A.P. 125, *132*
Lositskaya, T.V. 119, *132, 145*
Lowry, O.H. 32, *36*
Lozovaya, G.I. 271, *284*
Lozovaya, V.V. *366*
Lyman, H. 74, *90-1*
Lyons, J.M. 349, *355*
Lyttleton, J.W. 307, *314*

M
Mache, R. 111, *113*
Machold, O. 276, *284*
Makhovskaya, M.A. *366*
Makino, F. 31, *36*
Malashevich, A.V. *132*
Malyshev, O.G. 177-8, 184, *185*
Mamushina, N.S. 204, *208*
Mandel, M. *90*
Manning, J.E. 75-6, *90*
Marekha, L.N. 209, *215*
Margoliash, E. *91*
Margulies, M.M. 93, *103*
Margulis, L. 37, *41*
Markham, J.W. *48*
Marmur, J. 31, *36*
Marquardt, D.W. 241, *244*
Marré, E. 142, *145*
Marschall, O. *284*
Marsh, H.V.Jr. 119, 124, *132*
Marushige, K. *35*
Mathis, P. 239, *244*
Matorin, D.N. *118,* 263-4, *268*
Matrone, C. *132*
Matsuda, K. 29, *30*
Matsuka, M. 74, *90*
McCarthy, B.J. 29, *30*
McGowan, R. 80
McWilliam, J.R. 355
Mehlhorn, R.J. *355*
Meister, A. *284*
Melandri, A. *90, 145, 244, 261, 314, 362*
Mendel, G. 64
Menke, W. *284*
Menz, K.M. 333, *340*
Merezhkovskiï, K.S. 370
Mets, L.J. 53, 55-8, *62*
Metzner, H. *132, 175, 244, 261, 314*
Michel, J.-M. *314*
Mikelska, E.I. *103*

Mikul'ska, E. *8*
Miller, A.O.A. 101, *103*
Miller, P.L. 76, 89, *90, 201*
Milthorpe, F.L. 161, *166*
Mirsky, A.E. *36*
Mitrofanov, B.A. 363, *366*
Mitsuk, Z.I. *132*
Miyachi, S. 149, *157,* 171, *175*
Modolell, J. 76, *90*
Mohberg, J. 31, *36*
Mohr, H. 168-9, *175*
Mokronosov, A.T. 371
Monsi, M. 315, *340*
Monteith, J.L. 321, 323, *340*
Moore, R.L. 29, *30*
Moscarello, M.A. *36*
Moshkov, D.A. *184*
Moss, D.N. 333, *340*
Motorina, M.V. *339*
Moudrianakis, E.N. 289, *293*
Mühlethaler, K. 196, *200*
Mukhamadiev, B.T. 203, 205-7, *208,* 225, 231, *236,* 373
Müller, A.J. 189, *200,* 209, *215*
Müller, B. 80, *90, 314*
Müller, F. 247-8, *253*
Mullinix, K.P. 51-2, *62*
Munday, J.C.Jr. 255, *261*
Munns, R. 93, *103*
Murashige 210-3, 215
Murphy 162
Mushketik, L.S. 287
Mustafaev, A. *143*
Muzaffarova, S. 93-4, *103, 314*
Myers, J. 63, *89*

N
Nadler, K.D. 126, *132,* 133, *145,* 295, *301*
Nagy, Á. *244*
Nagy, Á.H. 243, *244-5, 261*
Nair, H. *314*
Nasyrov, Yu.S. 93-4, *103,* 133, 135, *143-5,* 189, *200-1,* 207, *208, 236,* 263, 268, 275, 282, *284,* 307, *314,* 330, *340,* 369-70
Nátr, L. 333, *339*
Nechaeva, E.P. 168, *175*
Negmatov, G. *143*
Nekrasov, L.I. 272-3, 277, *284*
Netrawali, M.S. 31, *36*
Neuman, J. 74, *90*
Nguen-Tkhyok *340*
Nichiporovich, A.A. *175,* 315-7, 320-1,

323-5, 328-9, 331-2, 338, *339-40*, 373
Nielsen, O.F. *145, 201, 285*
Nigon, V. *90*
Nikolayeva, G.N. *132*
Nilan, R.A. 135, *145,* 193, *200*
Nimmen, L. van *103*
Nishimura, M. 275-6, *284*
Noack, K. 295, *301*
Nobs, M.A. *339*
Noodén, L.D. 31, *36*

O
Odintsova, M.S. 1, *8,* 37, *41,* 93, *103*
Ogasawara, N. 171, *175*
Ogawa, T. *285*
Ogren, W.L. 172, *175*
Ohlenbusch, H. *35*
Okanenko, A.S. 287, *293, 366*
Olivera, B. *36*
O'Neal, D. 153, *157, 185*
Oparin, A.I. *41, 103,* 105, 109, *113*
Osawa, S. *62*
Osmarova, I.S. 169, *175, 269*
Osmond, C.B. *314,* 338, *339-40*
Ostapenko, L.A. *340*
Ostrovskaya, L.K. 137, *145,* 271, *284,* 287, *293*
Otaka, E. 52, *62*

P
Pacséry, M. *245*
Paech, K. *284*
Palade, G.E. *103*
Pap, E. 239
Paramonova, T.K. 119, *132, 145*
Park, R.B. 272-3, 277, 279, *284,* 310, *314*
Parker, J.H. *91*
Parnas, H. *103*
Parthier, B. 105, *113*
Paszewski, A. *103*
Patterson, V.D. *314*
Payer, H.D. 171, *175*
Pearce, R.B. 330, 333, *340*
Pearcy, R.W. *339*
Pennington, C.J.Jr. *89*
Persanov, V.M. 177
Petrenko, S.S. 287, *293*
Pham Than Ho 225
Philippovich, I.I. 106, 108-9, 111, *113*
Pichugina, N.G. *284*
Pigott, G.H. 41, *41*
Pinevich, V.V. 31-2, *36,* 369
Pirson, A. 173, *175*

Plaut, W. 1, *8*
Pochinok, Kh.N. *366*
Podchufarova, G.M. 119, 124, 129, *131*
Polishchuk, A.I. 287, *293*
Pollak, J.K. *314*
Pon, N.G. *185*
Popova, E.A. *132*
Porcile, E. *283-4*
Poucke, M. 173, *175*
Poyarkova, N.M. 167, 170, 172, *175*
Pratt, E.A. *89*
Pridham, J.B. *244, 301*
Pringsheim, F. 63, *90*
Pringsheim, O. 63, *90*
Prokof'eva, N.A. *284*
Provasoli, L. 63, *90*
Prudnikova, I.V. 119, *132, 145*
Pshenichnaya, A.K. *293*
Punnett, T. 307, *314*
Pupillo, P. 80, *90*
Putterman, G.J. *91*

R
Raaf, J. 31, *36*
Rabinowitch, E.I. 247, *253,* 278-80, *284*
Rackham, O. 165, *166*
Radunz, A. 272-3, 281, *284*
Radzhabov, H. 93
Raison, J.K. *145,* 349, 351-2, 354, *355*
Rajki, E. 358, *362*
Rajki, S. 357-8, *362*
Rakován, J.N. *244, 261*
Ramanis, Z. 43, 47, *48,* 52, *62,* 189, *201*
Rana, R. 209, *215*
Randall, R.J. *36*
Raps, S. *144, 283*
Rasoriteleva, E.K. *215*
Ratner, V.A. 295, *301*
Raveed, D. *285*
Rawson, H.M. 363, 365, *366*
Rawson, J.R. 67, *90*
Ray, D.S. 74-5, *90*
Rédei, G.P. 135, *144*
Reger, B.J. 79, *90, 145*
Reïngard, T.A. 287, 291, *293*
Rhoades, M.M. 66-7, *90,* 194-5, *201*
Richards, O.C. *90*
Rikhireva, G.T. *284*
Riley 162
Ris, H. 1, *8*
Rizzo, P.J. 31, *36*
Röbbelen, G. 134-5, *145,* 190, 196, *201,* 209, *215*

381

Robertson, D.S. 217-8, *223*, 239, *244*
Rodin, L.E. 315, 321, *340*
Rollin, P. 84, *90*
Romanova, A.K. 178, *185*
Rosebrough, N.J. *36*
Rosenberg, J.L. *156*, 272, *283*
Rosenberg, L.L. *89*
Rosmarin, M.N. 33, *36*
Ross, C. 162, *166*
Ross, Yu.K. 315, 321, 324, *340*
Roth, T.F. *314*
Rowley, J.A. 307, *314*
Roy, H. 289, *293*
Rubin, A.B. *118*, 263, *284*
Rubman, J. *91*
Rudoï, A.B. 125-6, *132*, *145*
Rumberg, B. *253*
Rusch, H.P. 31, *36*
Ryan, R.S. *90*
Ryzhkov, V.L. 135, 139, *145*

S
Saakov, V.S. *236*
Sadykov, A.S. 264, *268*
Saeki, T. 315, *340*
Sager, R. 43, 47, *48*, 52, 54, *62*, 74, *90*, 189, *201*, 234, *236*, 372
Sagromsky, H. 203, *208*, 247, 372
Salvador, G. 63, 67, *90*
Sanina, A.V. *200*
San Pietro, A. *145*, *314*, 315, 321, 324, 331, *340*
Sapozhnikov, D.I. *144*, 271-2, 274, *283-4*
Sauer, K. 277, 279, *284*
Savchenko, G.E. 119, *132*, *145*, 245, *261*
Schatz, A. *90*
Schiff, J.A. 63, 66-74, 76-84, 87-89, *89-91*, 119, 124, *132*, 295, *301*, 371
Schlender, K.K. 297, *301*
Schmid, G.H. 167, *175*, *284*, 307, *314*
Schmidt-Clausen, H.J. *314*
Schmidt-Mende, P. *253*
Schoffeniels, E. *174*
Scholze, M. 117, *118*
Schori, L. 75
Schötz, F. 198, *201*
Schulman, M.D. 79
Schwarting, A.E. 217-8, *223*
Schwartzbach, S.D. 81-3, *91*
Scott, N.S. 67, *91*, *103*, 168, *175*, *301*, 306, *314*
Seely, G.R. 274-5, *284*
Selivankina, S.Yu. 120, *132*

Sell, H.M. *301*
Semenova, G.A. *314*, 372
Semenyuk, I.I. 287, *293*
Semeyenova, G.A. *314*
Semichaevskiï, V.D. 137, *145*, 271, *284*
Semikhatova, O.A. 204, *208*
Semyonova, G.A. 43
Šesták, Z. *144-5*, 159-60, 162, *166*, *208*, *236*, *284*, 315, 321, 324, *340*, 369
Šetlík, I. 343-4, *347*
Sevchenko, A.N. *283*
Seybold, A. 247, 250, *253*
Shaw, E.R. *285*
Shcherbakova, I.Yu. *269*
Sherman, F. 79, *91*
Shibata, K. *244*, *301*, 358, *362*
Shibles, R.M. 330, *339*
Shields, C.M. *48*
Shipman, N.A. *91*
Shlyk, A.A. 119-23, 125-31, *131-2*, 133, *145*, 240, *245*, 255, *261*, 371
Shneyour, A. 349-50, 353-4, *355*
Shuvalov, V.A. *268*
Siegel, A. *30*
Siekevitz, P. *103*
Siggel, U. *253*
Simpson, G.M. 363, *366*
Sinclair, J. 29, *30*
Sinegub, O.A. 137, *144-5*, 271, 277-80, *283-4*
Sineshchekov, V.A. 239, *245*
Sironval, C. *166*, 307, 312, *314*
Sisakyan, N.M. 105, *113*
Sisler, E.C. 119, 124, *132*, 295, *301*
Skerra, B. *253*
Skoog 210-3, 215
Skott, N.S. (= Scott, N.S.) *314*
Slack, C.R. 149, 152, *157*, 170, 172, *174*, 177, *185*, 306, *314*, 371
Slatyer, R.O. *339*
Slavíková, D. 343, *347*
Slobodskaya, T.A. *340*
Smaïli, R.M. (= Smillie, R.M.) 305-6, *314*
Smillie, R.M. 67, *89*, *91*, 93, *103*, 135, *145*, 168, *174-5*, *201*, *236*, *285*, 296, *301*, 305, 308-10, *313-4*, 349, 354, *355*, 371, 373
Smirnov, A.F. *236*
Smit, J. (= Smith, J.H.C.) 133, *145*
Smith, G.M. *89*
Smith, H.J. 52, *62*
Smith, J.H.C. 136, *145*, 274-5, *284*, 295, *301*

Sokhibnazarov, Sh. 189
Solárová, J. 159
Solov'ev, K.N. *283*
Šormová, Z. 33, *36*
Spandar'yan, O.A. *113*
Spencer, D. 111, *113, 314*
Spirin, A.S. 32, *36*
Šponar, J. 32, 33, *36*
Sprey, B. 47, *48*
Staba 210
Staehlin, T. 55, *62*
Stanier, R.Y. 217-8, *223*
Steinemann, A. 276, *284*
Stern, A.I. 64, 68, 70, 81, 84, *91*
Stern, R. 93, *103*
Stewart, J.W. *91*
Stolbova, A.V. 43, 47, *48, 144*, 217-8, *223*, 227, 230-1, 233, *236, 372*
Strain, G.C. 51-2, *62*
Stránský, Z. 287, *293*
Strehler, B.L. 263, *268*
Strelkova, T.I. 276, *283*
Strnadová, H. 159, 162, *166*
Strogonova, L.E. *340*
Sturani, E. *145*
Stutz, E. 67, 79, *90-1*
Sueoka, N. 369
Suleïmanova, A.Yu. 363
Sullivan, C.Y. *339-40*
Surzycki, S.J. 43, *48*
Sveshnikova, I.N. *132*
Svetaïlo, E.N. *103, 113*
Swain, T. *244, 301*
Sweers, H.E. 255, *261*
Swift, H. *48*
Swinton, D. *89*

T
Tagawa, K. 203, 207, *208*
Tageeva, S.V. 43, *200*, 274-5, *283*, 307, *314*, 372
Takamatsu, K. 275-6, *284*
Takamiya, A. *301*
Takeuchi, M. 31, *36*
Takhtadzhyan, A.L. 198, *201*
Tamaki, M. *62*
Tanaka, K. *62*
Tarchevskiï, I.A. 363, *366*
Taylor, A.O. 307, *314*
Temper, E.E. 225, 227, 230, *236*
Teraoka, H. *62*
Tevini, M. 208, *208*
Tewari, K.K. 1, *8*, 9-10, 13-4, 24, 27-8, *30*, 31-2, 34-5, *36*, 93, *103*, 111, *113*, 369, 373
Thom, A.S. 161, *166*
Thomas, J.B. 277, 279, *284*
Thomas, J.R. 9-10, 24, *30*
Thomson, W.W. 355
Thornber, J.P. 272-3, *284*
Thorne, G.N. 363, *366*
Thorne, S.W. 125, *132, 314*
Thruni, F.N. *236*
Tichá, I. 159, 161, *166*
Tilney-Bassett, R.A.E. 93, *103*, 135, *144*, 295, *301*
Timofeeff-Ressovsky, N.V. 190, *201*
Timofeev, K. *118*
Timofeeva-Ressovskaya, H.A. *200*
Ting, I.P. 338, *340*
Tobin, N.F. *103, 314*
Togari, I. 315, 321-2, 324, *340*
Togasaki, R.K. *103*, 149, *157*
Tolbert, N.E. 173, *174*
Tolibekov, D. 272, *284*
Tongur, A.M. 105, *113*
Tracey, M.V. *284*
Trebst, A. 264, *268*
Troughton, J.H. *339*
Ts'o, P.O.P. *36*
Tsugita, A. 31, *36*
Tsujimoto, H.Y. *208*
Tsuzuki, J. 31, *36*
Tugarinov, V.V. 225, 231, 233, *236*
Tumerman, I.A. 277, 279, *284*
Turishcheva, M.S. 1, *8, 103*, 369

U
Ulugbekova, G. 93
Uribe, E. 289, *293*
Usiyama, H. *208*
Usmanov, P.D. 133, 135, 138, *143, 145*, 189-90, 193-5, *200-1, 208*, 264, 267, *269, 284*, 371
Usmanova, O.V. 189

V
Vakhidova, L.R. *144, 208, 236*, 271
Val'ter, G. *132*
Varner, J.E. *284*
Vasconcelos, A.C.L. 39, *41*, 55, *62*
Vasilenok, L.I. 287
Vasil'eva, N.G. *340*
Vater, J. *253*
Vatti, K.V. 190, *201*
Vavilov, N.I. 234

Vecher, A.S. 195, *201*
Vedenina, I.Ya. *185*
Velemínský, J. *145,* 209, *215*
Vernon, L.P. 272-3, 281, *285*
Vesitsky, A.Y. *145*
Vezitskiĭ, A.Yu. 119, *132, 145*
Viktorova, G.V. *201*
Virgin, H.I. 173, *175*
Vlasenok, L.I. 119, *132, 145*
Vlasova, M.P. 169, *175, 340*
Voldeng, H.D. 363, *366*
Volkmann, D. 74, *90*
Volkova, N.V., 287, *293*
Volovik, O.I. 287, *293*
Vorob'eva, I.P. 132, 277
Vorob'eva, L.M. *261,* 279, *285*
Voskresenskaya, N.P. 167-9, 173-4, *175,* 263, *268-9*
Vreese, A.M. de *103*

W

Waddington, G.H. 200, *201*
Wal, U.P. van der 277, 279, *284*
Walker, D.A. 178, *185*
Wallace, D.H. 330, 333, *339*
Wallace, R.H. 217-8, *223*
Walles, B. 135, *145,* 239, *245,* 307, *314*
Watson, D.J. 315, 329, *340*
Waygood, F.R. 111, *113*
Weatherbee, J.A. 87-8, *91*
Weibel, E.R. 196, *201*
Weier, J. 135, *145*
Weier, T.E. 108, *112*
Weikard, J. *253*
Weisblum, B. 76
Weissner, W. 307, *314*
Werz, G. 111, *113*
Wettstein, D. von 134-5, 137, *145,* 189, 193, 196, *201,* 271, 276, 278, *285,* 307, *314,* 371, 373
Wettstein, F. von 194-5, *201*
Whatley, F.R. *89,* 264, *268*
Whitfeld, P.R. 111, *113*

Widholm, J. 31, *36*
Wild, A. 251, *253*
Wildman, S.G. 1, *8,* 32, 34-5, *36,* 88, *90,* 93, *103,* 111, *113*
Williams, J.P. 1, *8, 36*
Wilson, A.T. 149, 151, *157*
Winkle, R.P. *236*
Witt, H.T. 247, 252, *253*
Witts, K.J. 329, *340*
Wollgiehn, R. 105, *113*
Wolstenholme, D.R. *90*
Woo, K.C. 308, *314*
Woodcock, C.L.F. 2, *8,* 111, *113*

Y

Yaginuma, N. *244*
Yakubova, M.M. 263, *269*
Yasniköv, A.A. 287, 291-2, *293*
Yokendo 315, 321-2, 324, *340*
Young, V.M.K. 295, *301*
Yukhananova, L.N. *144, 208,* 271-2, 278, 280, *283*
Yunusov, S.Yu. 194, *201*
Yurina, N.P. 37-8, *41*
Yusupova, G.A. *144, 208,* 271-2, *283, 285*

Z

Zacharias, I.M. 209, *215*
Zaĭtseva, N.A. 287, *293*
Zakharov, I.A. 227, *236*
Zalenskiĭ, O.V. 205, *208,* 231, *236, 268, 314,* 315, 321, *340,* 371, 373
Zeldin, B. 76, 80, 84, *89*
Zeldin, M.H. 70, *89, 91*
Zelitch, J. 315, 324, 331, *340*
Zen'kevich, E.I. 125, *132*
Zetsche, K. 93, *103*
Zhdan-Pushkina, S.M. 32, *36*
Ziegler, H. 80, *90,* 306, *314*
Ziegler, I. 80, *90, 314*
Zimmer, K.G. 190, *201*
Zucker, M. 167, *175*

SUBJECT INDEX

A
Acetylphosphate 292
Actinomycin D 94-9, 102, 121-2, 124-6, 128, 131, 141, 143
Adaptation of photosynthesis 357-62
Alpha-particles 193-4, 200
α-amanitin 51-2
δ-aminolevulinic acid 119, 129-31, 134, 296-9
δ-aminolevulinic acid (ALA) synthetase 124-5, 130, 134
Amyloplast 177, 184
Antheraxanthin 240
Antibiotics (cf. also Actinomycin D, Antimycin A, Aurantin, Chloramphenicol, Cycloheximide, Dexteromycetin, Erythromycin, Neomycin, Spectinomycin, Streptomycin, D-threo-chloramphenicol) 67-8, 94, 221, 231-2, 275, 372
Antimycin A 204-6, 208
Assimilation number 363-5
Aurantin 299-300
Autumnization 357-62
Auxotrophy 235
8-azaguanine 121, 125-6, 128
Azide, sodium 155

B
Bacteriochlorophyll 277-9
5-bromuracil 126, 128
Bundle sheath cells 156, 177, 179-82, 184, 308-10

C
Callus 209-11, 213
Calvin cycle 149-50, 170-1, 179
CAM see Crassulacean Acid Metabolism
C_3 and C_4 plants 152, 156, 169, 171, 173, 308, 310, 312-3, 321-3, 325, 329-30, 333, 336-7, 355
Canopy structure 316-28, 374
Carbon fixation pathways 149-56, 167-74, 177-84, 371
Carbonic anhydrase 155
Carotene (cf. also Lycopene) 217, 228-30, 234, 239-43, 249, 255-60, 274-6, 281
ξ-carotene 239-43, 255-7, 259-60
Carotenoid amount 235, 248-9
Carotenoid deficient mutants 239-44, 255-60, 372
Carotenoid photoconversion 168, 292
Carotenoid synthesis 69-73, 78, 135, 282
C_4 dicarboxylic acid cycle see Hatch & Slack cycle
Chemiluminescence 267
Chimeras see Chlorophyll chimeras
Chloramphenicol 77, 79, 83, 93-4, 120-1, 123, 130-1, 296-8
p-chloromercuribenzoate 127
p-chlorophenyldimethyl urea see CMU
Chlorophyll amount 163, 165, 169, 181, 199-200, 203-4, 206-8, 217, 228-30, 235, 241, 248-9, 260, 268, 364
Chlorophyllase 134
Chlorophyll a to b transformation 120, 125-8, 135
Chlorophyll b role in photosynthesis 247-52
Chlorophyll chimeras 209-15
Chlorophyll degradation 218, 239
Chlorophyll fluorescence 137-8, 140-2, 240, 255-60, 263-8, 272
Chlorophyll holochrome 134
Chlorophyll in photosynthesis 160, 163-5, 247-52, 363-5, 372
Chlorophyll in vivo forms 115-6, 133-43, 207-8, 241, 243-4, 255, 273-5, 280-1, 357, 359-62
Chlorophyll mutants see Mutants
Chlorophyll-protein-lipoid synthetic complexes 271-82, 372-3
Chlorophyll synthesis 68-70, 77-84, 119-31, 133-43, 282, 295-300, 306, 371
Chlorophyll synthetase 119, 125, 130
Chloroplast, agranal 156, 308-11
Chloroplast differentiation see Chloroplast ontogenesis . . .
Chloroplast dimensions 195-6
Chloroplast fractionation 109
Chloroplast fusion 44-7
Chloroplast gene, plastogene, extranuclear gene 43, 55, 60, 62, 134, 222, 233-4, 370, 372
Chloroplast ontogenesis, differentiation 2, 4, 63-89, 93-102, 305, 370, 374
Chloroplast origin 35, 37-41, 51-62, 64, 88, 369

385

Chloroplast ultrastructure 105-12, 127, 156, 169, 196-200, 242, 247, 252, 272, 305-13, 370-1
Chondriome 225
Chromatophore 277, 280-1
Chromosome number 195
Circular dichroism 240
CMU 74
CO_2 effect on photosynthesis 336-7
CO_2 in canopy 320-3
Compensation point 333
Conductance see Resistances to gas transfer
Coproporphyrinogen 134
Crassulacean Acid Metabolism 152, 156
Cycloheximide 77, 81, 83, 88, 93-4, 133, 143, 234
Cytochromes 77-9, 93, 234

D

DCMU 71-3, 86-8, 154-5, 203-8, 231, 260, 264-6, 373
Delayed fluorescence see Chlorophyll fluorescence
Detergents (cf. also Dodecylsulphonate, sodium, Digitonin) 110-1, 115, 279-80, 282
Dexteromycetin 123
Diaphorase 352
3,/3,4-dichlorophenyl/-1,1-dimethyl urea see DCMU
Digitonin 243, 291
Dinitrophenol 83
Diquat 352
Dithionite 258
Dithiothreitol 153
DNA amount 74-5
DNA catenated dimers 17, 27
DNA circular dimers 15-6, 18, 27
DNA cistrons 93
DNA de- and renaturation 12-3, 75, 372
DNA electron microscopy 2-8, 13-21
DNA length distribution 20-1
DNA melting profile 10, 12, 32-3
DNA membrane complex 1-8
DNA molecular weight 2, 9-21, 75-6
DNA, nuclear and chloroplast 9, 11, 29, 32-3, 35, 51, 76, 79, 84, 115-7, 306, 369, 374
DNA polymerase 76, 111, 234
DNA protein complex 31-5, 74, 369
DNA purification 10-1
DNA replication 226, 231-2, 369
DNA-RNA hybridization 9-10, 22-9
DNA-RNA ratio 109
Dodecylsulphonate, sodium 280
Dominance 343-4, 346, 373

E

Efficiency of photosynthetic radiant energy utilization 316, 320-5
Electron spin resonance 351
Embryonic test 189
Endoplasmic reticulum 108
Endosymbiont hypothesis 58-62, 88
Energy migration 280-1
Enzymes of carbon fixation see Glyceraldehyde 3-phosphate dehydrogenase, Malic enzyme, Phosphoenol pyruvate carboxylase, Phosphoribulokinase, Pyruvate kinase, Ribose-5-phosphate isomerase, Ribulose 1,5-diphosphate carboxylase, Triose phosphate dehydrogenase
Enzymes of chlorophyll synthesis (cf. also δ-aminolevulinic acid synthetase, Chlorophyllase, Chlorophyll synthetase) 134
Epistasis 343-4, 346, 373
Erythromycin 52-6, 58, 233
Ethionine 128
Ethyl methanesulfonate 53, 193-4, 200
Etioplast 2, 4, 102, 119

F

Ferredoxin 93, 234
Ferredoxin-NADP-reductase 234, 352-5
Ferulic acid 212
Flavins 168
Fluorescence of chlorophyll see Chlorophyll fluorescence
5-fluorouracil 121, 125-6, 128
Fraction I protein see Ribulose diphosphate carboxylase

G

Gametes 43-8, 372
Gamma-rays 193-4, 200, 263
Gene, chloroplast see Chloroplast gene . . .
Gene control 52-4, 60, 80, 119
Gene, extranuclear see Chloroplast gene . . .
Gene interactions, non-allelic 343-6
Gene, nuclear 55, 60, 62, 134-5, 143, 200, 233-5, 268, 370-1
Gene number 29
Gene transfer 60
Genome 43, 52, 58, 60, 67, 74, 76, 143, 194-200, 334, 372
Genome, plastid see Plastome

Genotype and phenotype 225, 344
Glyceraldehyde 3-phosphate dehydrogenase 93, 170-2, 178-9, 183, 306
Glycollate pathway 167, 173-4
Golgi apparatus 69
Granum 106, 108-10, 112, 156, 239, 242, 305-10, 313
Growth analysis characteristics 316-22, 324-6

H
5-haloiduracils see 5-bromuracil, 5-fluorouracil
Hatch & Slack cycle 170, 177, 183, 371
Hayman's six parameter model 343-6
Heteroplastid 135-6
Heterosis 332, 343-6, 373
Hill reaction 162-6, 169, 243, 251-2, 263-8, 308-10, 312, 350
Hydroxylamine 258

I
Intercellulars 160
α-iodoacetamide 127
Irradiance, effect on photosynthesis of 167-74, 204-5, 252, 309-10, 312-3, 322-5, 328, 330, 337, 365, 373
Irradiance, effect on plastid development of 70-4, 82-7

K
Kinetin 120, 123-5, 131, 214
Krebs cycle 151, 183

L
L_{opt} 316, 318, 326
LAI 318-21, 324, 326
Leaf area and thickness 163
Leaf area index see LAI
Light respiration see Photorespiration
Locus 54, 58, 73, 231, 371
Lutein 217, 228-9, 234, 248-9
Lycopene 239-42, 255-7, 259-60

M
Malic enzyme 177-81
Mendelian inheritance 43, 47, 54, 65-7, 73, 191, 221, 233, 239, 372
Methionine 128
N-methyl-N-nitrosourea 189-94, 200, 209
Microbody 72
Mineral nutrition and photosynthesis 335
Mitochondrion 58, 61, 67-9, 71-2, 74-6, 79-80, 83, 89, 159, 233-5
Mutants 44-7, 52-4, 56-7, 73, 115-7, 119, 133, 137-40, 177-84, 189-200, 203-8, 209-15, 217-23, 225-36, 239-44, 247-52, 255-60, 263-8, 371-2

N
NADP photoreduction 308, 350-5
Neomycin 233
Neoxanthin 217-8, 228-30, 234, 248-9
Net assimilation rate see Net photosynthetic productivity
Net photosynthetic productivity 318-20
Nitrosoethylurea 230, 232-3
Nitrosoguanidine 230, 232
12-nitroxide methyl stearate 352
Nuclear transformation 59
Nuclei fusion 44-6

O
Oxygen effect on photosynthesis 169, 172-3, 218, 251, 336-7

P
Paramylum 69, 71, 86
PEP carboxylase see Phosphoenol pyruvate carboxylase
Peroxisome 159
Phenotype see Genotype and phenotype
Phosphatase 287-93
Phosphoenol pyruvate carboxylase 149, 152-3, 170-4, 177, 336-7
Phosphoribulokinase 93, 178-9, 234
Photophosphorylation 162-6, 169, 174, 203-4, 206-8, 252, 287-93
Photorespiration 251-2, 322-3, 333
Photosynthates (cf. also Starch) 73, 151-6, 159-60, 177, 181-4, 206, 315, 328, 338, 363, 365, 371
Photosynthesis adaptation see Adaptation of photosynthesis
Photosynthesis and productivity see Yield
Photosynthesis, diurnal variation 364-5
Photosynthesis, effect of CO_2 on see CO_2 effect on photosynthesis
Photosynthesis, effect of irradiance on see Irradiance, effect on photosynthesis of
Photosynthesis, effect of light colour on 167-74
Photosynthesis, effect of mineral nutrition on see Mineral nutrition and photosynthesis
Photosynthesis, effect of O_2 on see Oxygen

effect on photosynthesis
Photosynthesis, effect of temperature on see Temperature effect on photosynthesis
Photosynthesis, ontogenetic changes of 159-66, 365
Photosynthesis, role of chlorophyll in see Chlorophyll in photosynthesis
Photosynthetic gas exchange see Photosynthetic rate
Photosynthetic matter accumulation see Photosynthetic rate
Photosynthetic rate 68, 70, 72, 78-9, 84, 159-66, 204-5, 251, 255-7, 259-60, 319-25, 343-6, 358
Photosynthetic unit 160
Photosystem I 93, 115, 137, 208, 243, 256-7, 259, 263-4, 271, 273, 281, 306, 308
Photosystem II (cf. also Hill reaction) 71, 93, 115, 177, 208, 243-4, 255-6, 259-60, 263-4, 273, 305-6, 308-10, 312-3
Phycoerythrin 310, 313
Phytochrome 84, 168, 173, 296, 299, 306
Phytoene 240
Phytofluene 240
Plastid replication 64-6
Plastogene see Chloroplast gene . . .
Plastoglobuli 196, 199
Plastome 43, 47, 52, 67, 115, 143, 225, 370
Plastoquinone 247
Polygenic systems 343-6, 374
Polyploidy 332, 374
Polyribosomes 102, 106, 110-2
Porphobilinogen 134
Precursors and intermediates of chlorophyll biosynthesis 134
Productivity of photosynthesis see Yield
Prolamellar body 68-9, 71, 85
Proplastid 2, 4, 63, 67-70, 72, 78
Protein complex with DNA see DNA protein complex
Protein synthesis 95, 102, 110-2, 121, 124, 140-3, 167-9, 231, 278, 296
Proteins, chloroplast 95, 117
Proteins, ribosomal 37-41, 52-62, 105
Protochlorophyll (ide) 84, 120-5, 129-31, 134-5, 240, 258, 295-300, 306
Protons 193-4, 200, 289, 291-2
Protoporphyrin 134
Protoporphyrinogen 134
Pyrenoid 44, 69, 71, 73, 85, 87-8
Pyridoxal phosphate 292
Pyruvate kinase 287, 290-2

Q
Quantasome 272-3, 279
Quantum requirement 324

R
r_a 159-61
r_i 159-60
r_m 159-61, 163-5
r_s 159-66
r_x 159-61
Regenerants 209-15
Resistances to gas transfer 159-66, 333, 337
Respiration 81, 83, 251, 320, 328, 330
Ribose-5-phosphate isomerase 178-9
Ribosomes (cf. also Polyribosomes) 37-40, 52-62, 67, 76, 81, 88, 93-4, 105-6, 117, 133, 143, 234, 370
Ribulose 1,5-diphosphate carboxylase 77-9, 81, 83-4, 87-8, 149, 152-3, 169-74, 177-9, 183, 234, 243, 333, 336-7
Rifampicin 95-6, 99-100, 234
RNA content 109
RNA polymerase 51-2, 111
RNA/protein ratio 38-9
RNase 101-2
RNA synthesis 94-5, 120-1, 124-5, 127-8, 131, 168
/ct/RNA 10, 94-5, 97-102, 370
mRNA 93-102, 142, 370
/r-/RNA cistrons 22-9, 93, 234
/r-/RNA fractionation 22, 55, 109
tRNA 93, 99

S
Sowing times and yields 358
Spectinomycin 234
Starch 44, 196, 199, 247, 250, 252
Stigma 44-5
Stomata 159-66, 250, 333
Streptomycin 54, 74, 76, 78, 81, 88, 231-3
Stroma lamellae 108-11, 312-3

T
Temperature effect on photosynthesis 349-55, 357-62, 365
D-threo-chloramphenicol 135
Thylakoid 44, 47, 68-9, 71, 85, 106, 108-9, 112, 197-9, 247, 252, 351-2, 373
Tissue culture 209-15
Transcription-translation system 94, 105, 143, 225-6, 234, 299
Triose phosphate dehydrogenase 77-81, 83
Trollein 217-8, 234

U
Uroporphyrinogen 134-5

V
Vacuole 199
Variegated leaf 139, 266-8
Vavilov's law 234
Violaxanthin 217, 228-9, 234, 248-9

W
Wind speed in canopies 321-2

X
Xanthophylls see Antheraxanthin, Lutein, Neoxanthin, Trollein, Violaxanthin
Xeromorphy 333
X-rays 193-4, 200

Y
Yield 249-50, 252, 315-39, 358, 373
Yield, biological and economic 317

Z
Zygote 43-8, 54, 64, 372

PLANT INDEX

A
Acetabularia 2
Alfalfa 330
Algae blue-green (*cf.* also *Anabaena, Anacystis, Lyngbya, Phormidium, Spirulina, Synechocystis*) 31, 35, 37-41, 58, 88
Algae green (*cf.* also *Acetabularia, Chlamydomonas, Chlorella, Eremosphaera, Scenedesmus*) 38, 137-8, 203-8, 225-36, 372
Algae red (*cf.* also *Porphyridium*) 308, 310, 313
Allium cepa see Onion
Anabaena 38-40
Anacystis 38-40
Antirrhinum 38, 115-7, 135
Arabidopsis 133, 135-6, 209, 211-4, 370-1
Artichoke 29
Atriplex 333
Avena see Oat
Avocado 349

B
Bacteria (*cf.* also Bacteria, photosynthetic) 1, 2, 6, 29, 32-3, 38-41, 52, 58, 76
Bacteria, photosynthetic (*cf.* also *Chromatium, Rhodopseudomonas*) 37-41, 137, 277-9
Banana 349
Barley 66, 120-3, 126, 129-30, 135, 168, 194, 247-52, 278, 296-300, 323, 349, 371-2
Bean 10-2, 24-7, 29, 38, 161, 330, 337, 349-50, 352
Beet see Sugar beet, Beet
Beet root 349
Beta see Sugar beet, Beet
Brassica oleracea var. *botrytis* see Cauliflower
Broad bean 167, 169-74, 335-6
Bushel baskets 322

C
Cauliflower 349
Cereals 317, 325, 327, 329-30, 363
Chlamydomonas 1, 43-8, 52-8, 62, 73-4, 138, 217-23, 225, 229-35, 369
Chlorella 38, 40, 73-4, 137-8, 171, 203-8, 225, 227-8, 230-1, 234-5, 251, 372-3
Christmas tree 322
Chromatium 38, 40, 280
Cotton 133, 138-42, 263-8
Crassulaceae 329
Cucumber 343, 349
Cucumis see Cucumber
Cynara see Artichoke

D
Dictyostelium 38

E
Equisetum 38
Eremosphaera 87
Escherichia 22, 38-40, 52, 69
Eucaryota 234
Euglena 63-88, 371

F
Ferns 168
Fungi 31, 38

G
Glycine see Soybean
Gossypium see Cotton
Grasses 322, 324, 355
Griffithsia 310

H
Helianthus see Sunflower
Hordeum see Barley

I
Ipomea batatas see Sweet potato

L
Lactuca see Lettuce
Lemna 326
Lettuce 9, 10-2, 14-5, 21, 24-9, 349-50, 355
Lycopersicon see Tomato
Lyngbya 38-40

M
Maize 4, 9, 10-2, 14-5, 21, 24-9, 38-9, 51-2, 61, 66-7, 73, 149-56, 167, 169-74, 177-84, 194, 239-44, 255-60, 308-13, 321-3,

325, 327, 334-8, 343-6, 349, 352, 371-2
Medicago see Alfalfa
Micrococcus 11
Mung bean 39
Musa see Banana
Mustard 173

N
Nicotiana see Tobacco
Nuphar 326

O
Oat 9, 14-6, 21, 25, 28-9, 330
Ochromonas 76
Oenothera 67, 198
Onion 29
Oryza see Rice

P
Pea 1-14, 17-9, 21-9, 31-5, 37-8, 40, 93-102, 108, 112, 135, 153, 194, 248, 287, 310-3, 349-51, 353-5, 370
Pear 349
Pelargonium 115-7
Persea see Avocado
Phaseolus see Bean
Phormidium 39
Pine 73
Pinus see Pine
Pirus see Pear
Pisum see Pea
Plectonema 38-40
Porphyridium 310
Potato 349, 371
Procaryota 225

R
Radish 76
Raphanus see Radish
Rhodopseudomonas 38, 40
Rice 329-31, 334
Rye 127-8

S
Saccharum see Sugar cane
Scenedesmus 74, 225, 227, 229-31, 234, 372
Secale see Rye
Sinapis see Mustard
Solanum tuberosum see Potato
Sorghum 321, 325, 349, 355
Sorgum see Sorghum
Soybean 330
Spinach 2, 9-12, 14-5, 21, 24-9, 149-51, 153
Spinacia see Spinach
Spirulina 38
Sugar beet, Beet 194, 323, 325, 329, 336-8, 351
Sugar cane 325
Sunflower 325
Sweet potato 349
Swiss chard 350
Synechocystis 38-40

T
Tobacco 34-5, 38, 343
Tomato 349-55
Triticum see Wheat
Typha 324, 326

V
Vicia see Broad bean
Victoria 326

W
Wheat 38, 194, 322, 327, 329, 331, 334, 349, 357-62, 364-5

Y
Yeasts 31, 79

Z
Zea see Maize